Normal
SURFACE AN[illegible]

NORMAL SURFACE ANATOMY

Bruce Keogh BSc, MB, BS **and Stephen Ebbs** MB, BS

Department of Anatomy, Charing Cross Hospital Medical School, London

Illustrated by Mike Duffy

Foreword by T W Glenister CBE, TD, MB, BS, DSc, PhD
Professor of Anatomy, Charing Cross Hospital Medical School, London

William Heinemann Medical Books Ltd
London

Acknowledgements

This book has been based on the concept of visual presentation with explanatory text. To produce such a book would not have been possible without the encouragement of Mr Robin Williams who laid the facilities of The Department of Medical Illustration at Charing Cross Hospital at our disposal.

In particular we wish to record our warmest gratitude and indebtedness to Mr Mike Duffy who specially prepared all the diagrams, and to Mrs Janey Shemilt who tirelessly prepared and photographed both the modelling sessions and radiograph illustrations.

We would like to thank Mr John Maxey, the student model, for his patience and endless co-operation during the sometimes protracted photography sessions.

We owe much to Dr John Palfrey and Dr Jean Ross of The Department of Anatomy at Charing Cross Hospital Medical School, not only for their encouragement and advice, but also for the thankless task of reading the preliminary text.

Although all the diagrams have been specifically drawn for this book it has not been easy to find appropriate radiographs and scans and we would therefore like to express our heartfelt thanks to those mentioned below for providing us with valuable contributions:

Kitty Clark (*Positioning in Radiography*, Vols. 1 & 2, 10th edition, Louis Kreel (Ed), Heinemann Medical Books Ltd)
X-rays: Pages 25, 39, 41, 43, 44, 61, 63, 65, 67, 69, 71, 95, 105, 109.
Scans: Pages 49, 85

Dr Brendon Conry
Dept. of Radiology
St Bartholomew's Hospital
X-rays: Pages 17, 35, 57, 77, 87, 91, 95, 97, 105.
Scans: Page 49.

Dr Robert Gibson and
Dr Oliver Hennessey
Dept. of Radiology
Hammersmith Hospital
X-rays: Pages 189, 191

Dr Tim Cox
Dept. of Radiology
Hammersmith Hospital
Carotid Angiograms: Pages 217, 219

Dr James McIvor
Dept. of Radiology
Charing Cross Hospital
X-rays: Pages 185, 249

Mr Binder
Consultant in Dental Surgery
Charing Cross Hospital
X-rays: Page 205

Dr Reginald Dewkes
Dept. of Nuclear Medicine
Charing Cross Hospital
Scan: Page 45

Cardiology Department
Charing Cross Hospital
X-ray: Page 33

Cardiology Department
St Bartholomew's Hospital
X-ray: Page 45

Finally, we would like to thank Dr Richard Barling and Mr Chris Jarvis of Heinemann Medical books for their patience and assistance.

First published in 1984
by William Heinemann Medical Books Ltd,
23 Bedford Square, London WC1B 3HH

ISBN: 0-433-18345-4

Printed in Great Britain by BAS Printers,
Over Wallop, Hampshire and bound by
Dorstel Press Ltd, Harlow

Foreword

That the human body should be seen by the student as a dynamic, developing, functioning and eventually ageing entity is crucial if the study of anatomy is not to be a dull, demanding, almost forlorn endeavour. Supplementing the study of the cadaver with substantial attention to the anatomy of the living subject thus becomes a *sine qua non* of a well designed, medically oriented anatomy course.

This new book by Bruce Keogh and Stephen Ebbs should help many to bring this proper perspective into their anatomical studies. The reader will soon discover that the well-illustrated text is not limited to the confines often attributed to the topic of 'surface anatomy'. With full use of illustrations produced with the aid of modern imaging techniques, the gap that so often existed between the study of human morphology as observed in the dead on the one hand, and the exploration of the appearances and structure of the living in clinical examination on the other, has been very effectively bridged. In fact, a proper understanding of clinical phenomena and the proper application of basic clinical skills can only be aided by the use of this book.

It is a matter of some satisfaction to me that the authors' enthusiasm for this subject was fired during the period they spent as Demonstrators of Anatomy in my Department. This Department has always had a tradition of emphasising 'living anatomy' and this orientation is due to the vision and enthusiasm of my predecessor in the Chair of Anatomy at this School, William J Hamilton, who did so much to pioneer and foster this approach to anatomy.

This book, produced by dedicated young teachers, should appeal to and help a wide range of students, both medical and paramedical. It deserves to do well and I commend it warmly.

T W Glenister
Charing Cross Hospital
Medical School
London

1984

Preface

The facts available to the student of topographical anatomy have changed little during the last 20 years. However, in the medically-orientated anatomy curriculum, the emphasis of teaching has shifted towards the clinical application of the subject. This has developed in three ways. First, the 'clinical application' of anatomy, facilitating the understanding of a disease process on anatomical grounds; second, the surface marking of structures on the body wall; and third, the appearance of the structures when they are 'imaged', usually in radiographs. It is on the latter two aspects that this book concentrates. The relevance of this becomes apparent when attempting to relate the topographical anatomy of the cadaver to the living patient.

We hope that this book will prove of use not only to medical students, but also to doctors studying for postgraduate qualifications, physiotherapists and other paramedical students of anatomy.

Bruce E. Keogh
Stephen R. Ebbs

1984

Contents

1. Measurements

Adult man

Hand span	10″	25 cm
Hand breadth	4″	10 cm
Thumb breadth	1″	2.5 cm
Finger breadth	$\frac{3}{4}$″	2 cm

PHYSIQUE

The physique of an individual can be classified into one of the following groups:

1 Hypersthenic (pyknic). The subject has a short wide trunk. The ribs lie near the horizontal so the sternum is at a high level and the subcostal angle is wide. The heart is broad and squat. The abdomen is broad above, with the stomach and transverse colon lying largely in the upper abdomen, both lying more transversely than in other types.

2 Sthenic
3 Hyposthenic } *intermediate grades.*

4 Asthenic (leptosomatic). The trunk is long and narrow with an acute subcostal angle, the ribs lying obliquely. The abdomen is longer than in the hypersthenic, and the transverse colon and stomach descend to a lower level. The asthenic often has decreased tone in the muscular system producing round shoulders and a sagging anterior abdominal wall.

APPROXIMATIONS

The surface markings of all structures are approximations and generalisations. The problems of accuracy become most difficult when considering the contents of the abdomen. Age, sex and posture all have marked effects, as do respiratory movements. More subtle changes occur as a consequence of variable tone in the musculature of the diaphragm, abdominal wall and the viscera.

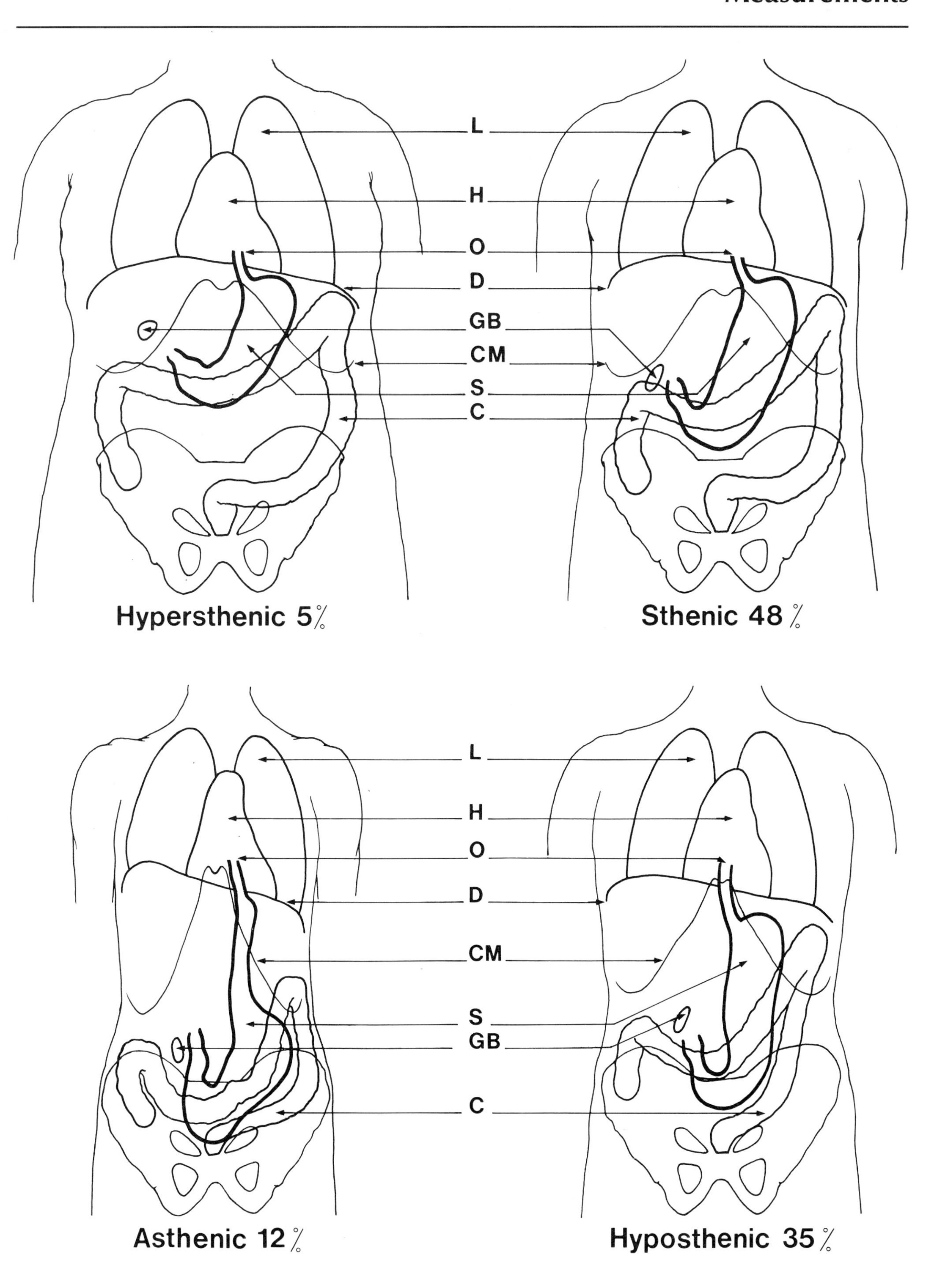

Physiques

L – Lungs. H – Heart. O – Oesophagus. D – Diaphragm.
CM – Costal margin. GB – Gall bladder. S – Stomach. C – Colon.

2. The Skin

The skin, or integument, is arguably the largest organ of the body. It covers the whole body and is continuous with the mucosal surfaces of the alimentary, respiratory and urogenital tracts. It also covers the lateral aspect of the tympanic membrane and, in modified form, constitutes the conjunctiva of the eye. The functions of the skin are myriad, but all relate to the fact that the skin forms the major interface between the individual and his environment.

FUNCTIONS

Protective. Defence against mechanical, thermal, osmotic and chemical trauma.

Immunological. It provides an effective physical barrier against microorganisms.

Thermostatic. Regulation of the blood supply to the skin, combined with control of the rate of loss of heat due to sweat production, enables precise control of heat loss.

Sensory surface. The skin contains a wide variety of somatic sensory nerve endings.

Metabolic. The vitamin D precursor 7-dehydrocholesterol is converted to cholecalciferol in the skin under the influence of ultraviolet light.

Social communication. The skin plays a vital role in non-verbal communication. Subtle, unconscious changes in either skin colour or contour can convey unmistakable messages. Vascular changes in the dermis affect skin colour according to emotion, and dermal or subdermal muscular effects may alter skin contour, e.g. frowning, smiling, goose-pimples. In addition, the skin secretes pheromones for the purposes of sexual attraction.

STRUCTURE

Skin consists of two main layers—the dermis and epidermis. The predominance of one or other layer markedly affects skin texture. The skin of the flexor surfaces has a thinner dermis and is therefore softer than extensor skin. It also tends to be more sensitive and has fewer hairs than the skin of extensor surfaces. Palmar and plantar skin, in addition to being bound down firmly to underlying structures for the purposes of grip, also have a greatly thickened epidermis for mechanical protection. However, the skin on the dorsum of the hand or foot is less bound down and may easily be pinched away from underlying structures.

Lines of Cleavage

The dermis contains multiple bundles of collagen and elastic tissue arranged in specific directions. If a conical object is driven through human skin, the bundles separate longitudinally and return together when the object is withdrawn. Thus, any incision made along the direction of the connective tissue bundles will be less damaging to the dermis and result in a more sightly scar than a disruptive transversely directed incision.

The lines of direction of the dermal connective tissue were plotted by Langer in 1861 and are shown in the diagram.

PIGMENTATION

Skin colour is due to a combination of five main pigments—melanin, melanoid, carotene, haemoglobin and oxyhaemoglobin. Melanin and melanoid are brown, carotene is yellow to orange, haemoglobin is purple and oxyhaemoglobin is red. The latter two are effective by virtue of the fact that they circulate through the superficial venous plexuses of the skin. Thus, changes in vascularity and/or oxygenation, either locally or generally, can alter skin colour, e.g. the pallor of fear or the dusky red complexion of anger. The quantity of melanin, melanoid and carotene varies genetically between individuals, and topographically and chronologically within the same individual. The degree of melanisation in any area depends on the number and/or activity of the local melanoblasts. The number of melanoblasts varies from 800–2000/mm^3, with maximum numbers in the mucus membranes, penis, face and limbs. Darker races have the same number of melanocytes but they produce more pigment. The dark pigment serves to protect the dermis and the rapidly mitosing basal layer of the epidermis from the harmful effects of excessive ultraviolet radiation. If normal skin is exposed to u.v. radiation, the melanin already present darkens and, later, the production of melanin is increased. Over a prolonged period there is an increase in the number of melanocytes.

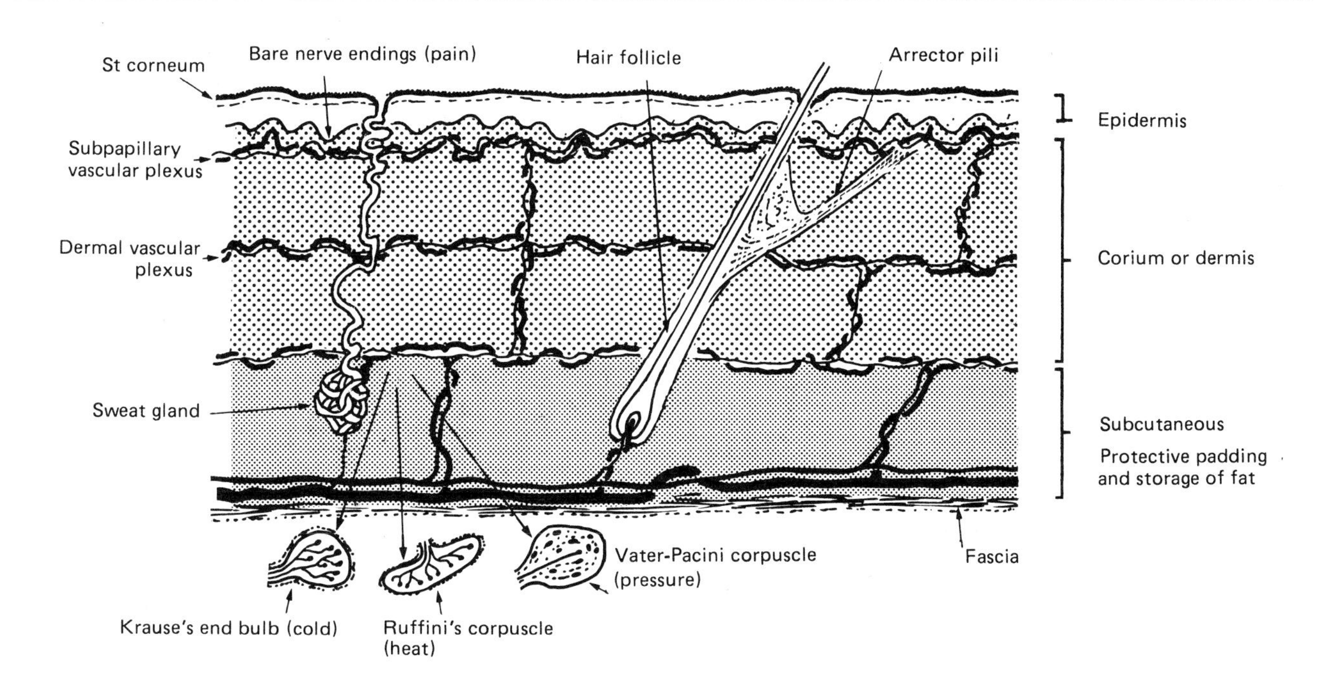

Cleavage lines

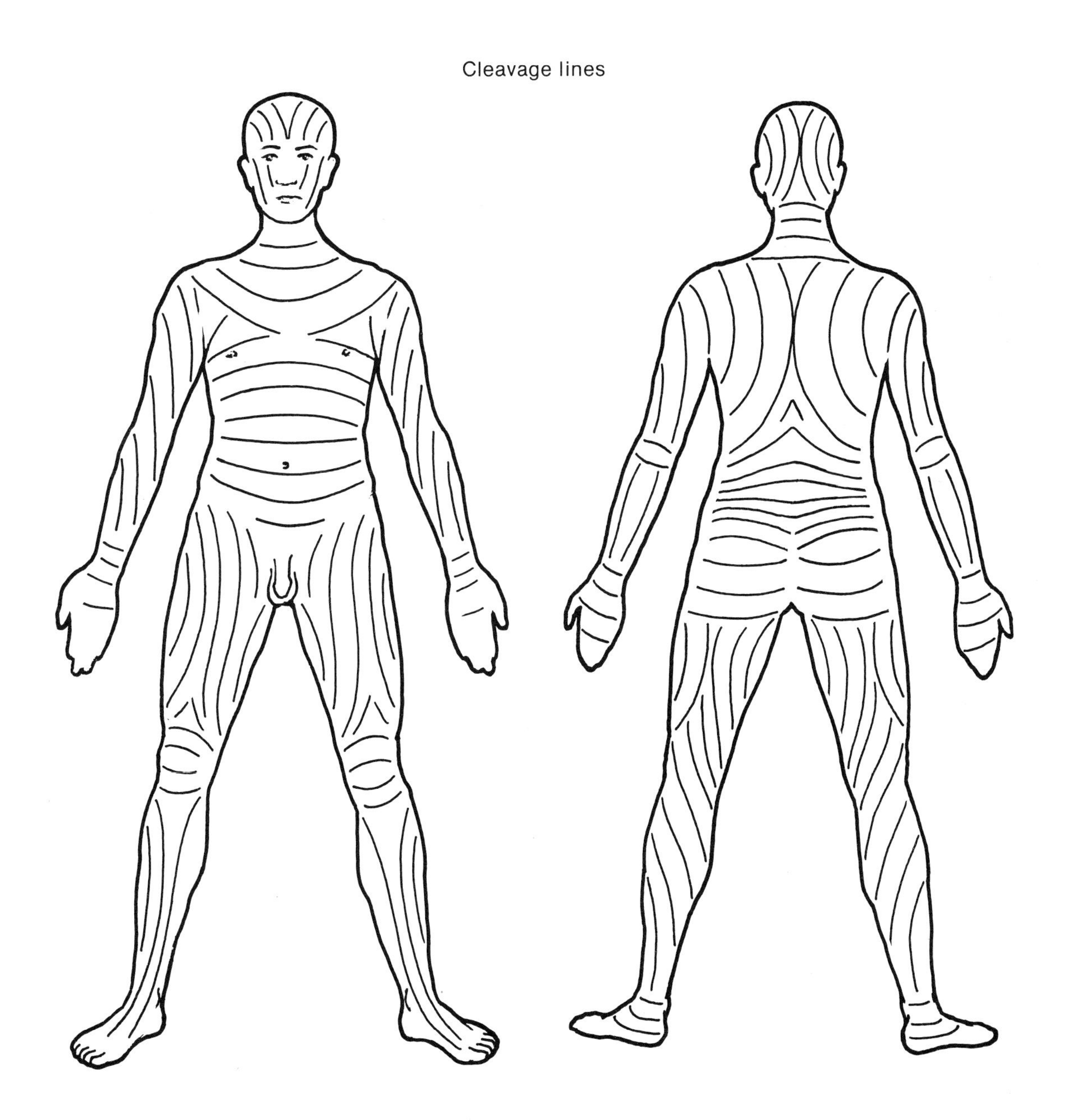

3. The Hair

DISTRIBUTION

Hair is present over most of the body except the palms and the soles, the umbilicus, the glans penis, the inner surface of the prepuce and the inner surfaces of the clitoris and labia majora and minora.

At puberty, several differences in the hair pattern between the sexes emerge. The male hair pattern is initiated with the appearance of pubic hair followed around one year later by axillary and facial hair and then by chest hair in some cases, and frontoparietal recessions in scalp hair. The female does not develop facial or chest hair and the pattern of pubic hair is different. Pubic hair in the male extends in a triangular fashion to the umbilicus. In the female, however, the hair lies in the form of an inverted triangle with the horizontal base lying approximately level with a line passing through both anterior superior iliac spines. In addition, male pubic hair often extends back into the gluteal cleft, but this is not the case in the female.

HAIR GROWTH

In mid-fetal life, almost the entire body is covered with fine, downy hair known as *lanugo* or primary hair. By birth this hair has been shed and replaced by *vellus* or secondary hair which remains over most areas of the body, but is replaced by *terminal hair* over the scalp, in the axillae, over the pubic regions and, in the male, over the face and chest.

Hair growth is cyclical in nature. The growth phase is followed by a phase of regression when the lower part of the follicle degenerates and the hair becomes detached from the papilla. The hair bulb then passes into a resting phase. When the new growth phase commences, the new hair pushes the old, loose hair out of the follicle. In humans, unlike some mammals, the growth phases of adjacent follicles are asynchronous, so that at any one time there are numerous follicles at all stages of the cycle.

In the scalp the growth phase is 3–4 years with a resting phase of 3–4 months. In other parts of the body the growth phase is measured in months with long resting periods.

Hair growth appears to vary with texture. Thus fine hair grows at about 1.5 mm per week and coarse hair at about 2.2 mm per week. Greying hair is due to both air bubbles in the hair shaft and diminished production of melanin at the hair bulb.

CLASSIFICATION

Hair type varies considerably between the races and has been classified as follows:

Heliotrichous. Helical or spiral hair as found in Negroids. In helical hair the loops are of constant diameter, whereas in spiral hair the loops diminish in diameter from the scalp outwards.

Leiotrichous. This is straight hair as seen in the Mongoloid races.

Cynotrichous. This hair may be straight, curly or wavy as seen in Caucasians.

Hair

Heliotrichous

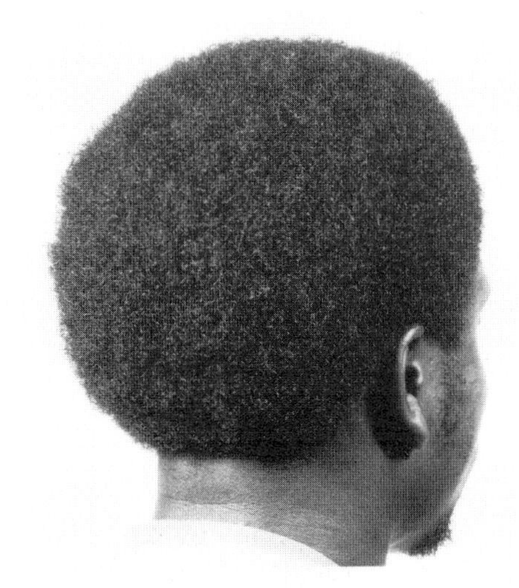

Leiotrichous

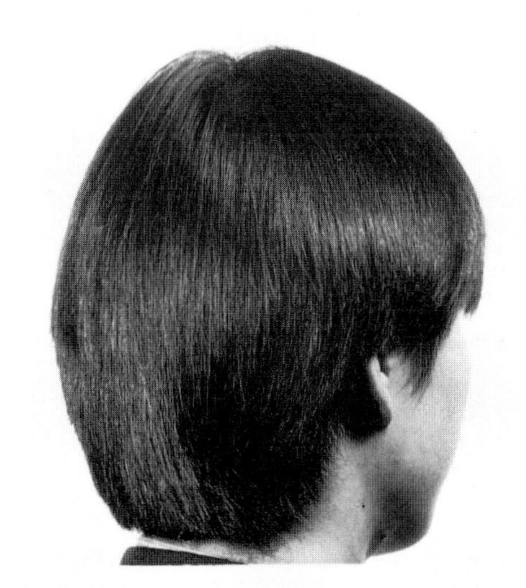

Cynotrichous

4. The Sweat Glands and Nails

SWEAT GLANDS

Sweat glands are simple tubular glands concerned with temperature control. The evaporation of sweat from the body surface extracts the latent heat of evaporation from the skin and thus helps to cool the body surface. In the human there are two types of sweat glands, the eccrine gland producing a watery secretion and the apocrine gland producing a thicker secretion. It is now thought that both types of gland secrete by the merocrine method.

Eccrine Glands

These glands are larger in areas of profuse sweat production such as the axillae and groin. There is also a greater number of glands, regularly spaced, in the skin of the palms and soles. There are relatively few glands over the neck and back.

Apocrine Glands

These are found mainly in the axilla, circumanal and genital region, the eyelids and areola. They are closely associated with hair and their ducts often open into the distal ends of hair follicles.

In females, these glands undergo involutional changes during the menstrual cycle. They are modified in the external auditory meatus to produce ceruminal wax.

NAILS

The nail plate [1] is a semitransparent, elastic, keratin plate.

The nail root [2] is the proximal part of the nail plate.

The nail fold [3] is a fold of skin overlapping the nail root.

The eponychium or cuticle [4] is a distal prolongation of the stratum corneum of the nail fold.

The nail wall [5] is a fold of skin overlapping each lateral nail border.

The nail bed [6] is a modified epidermis lacking the stratum granulosum. Proximally, beneath the root of the nail, it is concerned with nail growth. Distally, it merely provides a surface for the growing nail to glide over. Beneath the nail the corium is thick and vascular. It is thrown into longitudinal folds which are responsible for the longitudinal ridges seen on the nail plate. The vascularity of the corium accounts for the pink colour of the nail bed which may become pale in anaemia. Near the root, the longitudinal ridges are less prominent, the vascularity is reduced and the nail tissue is more opaque producing the *lunule.*

The lunule [7] is most obvious in the thumb but becomes progressively less visible until in the little finger it is entirely hidden beneath the nail fold.

Nails grow faster in the fingers than the toes—the replacement time being around six months in the fingers and one year to 18 months in the toes, a fact to be borne in mind when treating fungal infections of the nails with systemic agents.

In the hand, growth is quickest in the longest digit and slowest in the little finger. Growth is also faster in summer than in winter.

Sweat glands

Eccrine

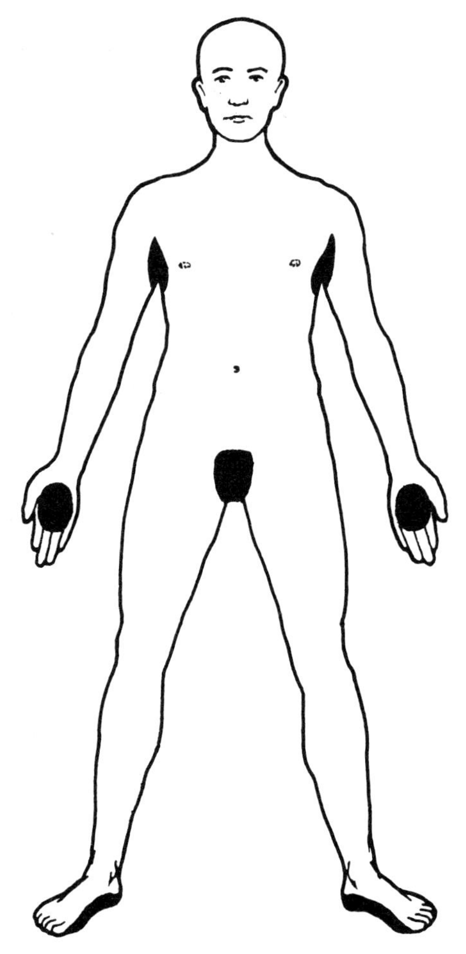

Axillae
Groin
Palms and soles

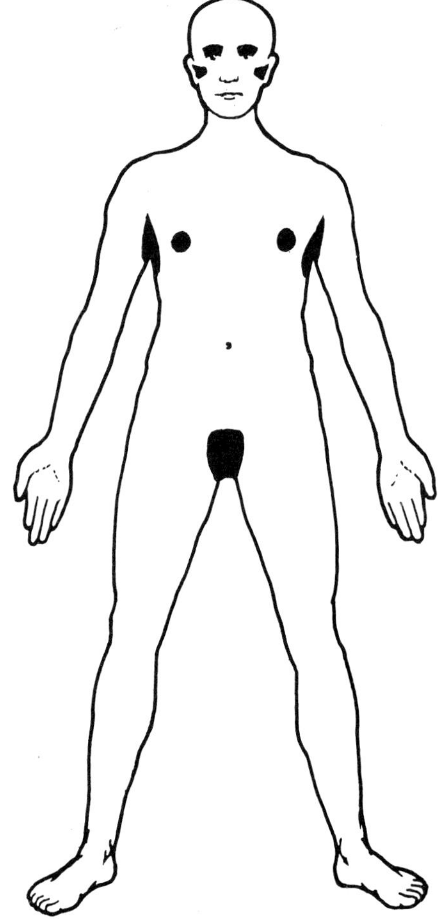

Eyelids
External acoustic meatus
Axillae
Areolae
Groin

Nails

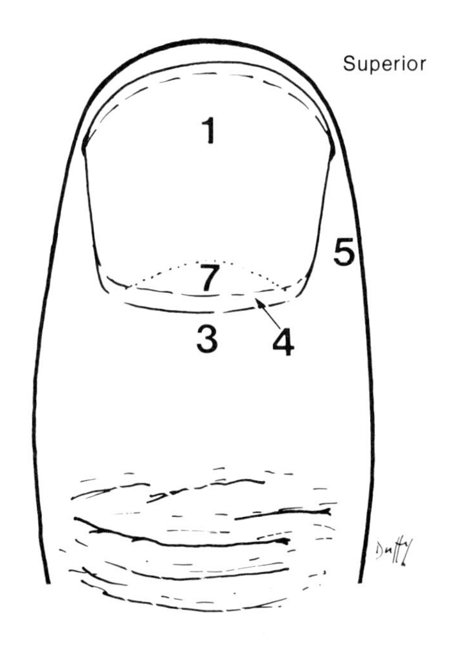

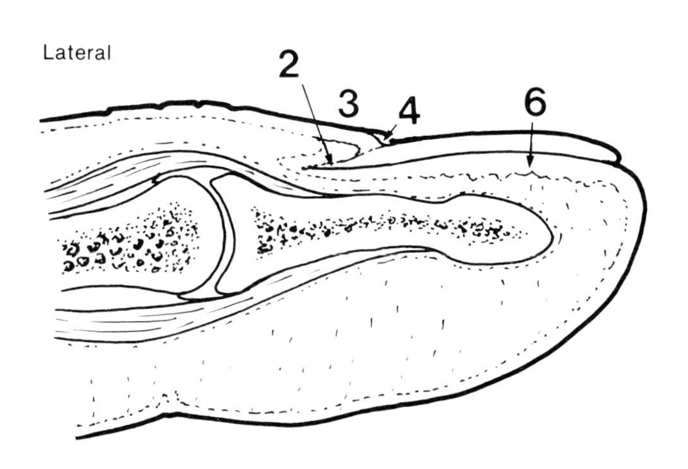

5. The Back

SURFACE FEATURES

Median furrow [**1**]. This runs from the external occipital protuberance to the natal cleft. It is shallow in the lower cervical region and deep in the mid-lumbar region. Inferiorly, it widens into a triangle whose basal angles are the dimples which overlie the posterior superior iliac spines, and the apex of which lies at the upper end of the natal cleft, over the third sacral spinous tubercle.

Splenius capitis [**2**]. The muscle which floors the apex of the posterior triangle of the neck. Its bulk becomes most obvious when the neck is extended against resistance.

Trapezius [**3**]. A flat triangular muscle which, with its partner, produces a trapezium-like shape with the lateral angles at the shoulders, the superior angle at the external occipital protuberance, and the inferior angle at the spine of T12.

Teres major [**4**]. A thick band of muscle between the inferior angle of the scapula and the humerus which, with the latissimus dorsi, forms the posterior wall of the axilla.

Latissimus dorsi [**5**]. The lower, or lateral border, produces an elevation along a line drawn from the middle of the iliac crest to the posterior fold of the axilla; this can be demonstrated when the arm is adducted against resistance.

Sacrospinalis [**6**]. Part of the erector spinae group of muscles, it extends for one hand's breadth on either side of the midline between the iliac crest and the twelfth rib. The bulk of the muscle may be demonstrated by extending the vertebral column against resistance.

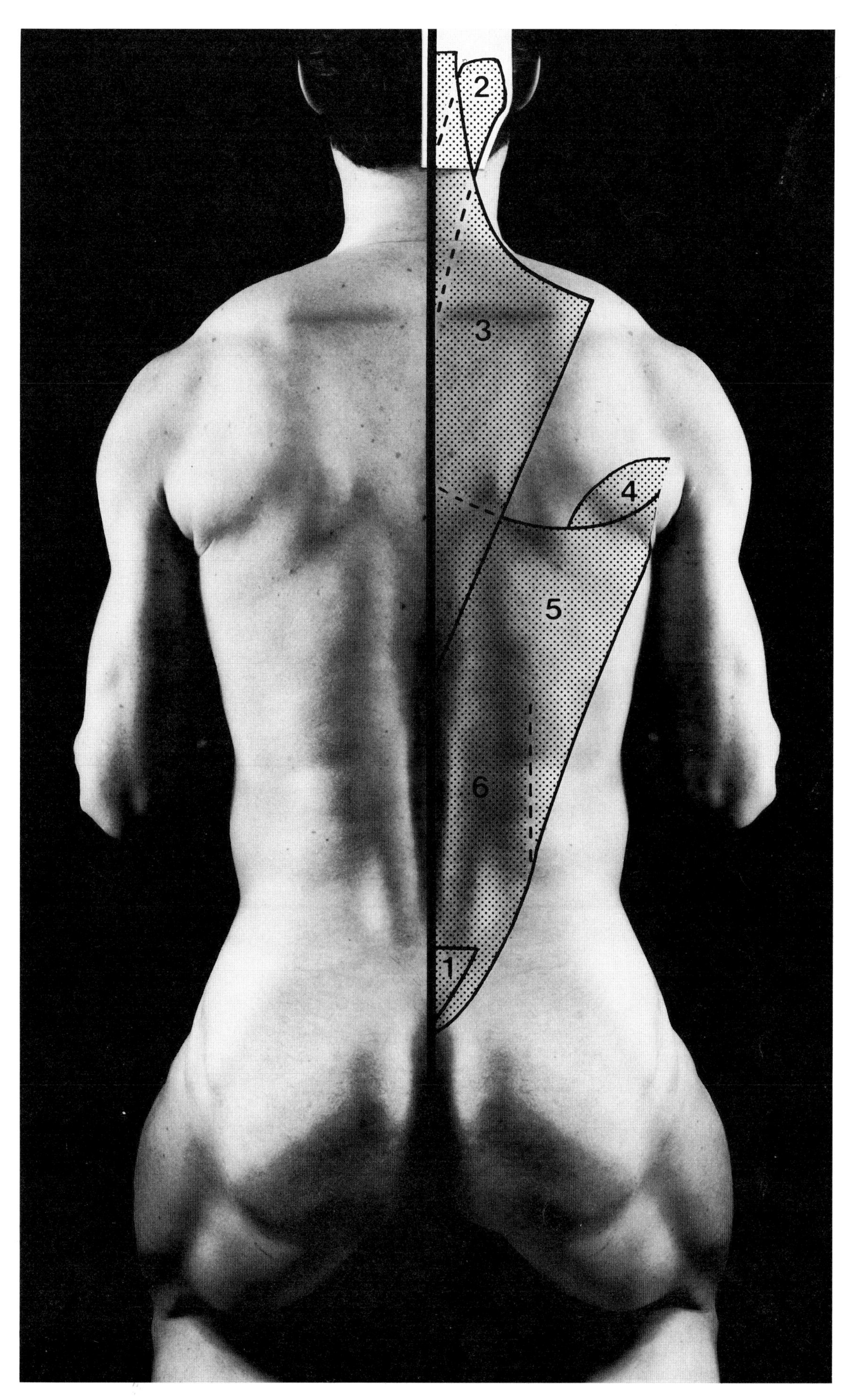

Surface features of the back

BONY LANDMARKS

Cervical

The transverse process of the atlas [**1**] can be felt in the hollow below the external ear.

The spine of C2 [**2**] is the first bony prominence encountered when the finger is drawn downwards in the midline. This lies 5–7 cm below the external occipital protuberance.

The spine of C7 [**3**] is the upper of the two prominences found in the root of the neck, the lower being the spine of T1.

The tips of the cervical transverse processes lie approximately 2.5 cm from the midline.

Thoracic

The thoracic spines slope downwards so that their tips lie at a lower level than the corresponding bodies. The spines can usually be felt but not seen.

In the upper and lower thoracic regions, the tips lie opposite the upper part of the body of the vertebra below; in the mid-thoracic region they lie opposite the lower part of the vertebra below.

The spine of T1 [**4**] is the lower of the two prominences in the root of the neck.

The spine of T3 [**5**] is at approximately the same level as the spine of the scapula.

The spine of T7 [**6**] is approximately opposite the inferior angle of the scapula.

The spine of T12 [**7**] lies about midway between the level of the inferior angle of the scapula and the level of the upper border of the iliac crest.

Lumbar

The tip of the spine of each lumbar vertebra lies opposite the lower part of its own body.

The supracristal plane [**8**] is the plane passing through the summits of the iliac crests and between the spines of L3 and L4.

The space between the laminae of L4 and L5 [**9**] lies about 2.5 cm below this plane. Here the subarachnoid space lies 5 cm from the skin in an adult, but only 2.5 cm in a child.

The tips of the lumbar transverse processes lie about 5 cm from the midline.

Sacral

The sacral spines are variably fused to form the median sacral crest which is raised into four spinous tubercles.

The second spinous tubercle lies above the lower ends of the dura, arachnoid, subdural and subarachnoid spaces, and is midway between the posterior superior iliac spines.

The posterior superior iliac spines [**10**] are indicated by dimples above and medial to the buttock, two or three finger breadths from the midline.

The third spinous tubercle [**11**] lies at the upper end of the natal cleft and corresponds to the level of the upper end of the rectum.

The posterior sacral foraminae are approximately 2.5 cm from the median plane and 2.5 cm apart.

Coccyx

The coccyx lies deep to the natal cleft with its lower end 1 cm from the anus.

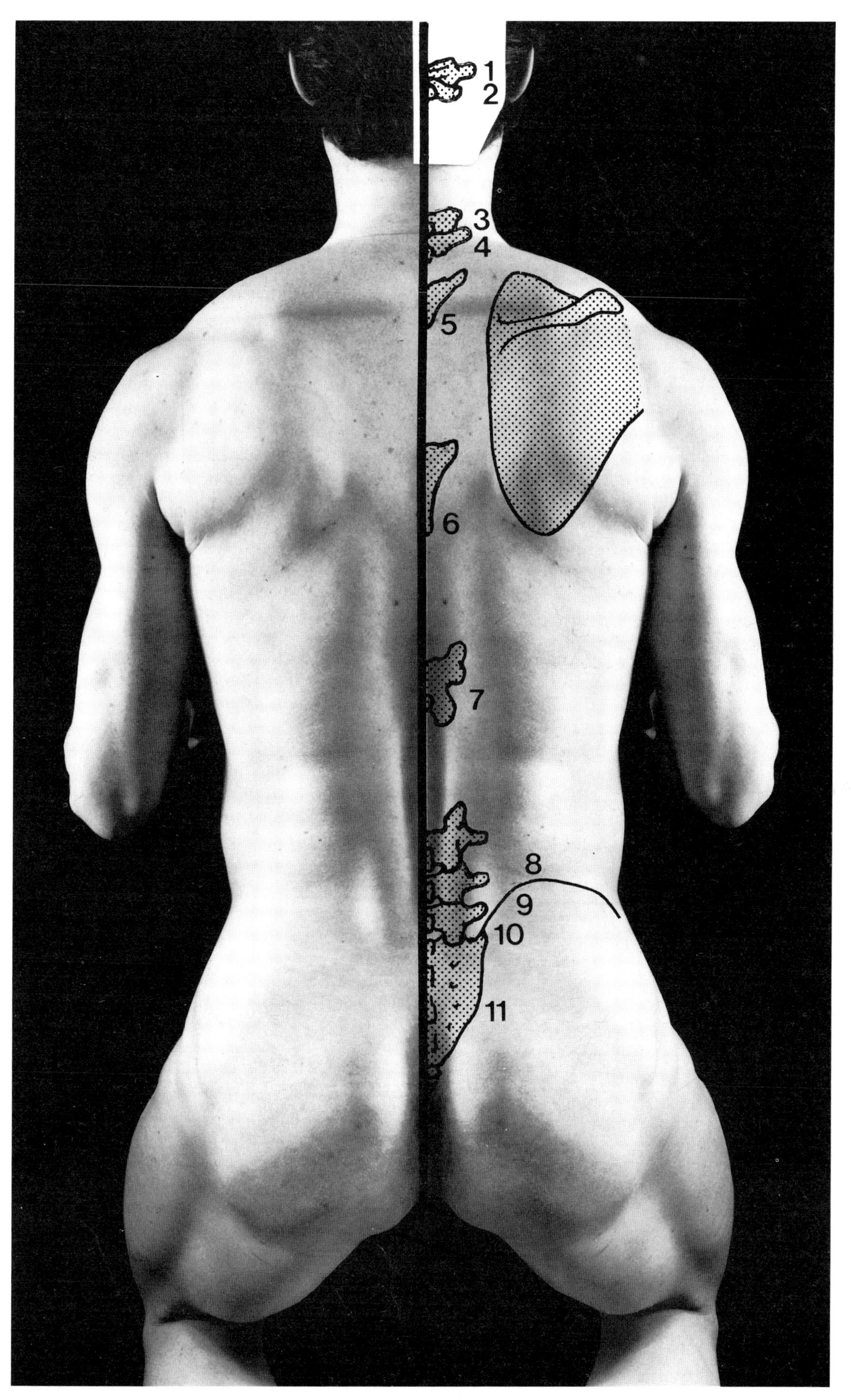

Bony landmarks of the back

CURVES

Sagittal Curves

Cervical [**1**]. This curve is convex forward and extends from the first cervical vertebra to the second thoracic. This develops as the child holds its head up. The most anterior vertebra of this curve is C6.

Thoracic [**2**]. This is concave forward and extends between T2 and T12. It is part of the primary curve, being present *in utero* when the fetus assumes a C-shaped posture. The upper portion of the curve may be exaggerated to produce the condition of 'round shoulders'. The most posterior vertebra of the thoracic curve may be T8, T9 or T10.

Lumbar [**3**]. Convex forward, it extends from T12 to the lumbosacral prominence (where the anterior surface of the fifth lumbar vertebra meets the anterior surface of the sacrum). The curve is more pronounced in women. The most anterior vertebra is L4.

Pelvic [**4**]. The anterior surfaces of the sacrum and coccyx are generally concave and face forwards and downwards. This is part of the primary curve.

Lateral Curve [5]

The upper portion of the thoracic curve usually has a slight convexity towards the hand most commonly used by the subject.

Abnormal Curves

Kyphosis. An excessive curvature of the vertebral column, convex posteriorly, producing, for example, 'round shoulders'.

Lordosis. An excessive curvature of the vertebral column, convex anteriorly, most often seen as an exaggerated lumbar curve.

Scoliosis. An abnormal lateral curvature of the vertebral column, convex to the left or right, usually in the thoracic region.

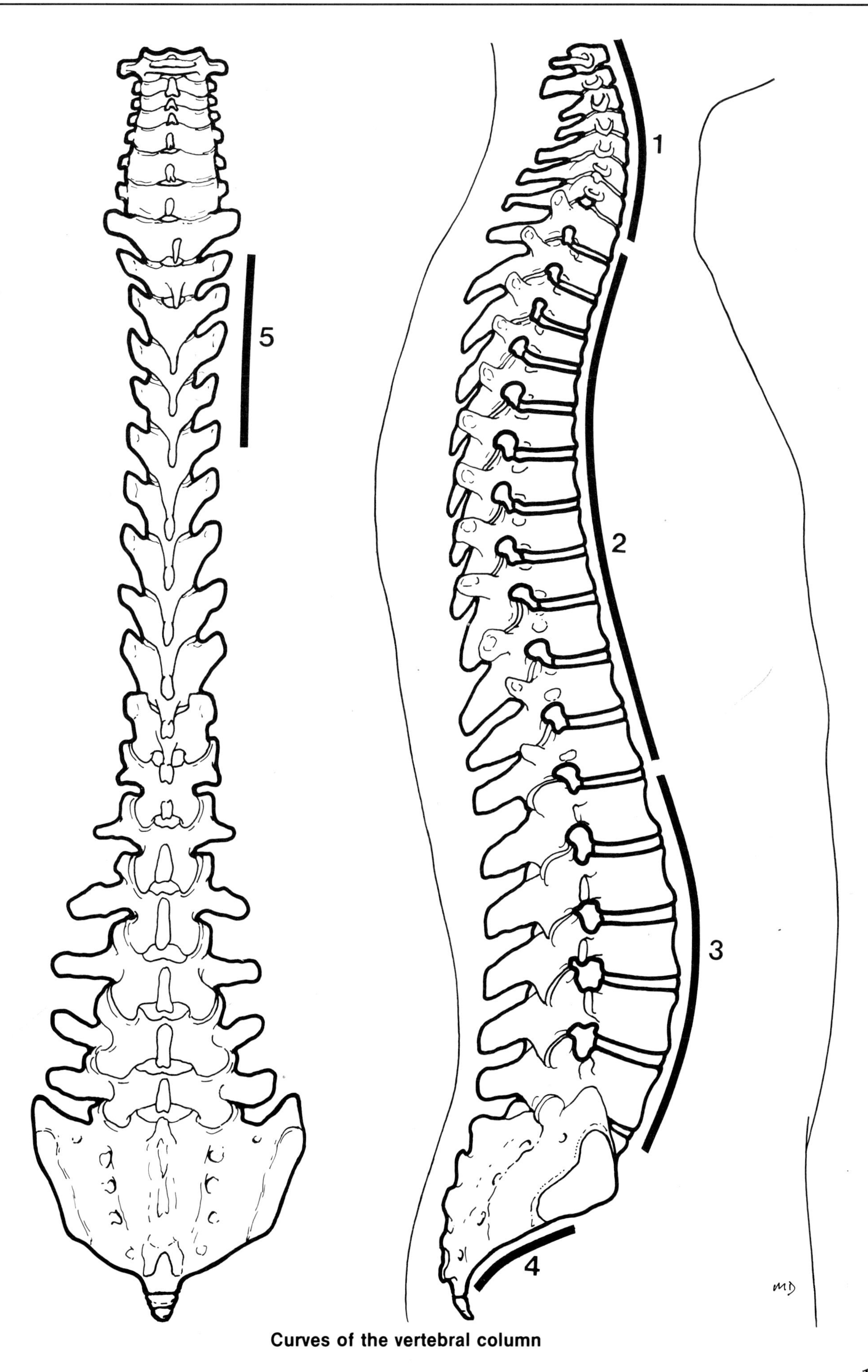

Curves of the vertebral column

SPINAL CORD

The cord starts at the foramen magnum which, in the flexed neck, lies midway between the palpable bony prominences of the external occipital protuberance and the spine of C2. In the child, it ends opposite the space between the spines of L3 and L4 and, in the adult, opposite the upper border of the spine of L2.

The nerve roots are attached 0.25 cm from the midline.

Roots	*Lie between*
C1–C8	Foramen magnum and C6 spine
T1–T6	C6–T4 spines
T6–T12	T4–T9 spines
L,S,C	T10–L1 spines

The enlargements of the cord can also be marked.

	Nerve level	*Vertebral level*	*Spinous process level*
Cervical enlargement	C5–T1	C4–C7	C3–C6
Lumbar enlargement	L2–S2	T10–L1	T9–T12

The sympathetic trunks and ganglia lie approximately 2.5 cm from the midline.

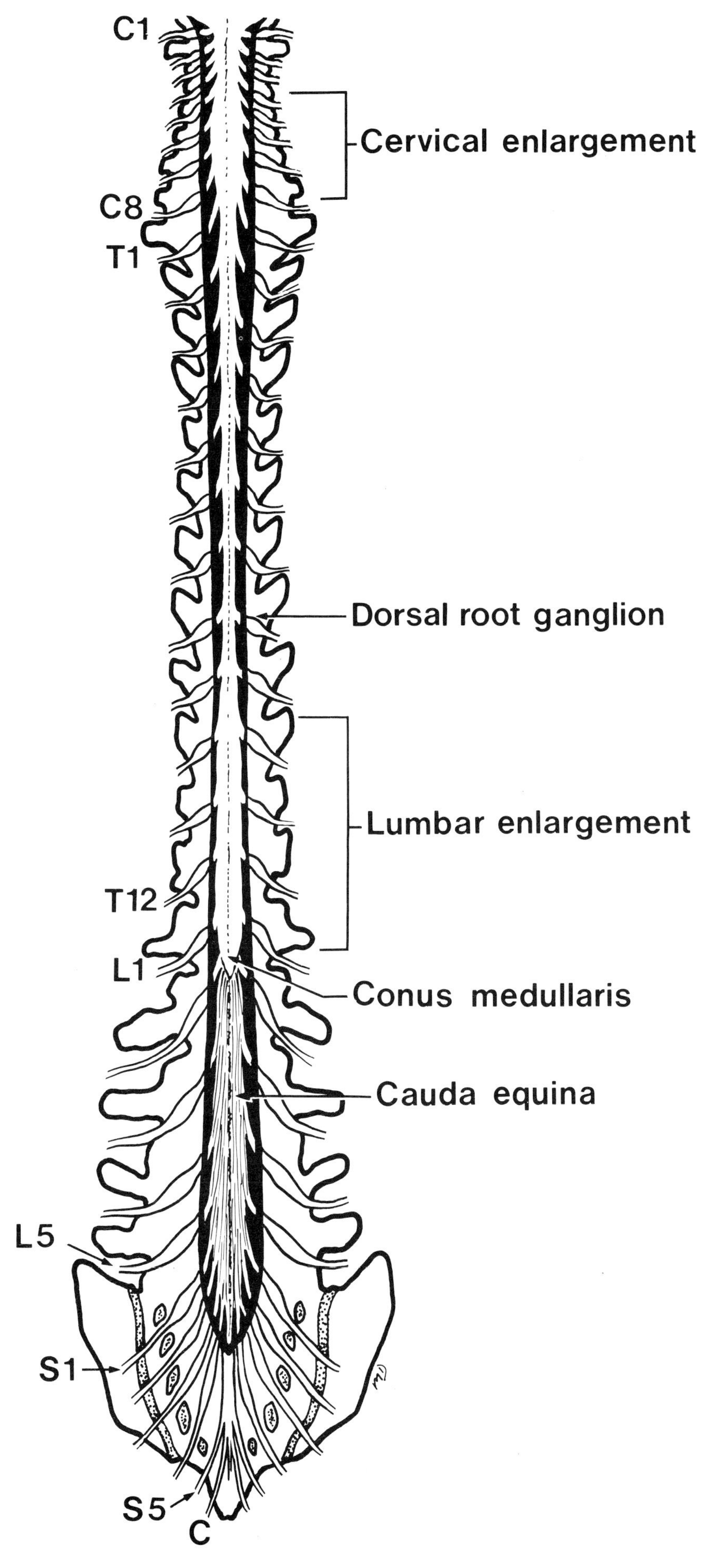

The spinal cord

RADIOLOGY OF THE BACK

The Vertebral Column

The bodies of the vertebrae appear as rectangles on a radiograph. They are mainly composed of cancellous bone whose trabeculated arrangement can be seen. The most obvious trabeculae run vertically to meet the stress of the body weight. The bodies are outlined by a cover of cortical bone approximately 1 mm thick. The upper and lower surfaces of the bodies, the vertebral end plates, may be indented by shallow protrusions of the intervertebral discs. These are known as Schmorl's nodes, but are of limited clinical significance.

The intervertebral discs are radiolucent and occupy the spaces between the vertebral end plates. In the cervical and thoracic parts of the spine they are all of similar height, about 5 mm. In the lumbar spine their height increases progressively down the column, except that the disc at the lumbosacral junction is narrower.

The pedicles are best seen in an antero-posterior view, except in the cervical spine where oblique views are necessary. In the thoracic and lumbar regions they produce oval shadows which lie over the lateral aspect of the bodies. An abnormality arising within the confines of the vertebral canal may lead to a widening of the interpeduncular distance.

The transverse processes can usually be recognised without difficulty except in the thoracic region where they are masked by the ribs.

The spinous processes produce central, vertically-aligned ovals when seen in the frontal view. In the lateral view, the size of the vertebral canal is indicated by the distance between the back of the body of the vertebra and the line of cortical bone at the anterior end of the spinous process of the same vertebra.

The sacro-iliac joints are viewed obliquely by the antero-posterior film. The more lateral lines of density represent the anterior and the more medial the posterior part of the joint cavity.

Contrast Studies

Normal nervous tissue is radiolucent, but the outlines of the cord and spinal nerves may be visualised by the introduction of contrast medium into the subarachnoid space.

Examination of the cord in this way is called myelography, and a myelogram may be produced by the use of an oily contrast medium. A water-soluble medium is less irritant and allows the visualisation of the nerve roots in a radiculogram.

Paravertebral Shadows

Abnormalities in the soft tissue shadows lying alongside the vertebral column may occur in a variety of conditions. In the lumbar region the abnormality will, if large enough, displace the psoas opacity. In the thoracic region the shadow of the abnormality is usually the first soft tissue change noted. In the cervical region the pharyngeal air translucency may be pushed forwards.

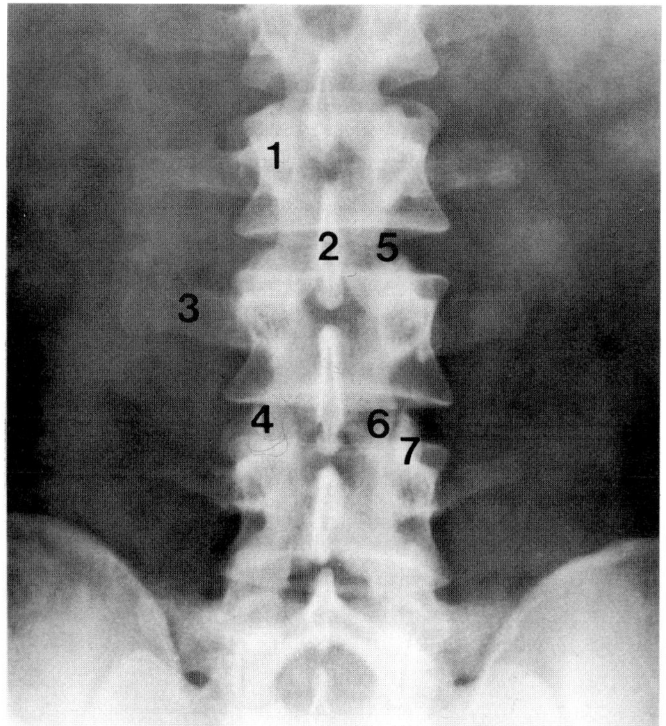

Frontal view

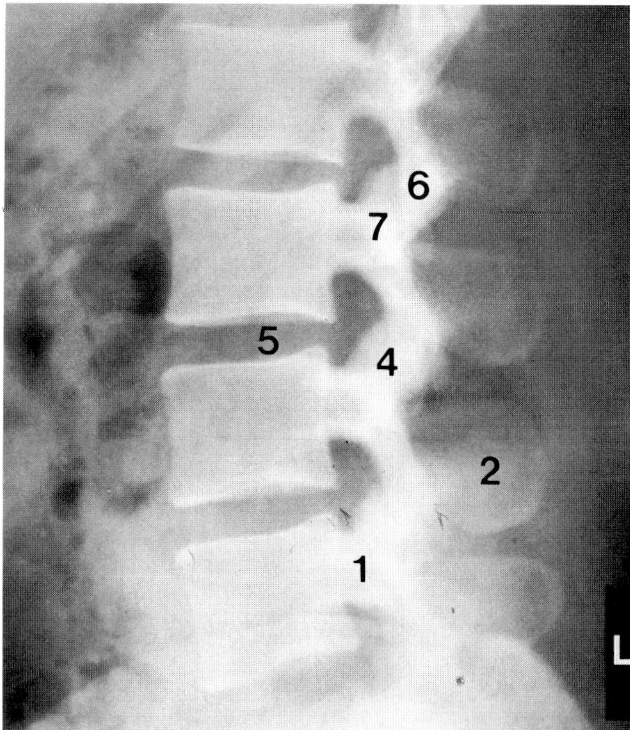

Lateral view

The lumbar spine

1. Pedicle
2. Spinous process
3. Transverse process
4. Apophyseal joint
5. Disc space
6. Inferior articular facet
7. Superior articular facet

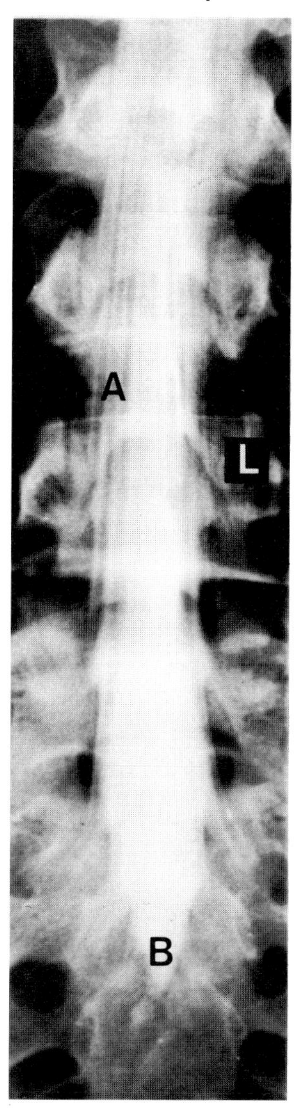

Normal radiculogram

A. Nerve roots forming cauda equina
B. Lower end of subarachnoid space

6. The Thorax

THORACIC WALL

Thoracic Vertebrae

See section on The Back (p. 10).

Sternum

The suprasternal notch [1] lies between the clavicles and the upper border of the manubrium, at the level of the lower border of the body of T2, (or the space between the first and second thoracic spines).

The manubrium [2] measuring about 4 cm from above downwards, lies at the same level as the bodies of T3 and T4, in front of the arch of the aorta.

The sternal angle of Louis [3] which lies 4 cm below the suprasternal notch, level with the lower border of the body of T4 (or the space between the third and fourth thoracic spines), marks the position of the manubriosternal joint.

The body of the sternum [4] is about 10 cm long and occasionally shows evidence of its four developmental segments, or sternebrae, in the form of palpable transverse ridges. It lies in front of the heart and the fifth to eighth thoracic vertebral bodies.

The xiphisternal joint [5] at the level of the body of T9, may sometimes be seen as a transverse ridge at the upper part of the epigastric fossa.

Clavicles

The clavicles [6] produce sinuous horizontal ridges visible at the junction of the neck and thorax.

The infraclavicular fossa [7] lies below the middle third of the clavicle.

Costal Cartilages

The cartilages increase in length from the first to the seventh and then successively shorten. The costochondral junctions lie on a line from a point 5 cm from the midline at the level of the sternal angle to 2.5 cm behind the lowest part of the tenth cartilage.

The first cartilage [8] about 2.5 cm long, lies below the medial end of the clavicle.

The seventh cartilage [9] is the lowest to articulate directly with the sternum.

The upper end of the linea semilunaris [10] at the lateral border of the rectus abdominis, crosses the tip of the *ninth cartilage* about halfway along the costal margin.

The tenth cartilage [11] forms the lowest part of the costal margin when viewed anteriorly, it lies at the level of the body of L3.

Ribs

To number the ribs:
Anteriorly, count downwards from the second costal cartilage which articulates at the level of the sternal angle.
Posteriorly, the medial end of the spine of the scapula overlies the fourth rib; the eighth rib lies just below the inferior angle of the scapula; the eleventh rib is the lowest rib crossed by the lateral margin of latissimus dorsi.

The posterior angles of the ribs are marked by a line drawn from 4 cm lateral to the spine of C7 to where the twelfth rib is crossed by the sacrospinalis.

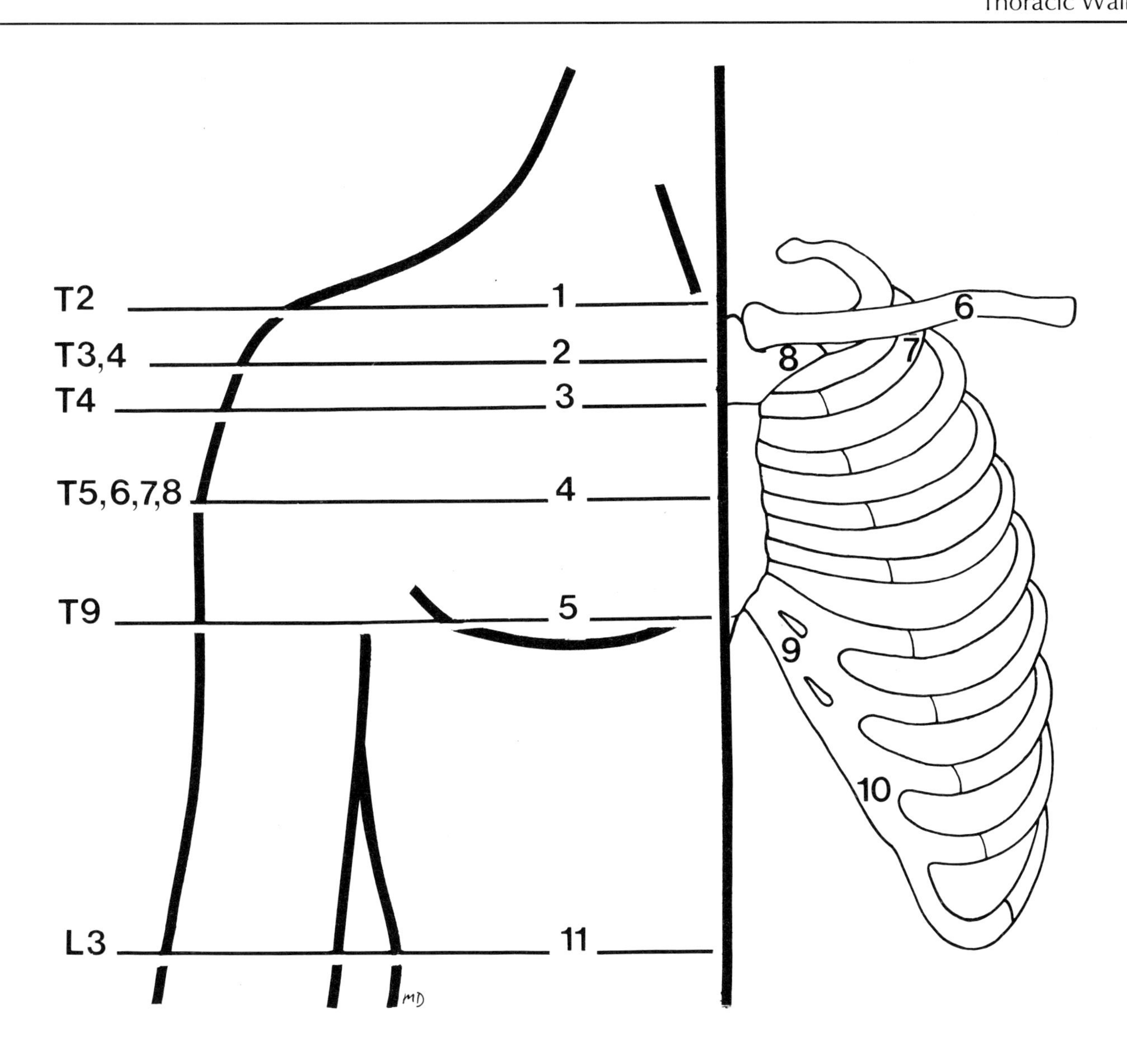

Skeletal elements of the thorax

MUSCLES

Pectoralis major [**1**]. In the male, the lower border forms a curved line leading to the anterior wall of the axilla and acts as a guide to the fifth rib and cartilage. In the female this border is largely masked by the breast.

Serratus anterior [**2**]. Covers the lateral thoracic wall below pectoralis major.

Pectoralis minor [**3**].

Subclavius [**4**].

These muscles are covered in greater detail under the section on the shoulder (*see* p. 116).

Only a narrow strip about 2 cm wide, down the middle of the sternum, is not covered by muscle.

The external intercostals lie between the ribs extending anteriorly as far as the costochondral junctions, where they are replaced by the external intercostal membrane, and posteriorly to the tips of the transverse processes.

The internal intercostals lie between the ribs as far back as the posterior angles where they are replaced by the internal intercostal membrane. Anteriorly they reach the lateral border of the sternum.

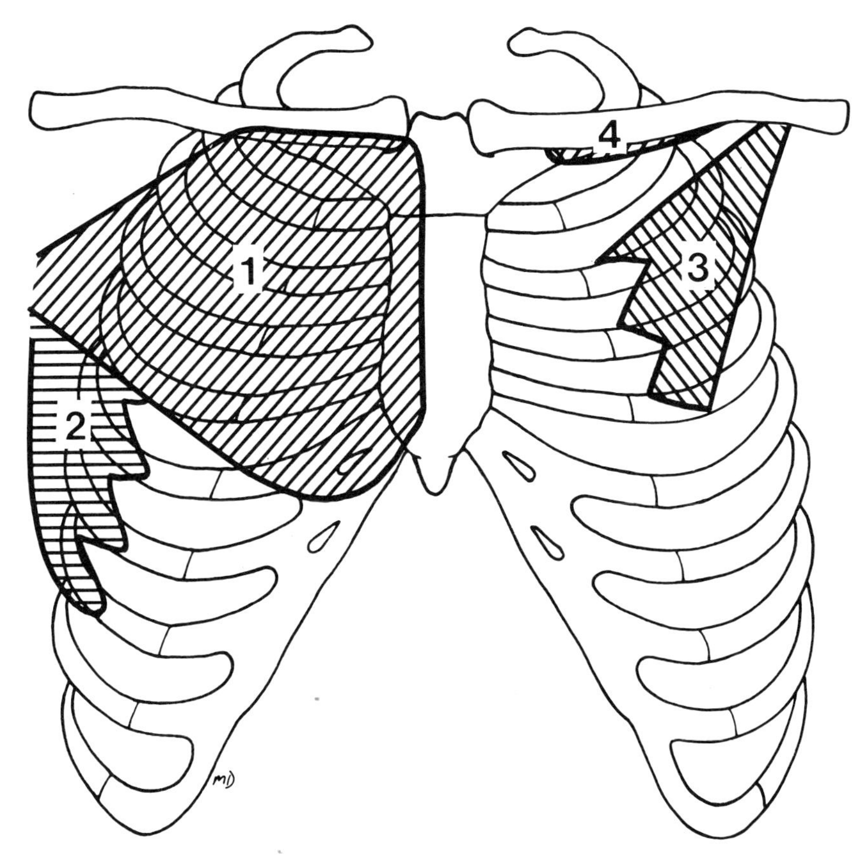

Muscles of the anterior thoracic wall

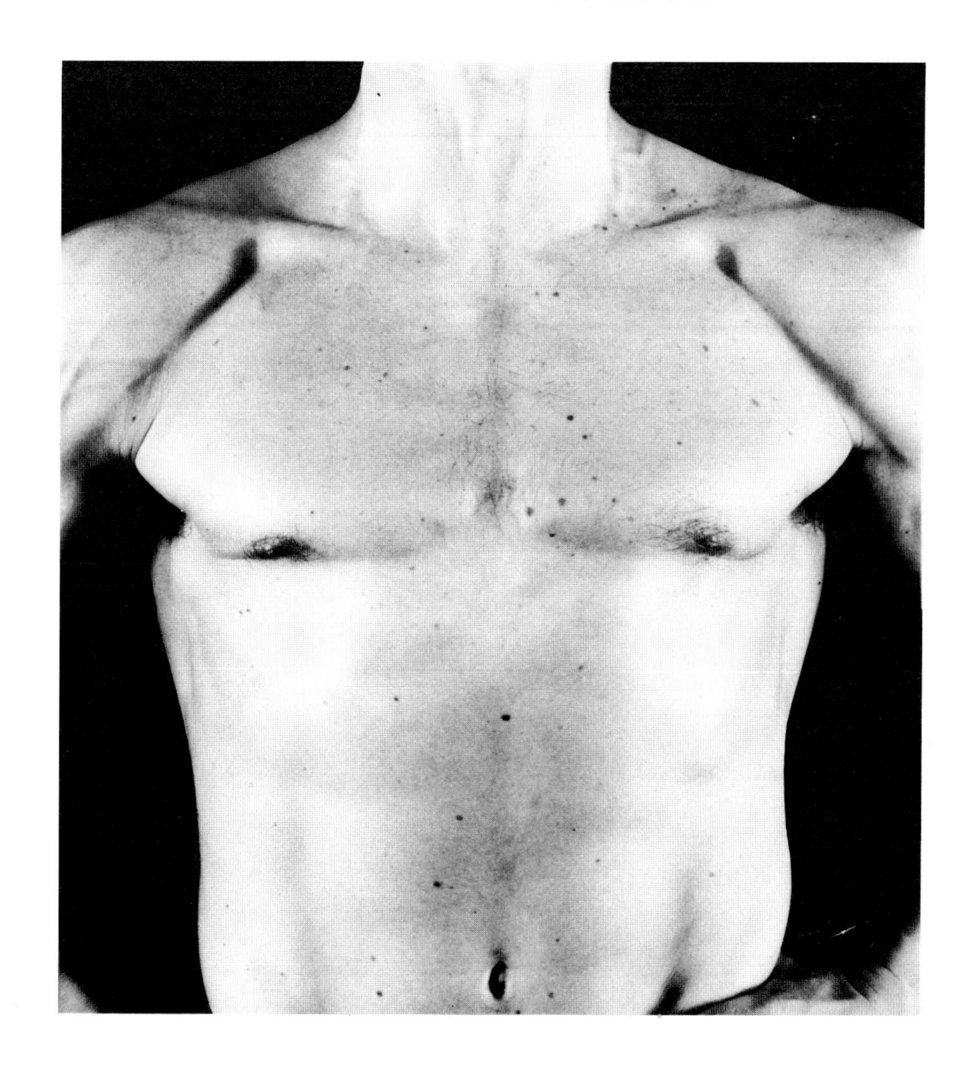

THE BREASTS

The male breast remains poorly developed throughout life. The nipple is usually situated over the fourth intercostal space, 10 cm from the midline.

The female breast develops to a variable degree during and after puberty, but its circular base occupies a fairly constant area over the second to sixth ribs, from the lateral border of the sternum to the mid-axillary line. The upper outer portion of the base is extended into the axilla as the axillary tail.

The nipple [**1**] projects just below the centre of the breast. Approximately 15–20 lactiferous ducts open on its flattened tip.

The areola [**2**] surrounds the nipple. In the nulliparous, it is pink in colour but changes to a dark brown during pregnancy. After lactation the colour lightens but never completely returns to pink.

The sebaceous glands [**3**] of the areola also enlarge during pregnancy to form Montgomery's tubercles.

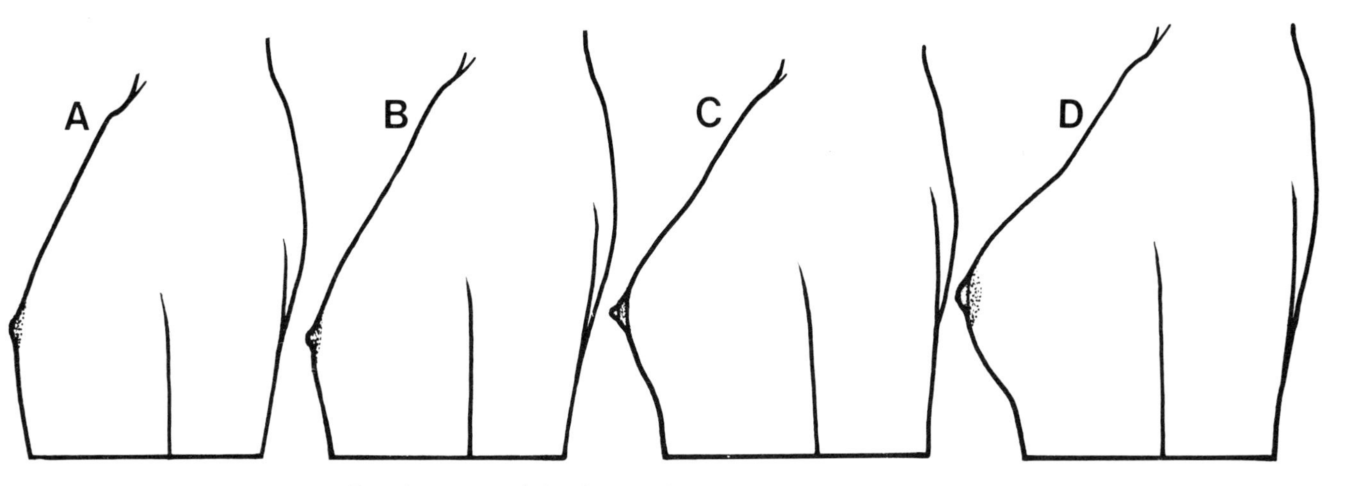

Development of the female breast

A – Pre-puberty, only the nipple is elevated
B – Pubertal development of the breast bud
C – The areola and nipple project above the breast
D – Adult form with projecting nipple and recessed areola

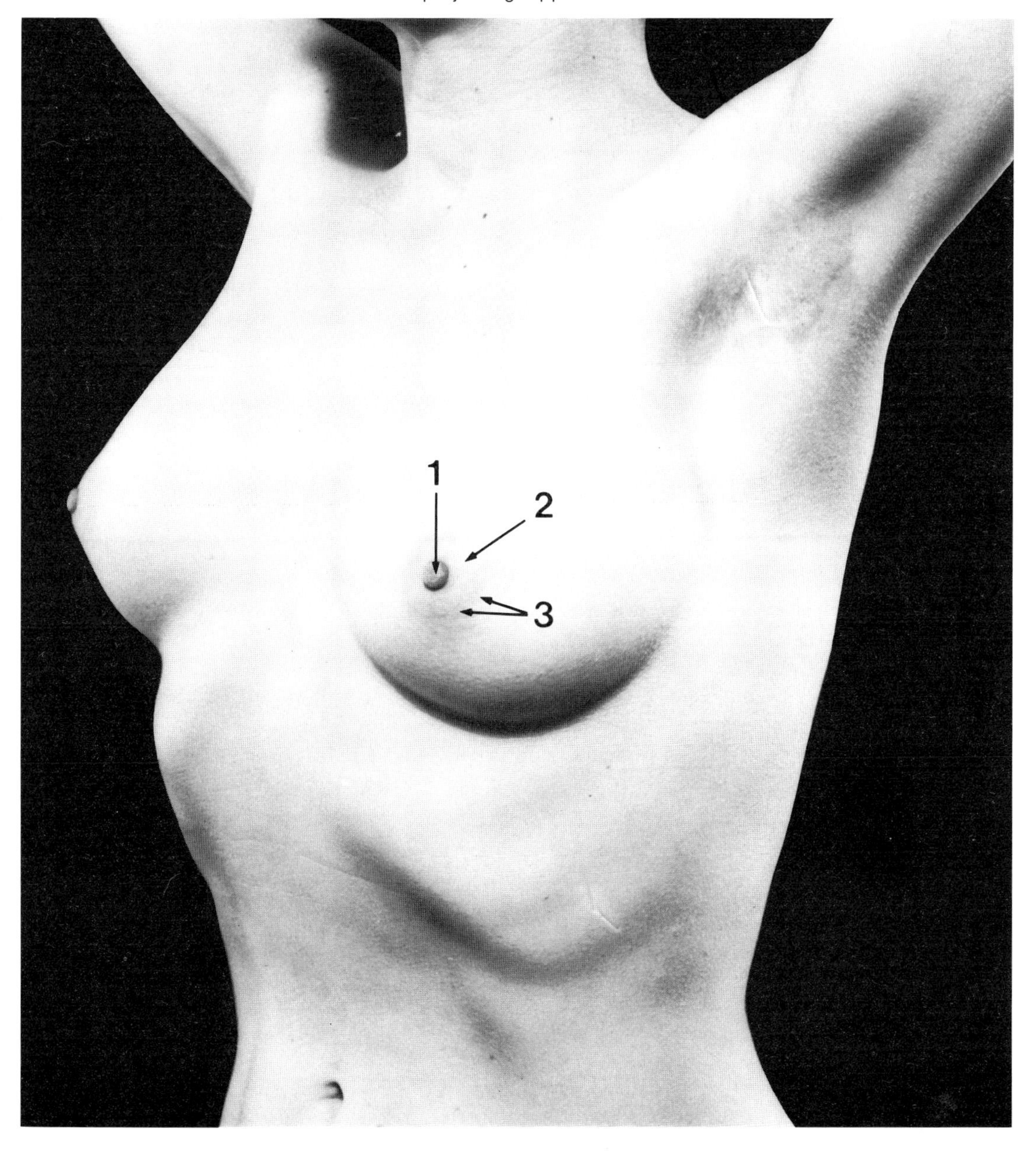

The normal female breast

THE TRACHEA

The trachea begins at the lower border of the cricoid cartilage, at the level of C6. It is approximately 10 cm long with an external diameter of 2 cm in the adult male, lying in the median plane and running almost vertically. A little to the right of the midline it divides into the two main bronchi. The bifurcation, at the level of the upper border of T5 in the cadaver, may be pulled downwards by as much as 2.5 cm during inspiration in the living.

THE DIAPHRAGM

The dome-shaped diaphragm separates the thorax from the abdomen. It extends from the attachment of the right crus, which reaches as far down as the disc between L3 and L4, to the highest point on the right dome which, in deep inspiration, is at the level of the body of T10.

The levels at which structures pass through or around the diaphragm are clinically important.

The aperture for the inferior vena cava lies at the level of the disc between T8 and T9, about 2.5 cm to the right of the midline. The inferior vena cava is accompanied by branches of the right phrenic nerve.

The oesophageal aperture is at the level of the body of T10. It also transmits the anterior and posterior vagal trunks, the oesophageal branches of the left gastric artery and the communications between the oesophageal and gastric veins. It lies less than 2.5 cm to the left of the midline.

The aortic aperture lies in front of the lower part of the body of T12, just to the left of the midline. The thoracic duct also runs through it. The azygos vein may enter the thorax through a variety of routes including the aortic aperture, but often runs lateral to or through the right crus. The left ascending lumbar vein leaves the abdomen in a similar fashion, usually passing through the left crus.

The branches of the left phrenic nerve pierce the left dome in a position comparable to the opening for the inferior vena cava.

The sympathetic trunks run behind the medial arcuate ligament.

The three splanchnic nerves all pierce the crura.

The subcostal nerves and vessels run in the subcostal groove of the twelfth rib behind the lateral arcuate ligaments.

The seventh to eleventh intercostal nerves and vessels pass between the digitations of the diaphragm as they run around the body wall.

The superior epigastric vessels run between the sternal and costal attachments of the diaphragm to enter the rectus sheath.

THE PHRENIC NERVES

The course of the phrenic nerve in the neck can be represented on either side by a line starting on the sternomastoid at the level of the laryngeal prominence, and running down to reach the clavicle between the clavicular and sternal heads of that muscle.

In the thorax, the course differs on the two sides and the surface marking is complex. It may be described by reference to the surface markings of those structures which relate to the nerves.
On the right side these are:

- the right brachiocephalic vein
- the superior vena cava
- the right border of the pericardium
- the terminal part of the inferior vena cava.

On the left, the line extends from the sternal end of the left clavicle down to the left edge of the manubrium in the first intercostal space, and then vertically across the second costal cartilage and along the left border of the heart.

The surface markings of all these structures are given in the relevant sections.

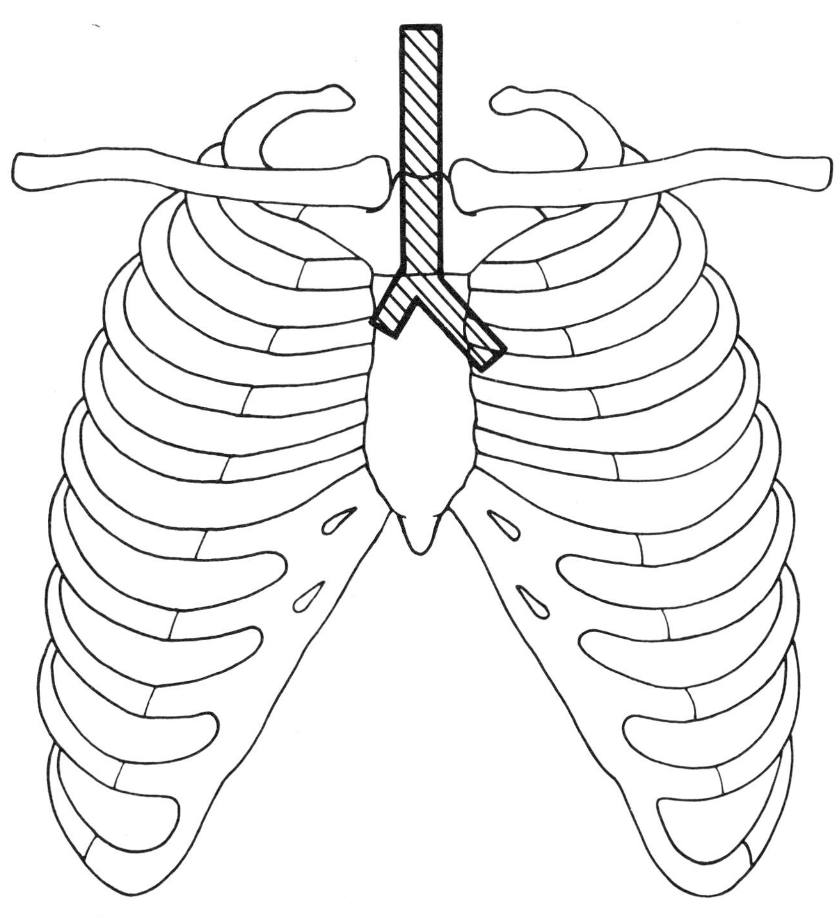

Surface projection of the trachea and main bronchi

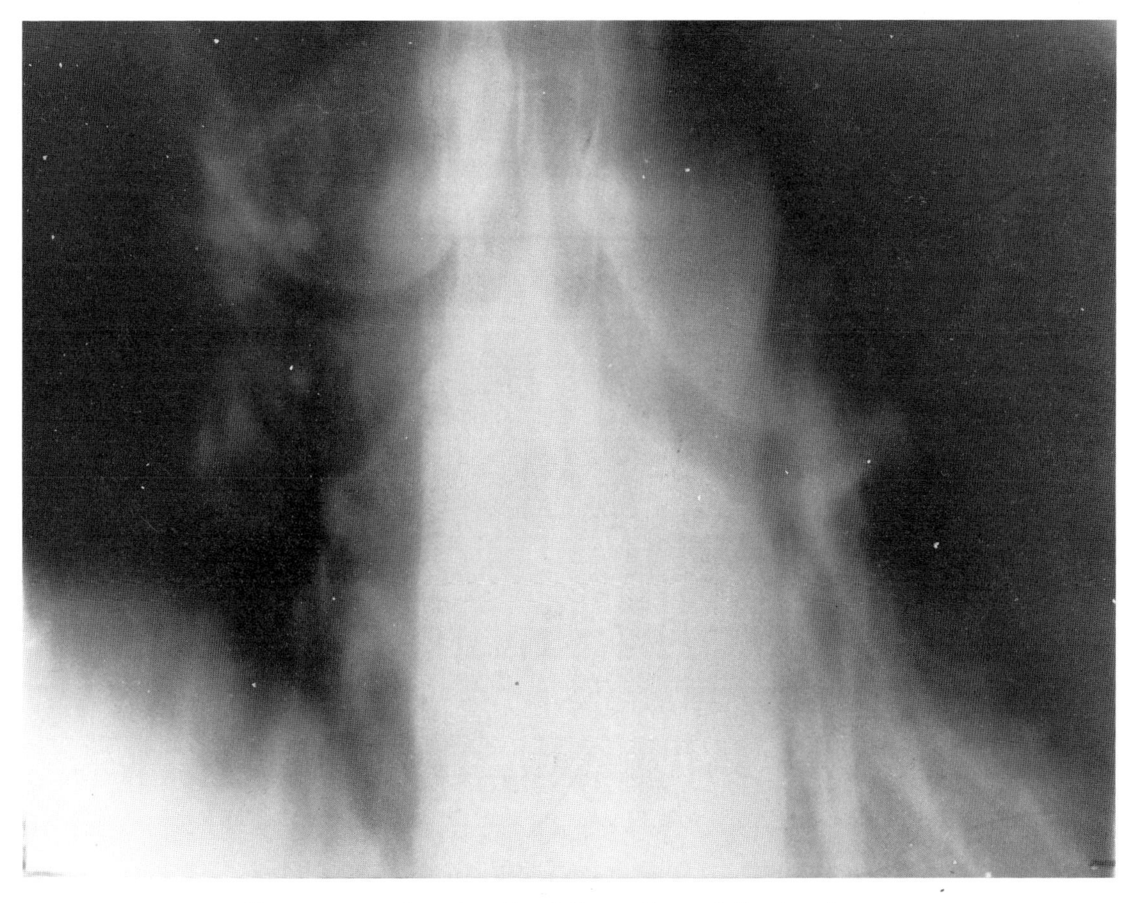

Tomogram showing the bifurcation of the trachea

THE PLEURA

The cervical pleura extends into the neck covering the summit of the lung. When viewed from the side it follows the line of the first rib but, when seen from in front, the limit can be marked as a curved line, convex upwards, from the junction of the medial and middle thirds of the clavicle to the sternoclavicular joint. The apex lies about 2.5 cm above the clavicle and the lung and pleura may be injured here by a penetrating wound.

The lines of junction of the mediastinal pleura, which clothes the interpulmonary region, and the costal pleura, which covers the inner aspect of the chest wall, run down behind the sternum. They are marked by starting at the sternoclavicular joints and extending a line from each to a point in the midline at the sternal angle. The lines are continued downwards in the midline to the level of the fourth cartilages. On the right, this is continued as a straight line to the right edge of the xiphisternal joint but, on the left, it arches out to run along the lateral border of the sternum to reach the lower edge of the left sixth chondrosternal joint.

The lower limit of the costal pleura continues laterally crossing the eighth rib in the midclavicular line; the tenth rib in the mid-axillary line; and the twelfth rib at the lateral border of the sacrospinalis, to reach a slightly higher level opposite the spine of T12.

A small area of pleura is exposed in the costovertebral angle, below and medial to the twelfth rib, behind the upper pole of the kidney. Here it lies in danger, notably from a loin incision to expose the kidney.

The pleura marks the limit of expansion of the lungs. In quiet respiration, the lower border of the lung is about 5 cm above the lower border of the pleura. The resulting slit-like cavity between the costal and diaphragmatic pleurae is the costodiaphragmatic recess. A similar situation exists on the left side behind the sternum and costal cartilages where the lung falls short of the pleura, producing the costomediastinal recess.

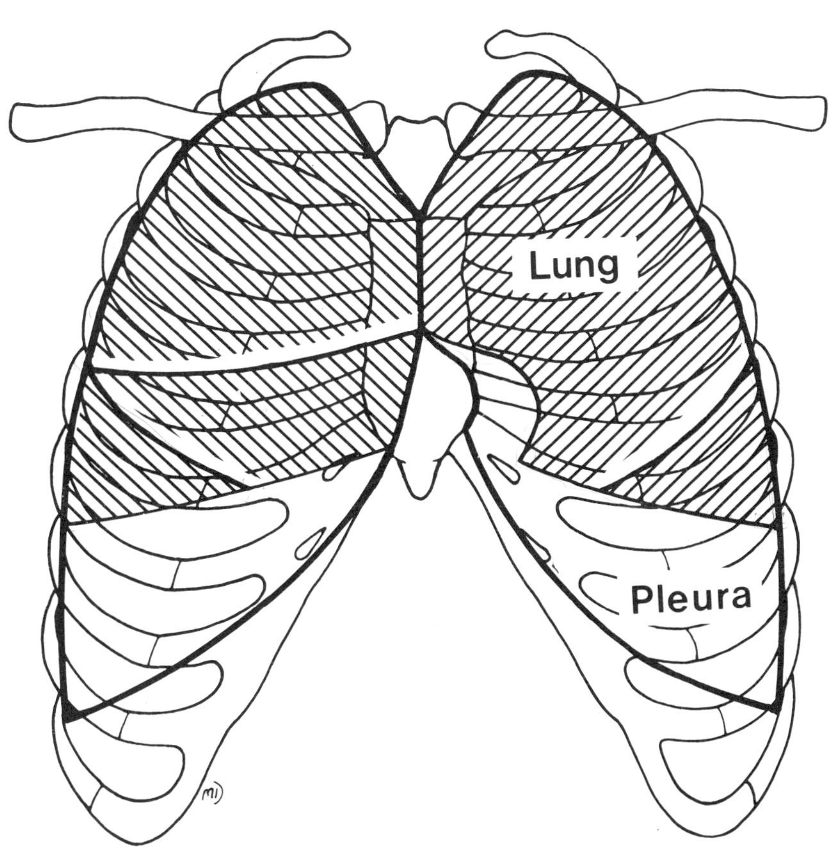

Surface projection of the pleura

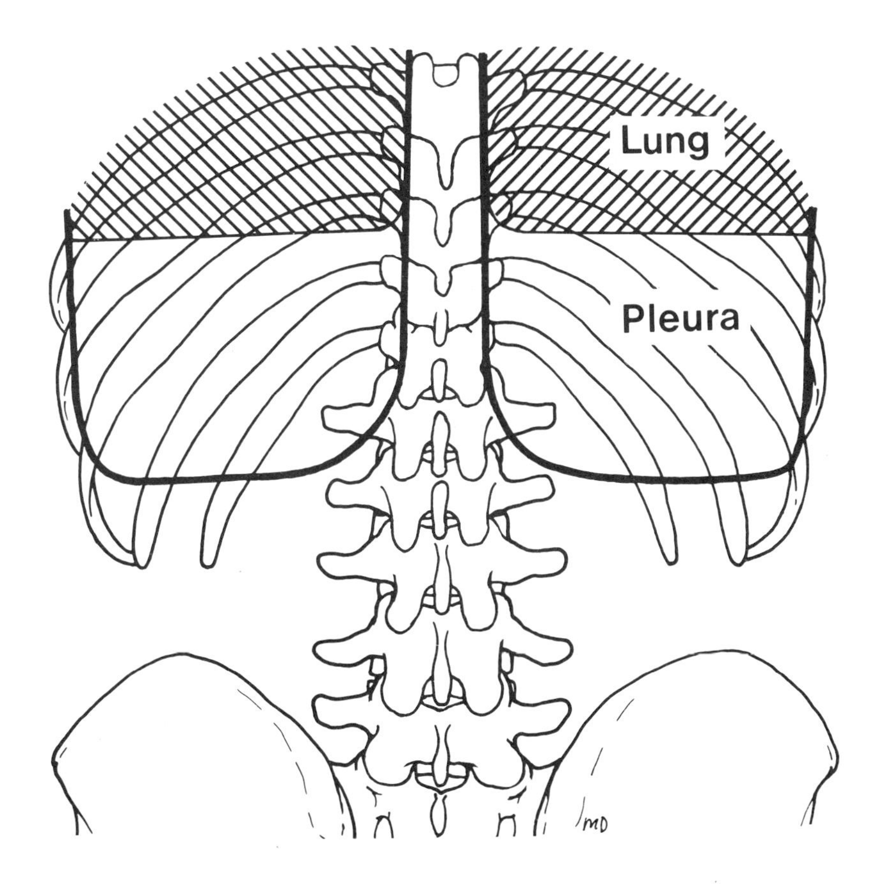

THE LUNGS

The surface marking of the apices of both lungs correspond to those of the cervical pleura. Similarly, the anterior border of the right lung corresponds to the right junction of costal and mediastinal pleurae down to the level of the sixth chondrosternal joint. The anterior border of the left lung curves away laterally from the line of pleural reflection beginning at the level of the fourth costal cartilage. This produces the cardiac notch, the most lateral aspect of which lies 3.5 cm from the edge of the pleura, i.e. the lateral border of the sternum, before the lung curves down to lie behind the sixth left costal cartilage 4 cm from the midline.

The inferior borders of the lungs are represented by lines running from the lower ends of the anterior borders and crossing the sixth rib in the mid-clavicular line; the eighth rib in the mid-axillary line; and the tenth rib 2.5 cm from the median plane. The inferior border moves only a short distance in quiet respiration but may move up to 7.5 cm in deep respiration.

The posterior border can be represented by a line drawn from the posterior end of the inferior border to a point 2.5 cm lateral to the spine of C7.

The hili of the lungs lie at the level of the second, third and fourth costal cartilages, parallel and 2.5 cm lateral to, the borders of the sternum. Posteriorly, they can be represented as vertical lines lying 5 cm from the midline at the levels of the spines of T4, T5 and T6.

The left oblique fissure corresponds to a line which starts 2.5 cm to the left of the interval between the spines of T3 and T4, and runs to the fifth intercostal space in the mid-axillary line, and then cuts the inferior border 7.5 cm from the midline.

The right oblique fissure has a similar course but starts opposite the fourth thoracic spine.

Both oblique fissures can be approximated by the medial borders of the scapulae when the arms are fully abducted.

The transverse fissure of the right lung is represented by a line leaving the oblique fissure in the fifth intercostal space in the mid-axillary line, running forwards along the lower border of the fourth cartilage to meet the anterior border of the lung.

Unusually an azygos lobe is formed by the azygos vein hanging down in a double fold of parietal pleura and separating the medial part of the right upper lobe.

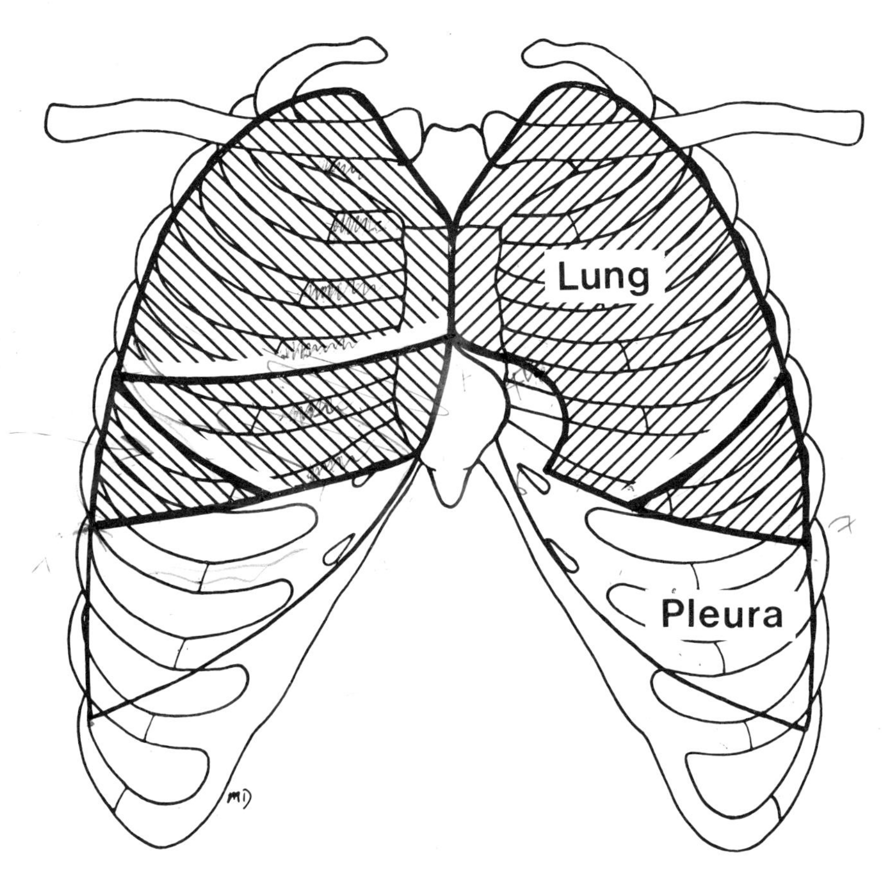

Surface projection of the lungs

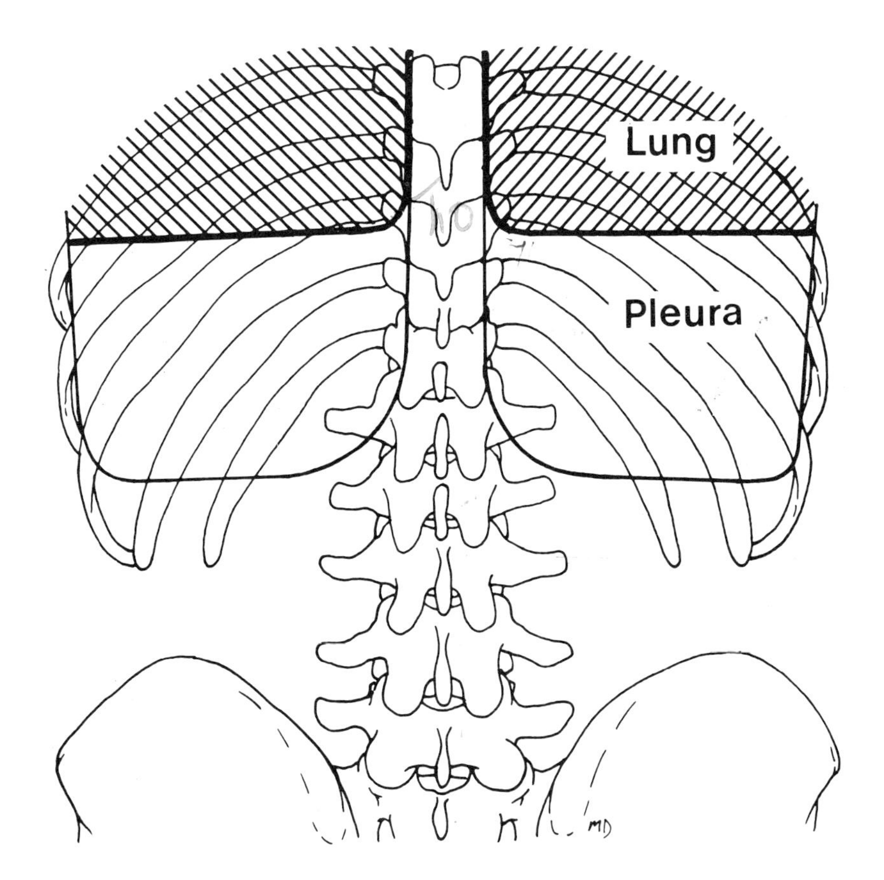

THE HEART

The heart rests on the diaphragm and is covered anteriorly by the body of the sternum and the third to sixth costal cartilages of both sides.

The base

Formed by both atria, this lies in front of the bodies of the middle four thoracic vertebrae, at the level of the fifth, sixth and seventh thoracic spines.

The sternocostal surface

This is not completely overlapped by the lungs and pleura, and is a region of confusing terminology.

The exposed area of pericardium, not covered by pleura, lying over the right ventricle is known as the *bare area*.

The area of complete cardiac dullness, or superficial cardiac projection, is the larger area not covered by lung. It can be marked by starting at:

Point A, in the middle of the sternum at the level of the fourth costal cartilage, and running a straight line to

Point B, the right edge of the xiphisternal joint, then extending a line to the left horizontally to

Point C, the left sixth costal cartilage, 4 cm from the midline.

The line returning to point A is markedly curved to the left and reaches 3.5 cm from the sternal edge.

The deep projection of the heart maps out the whole of the sternocostal surface. It is represented by gently convex lines joining the following four points:

The upper edge of the third right costal cartilage [**1**], 1.2 cm from the sternal border.

The middle of the right sixth chondrosternal junction [**2**].

The fifth left intercostal space [**3**], 9 cm from the midline. This represents the apex beat, the lowest and outermost point of definite cardiac pulsation. The apex beat varies in position and may lie from the fourth intercostal space above to the sixth costal cartilage below, and between 6 to 10 cm from the midline.

The lower margin of the left second costal cartilage [**4**]. 1.2 cm from the sternal edge.

These lines map out the four borders of the heart.

The right border is formed by the right atrium. The vena cavae enter the heart at the upper and lower ends of this border.

The lower border is mainly right ventricle, but includes 1.2 cm of left ventricle at the apex. It lies at the level of the xiphisternal joint.

The lower four-fifths of the left border are formed by the left ventricle and the upper fifth by the left atrium.

The upper border is formed by both atria. The superior vena cava enters the heart at its right end; the ascending aorta crosses the middle of this border; the pulmonary trunk bifurcates just above the left end and the right pulmonary artery runs to the right just above the border.

Chambers

The chambers of the heart can be marked out by drawing:

The coronary sulcus [**5**] separating the atria and the ventricles, from the upper medial end of the third left costal cartilage to the middle of the right sixth chondrosternal joint.

The anterior interventricular sulcus [**6**] from the third left intercostal space 2.5 cm to the left of the midline to a point 1.2 cm medial to the apex.

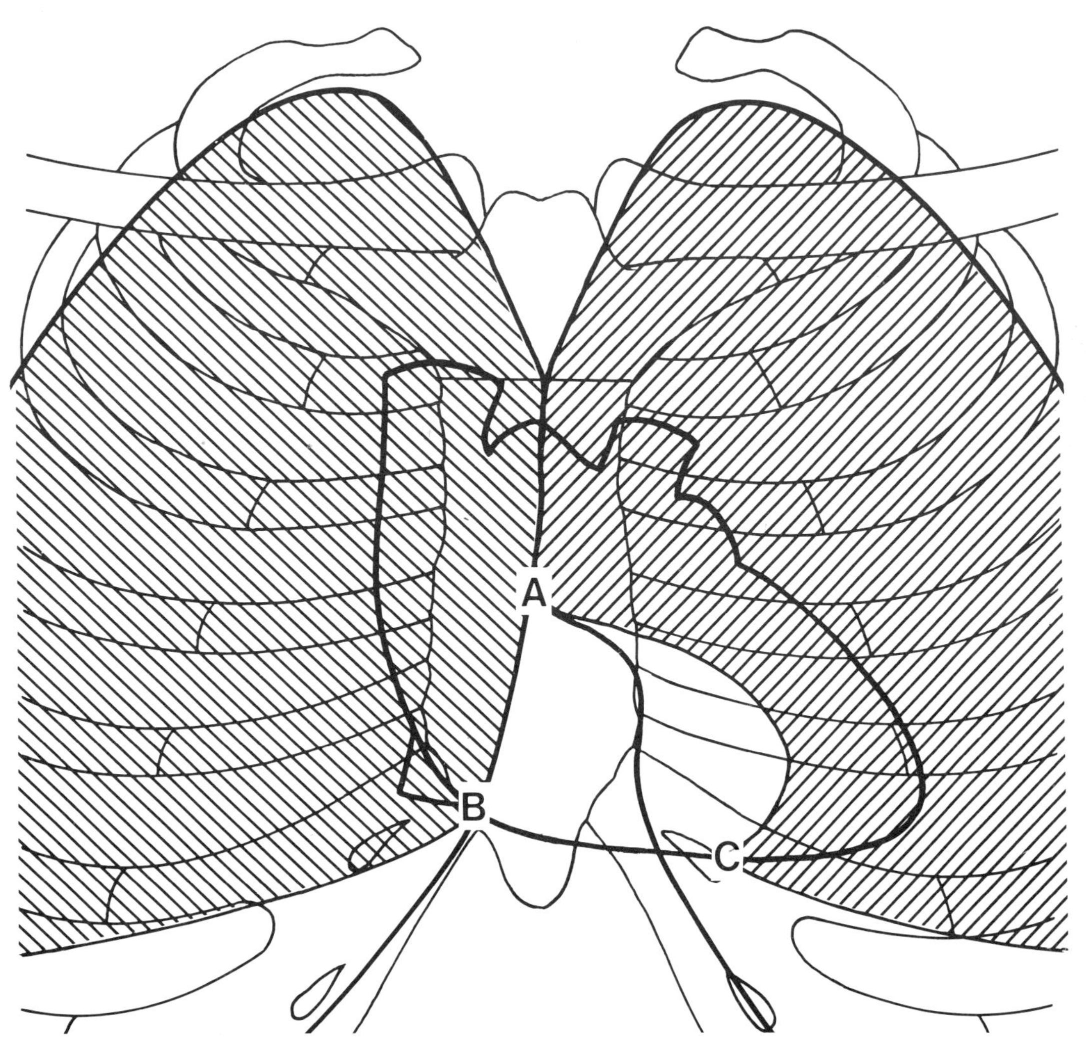

The sternocostal surface of the heart in relation to the lungs and pleura

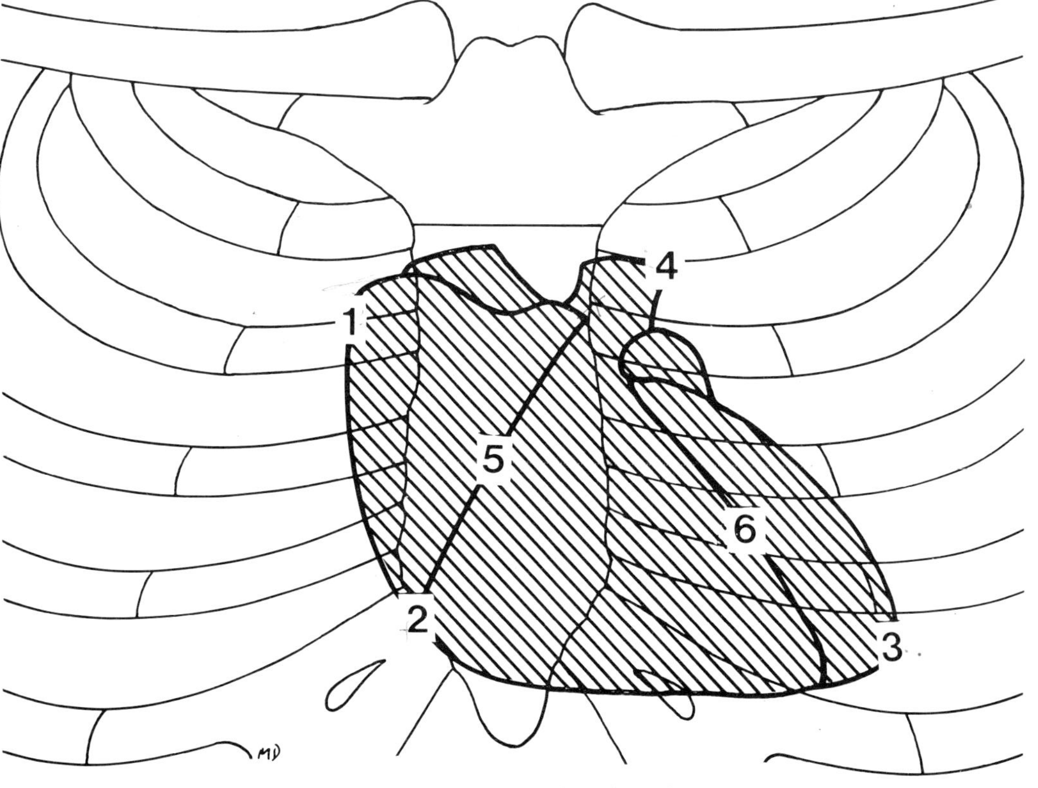

The sternocostal surface of the heart showing the coronary and anterior interventricular sulci

Orifices

Superior vena cava [**1**]. 2 cm wide, horizontal, centred on a point at the level of the third right costal cartilage, 1.5 cm from the midline.

Inferior vena cava [**2**]. 2 cm wide, horizontal, centred on a point at the level of the sixth right costal cartilage, 2.5 cm from the midline.

Pulmonary [**3**]. 2.5 cm wide, horizontal, centred on a point at the third left chondrosternal joint.

Aorta [**4**]. 2.5 cm wide, oblique, running down and right, starting from the medial end of the third left intercostal space.

Mitral [**5**]. 3 cm wide, oblique, running down and right, starting opposite thc fourth costal cartilages and lying wholly beneath the left side of the sternum.

Tricuspid [**6**]. 4 cm wide, almost vertical, centred at the level of the fourth intercostal space just to the right of the midline.

The normal sounds of valves closing, and abnormal sounds produced by turbulent flow, are conducted 'downstream' by the flow of blood. Remembering the surface markings of the valves and the direction of the blood flow, it follows that in particular regions certain parts of the heart can be heard better than others.

The aortic area [**A**] is at the sternal edge in the second right interspace.

The pulmonary area [**P**] lies in the second left intercostal space at the sternal edge.

The tricuspid area [**T**] lies at the left lower end of the sternum.

The mitral area [**M**] lies at the apex.

The Fibrous Pericardium

This has the same surface marking as the sternocostal surface of the heart except that the pericardium extends up to the second costal cartilage on the right.

The Cardiac Plexi

The deep cardiac plexus lies on the bifurcation of the trachea behind the body of the sternum, between the levels of the second and third costal cartilages, at the level of the upper border of T5.

The superficial cardiac plexus lies below the aortic arch, above the bifurcation of the pulmonary artery and beneath the sternal end of the second left costal cartilage.

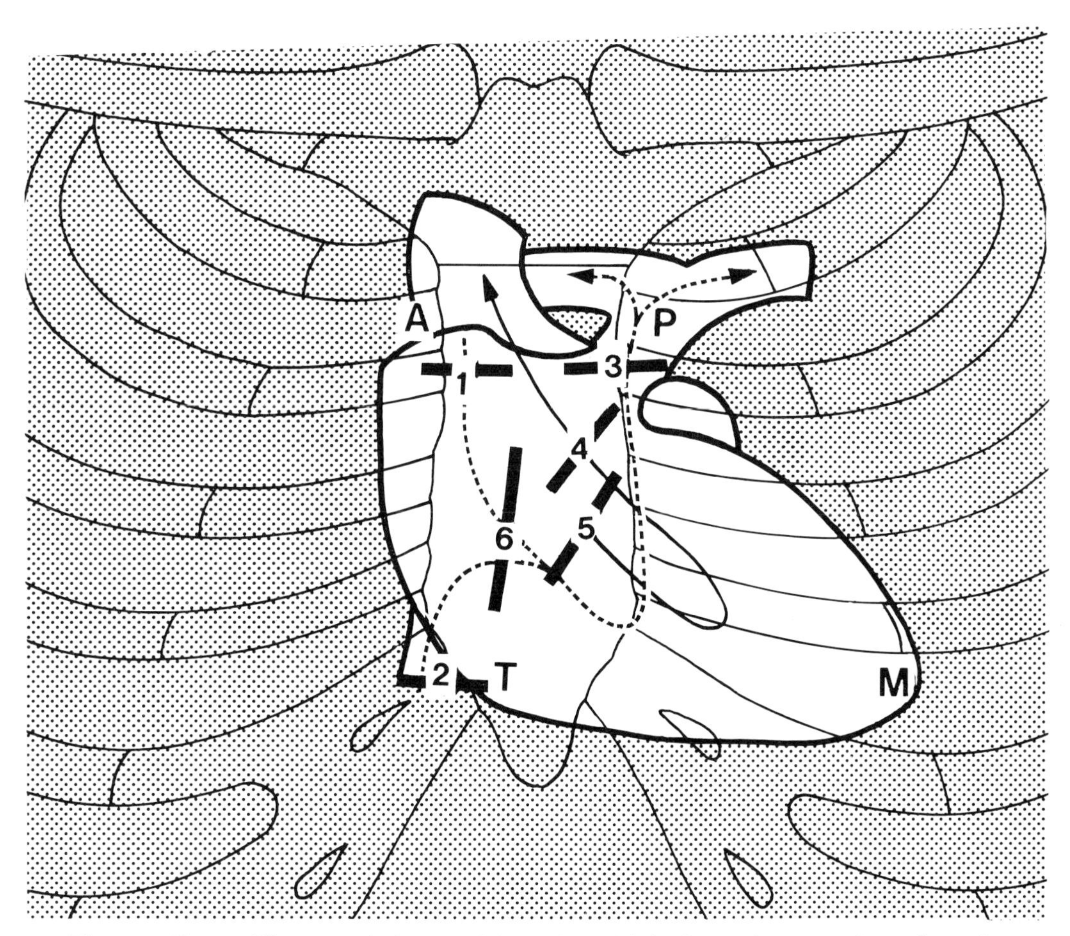

The cardiac orifices and the positions in which the valves are best heard

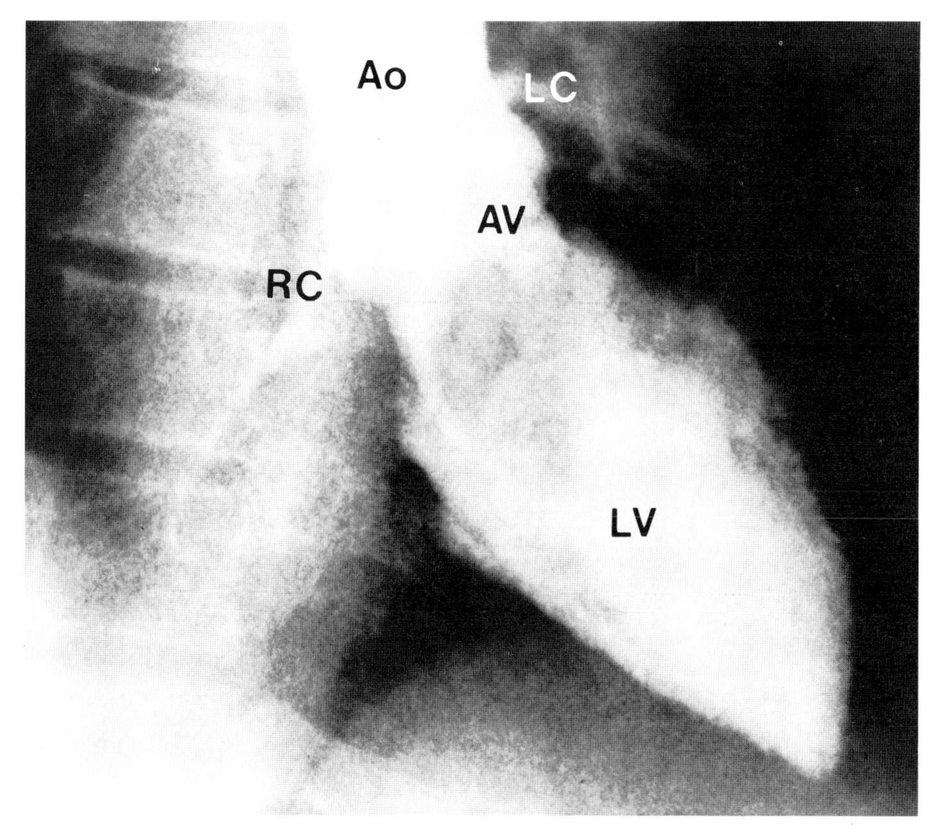

Left heart catheterisation

A catheter has been introduced via a large peripheral artery into the left ventricle (LV) and contrast medium is being injected. The contrast is passing through the aortic valve (AV) into the ascending aorta Ao. The right (RC) and left coronary (LC) arteries can be seen

ARTERIES

The pulmonary trunk [**1**]. The pulmonary trunk can be represented by a band 5 cm long and 2.5 cm wide. It runs from the pulmonary orifice, behind the third left chondrosternal joint, upwards and to the left, to reach the second left costal cartilage 1.2 cm from the sternal edge where it bifurcates.

The left pulmonary artery runs from the bifurcation of the trunk for 2.5 cm to the left.

The right pulmonary artery runs to the right from the bifurcation between the sternal ends of the second left and right costal cartilages.

The aorta [**2**]. The ascending aorta runs from the aortic orifice at the medial end of the third left intercostal space, up to the second right chondrosternal joint. The arch continues above the right side of the sternal angle and then turns down to lie behind the second left costal cartilage.

The descending aorta runs down from behind this cartilage, gradually moving across to reach a point just to the left of the midline, 9 cm below the xiphisternal joint where it enters the abdomen.

The innominate artery [**3**]. The innominate artery starts at a point deep to the centre of the manubrium and passes up to lie beneath the right sternoclavicular joint.

The left common carotid artery [**4**]. This can be represented as a line running from just to the left of the centre of the manubrium to the left sternoclavicular joint.

The left subclavian artery [**5**]. The left subclavian artery can be projected as a line from close to the middle of the left margin of the manubrium to the left sternoclavicular joint.

The internal thoracic arteries [**6**]. The internal thoracic arteries begin 2 cm above the clavicles between the clavicular and sternal attachments of the sternomastoid muscles. They run down to lie behind the second costal cartilages, 3 cm from the midline, 1.2 cm from the sternal border, and from there vertically down to the sixth costal cartilages. The arteries then divide into their musculophrenic and superior epigastric branches.

The intercostal neurovascular bundles [**7**]. These lie in the costal grooves on the lower borders of the ribs. The vein lies above the artery, which in turn lies above the nerve. The collateral branches of the arteries originate just lateral to the angles of the ribs, and then pass downwards to run along the upper border of the rib below.

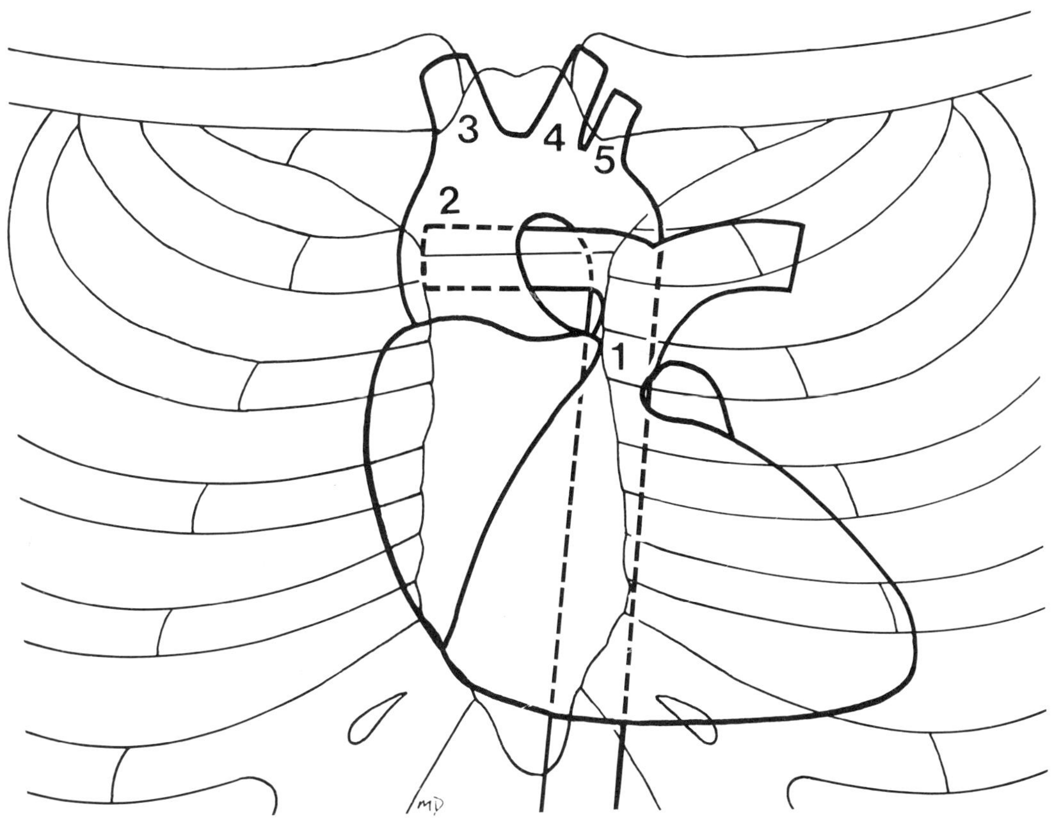

Surface projection of the major thoracic arteries

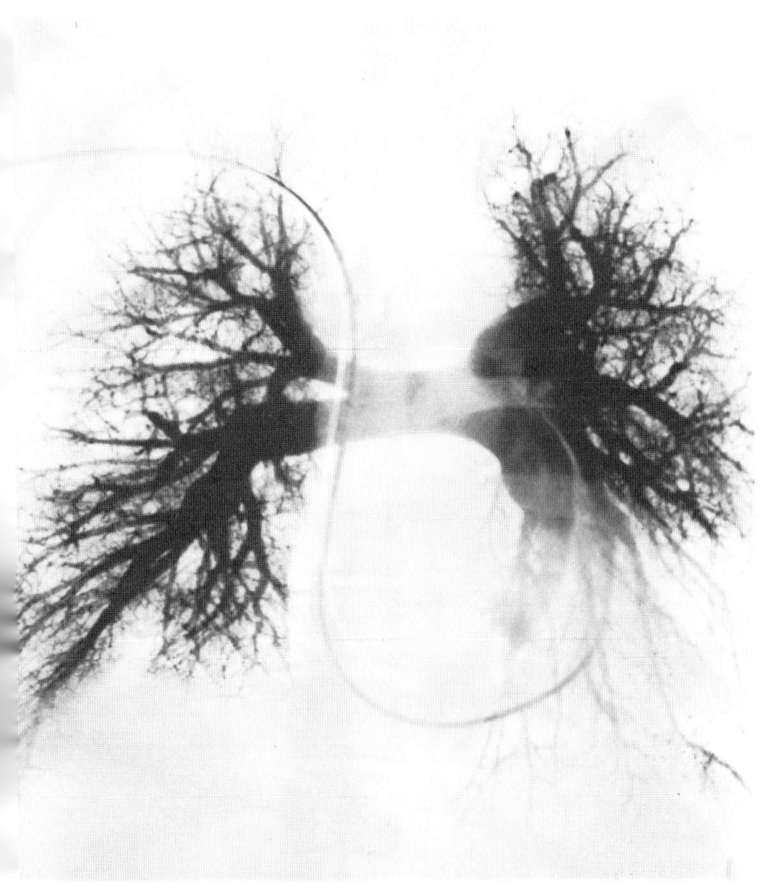

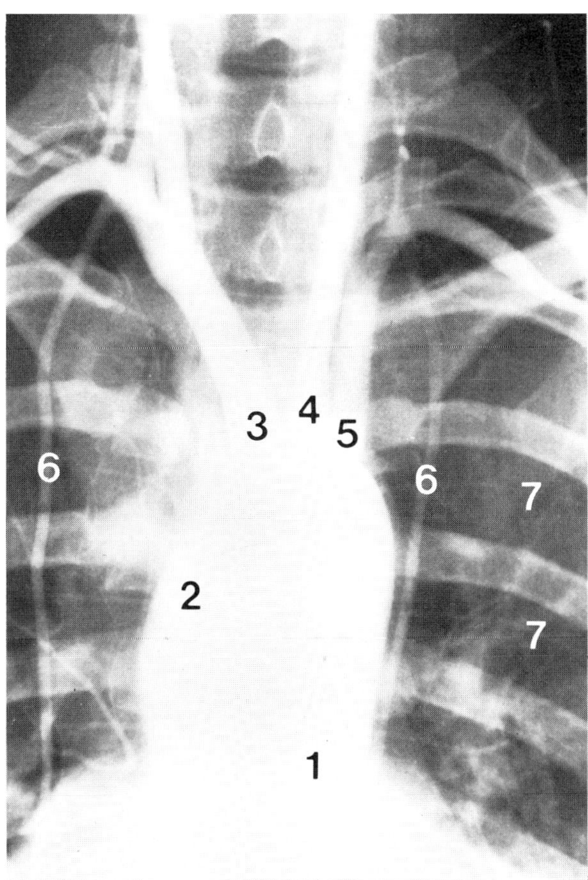

pulmonary angiogram. A catheter is threaded through the ·nous system and the right heart and contrast injected into the ılmonary vasculature.

An arch aortogram, showing contrast being injected into the arch of the aorta and passing into the major arteries.

VEINS

The right brachiocephalic vein [**1**]. This commences behind the sternal end of the right clavicle and then runs down to the lower edge of the first right chondrosternal joint where it unites with the left brachiocephalic vein to form the superior vena cava.

The left brachiocephalic vein [**2**]. Commencing behind the sternal end of the left clavicle, it then runs behind the left sternoclavicular joint before meeting the right vein behind the first right chondrosternal joint.

The superior vena cava [**3**]. This is formed by the union of the brachiocephalic veins at the lower edge of the first chondrosternal joint. It runs directly down to pierce the pericardium behind the second right cartilage, and enters the heart at the level of the third cartilage.

The inferior vena cava [**4**]. The inferior vena cava has only a short course in the thorax. It lies in the right cardiodiaphragmatic angle and enters the heart behind the sixth right costal cartilage.

The azygos vein. Beginning to the right of the midline, 10 cm below the level of the xiphisternal joint, it ascends vertically to the level of the body of T5, then curves forwards to enter the superior vena cava at the level of the second right costal cartilage.

THE THORACIC DUCT

The duct starts 10 cm below the xiphisternal joint, 2.5 cm to the right of the midline and then ascends vertically to the level of the sternal angle where it crosses obliquely to a similar position on the left of the midline. It continues to climb through the thorax and the neck to a point 2.5 cm above the clavicle, 2.5 cm to the left of the midline. It arches downwards to run between the two heads of the left sternomastoid muscle and enters the venous system, usually at the junction of the left jugular and subclavian veins.

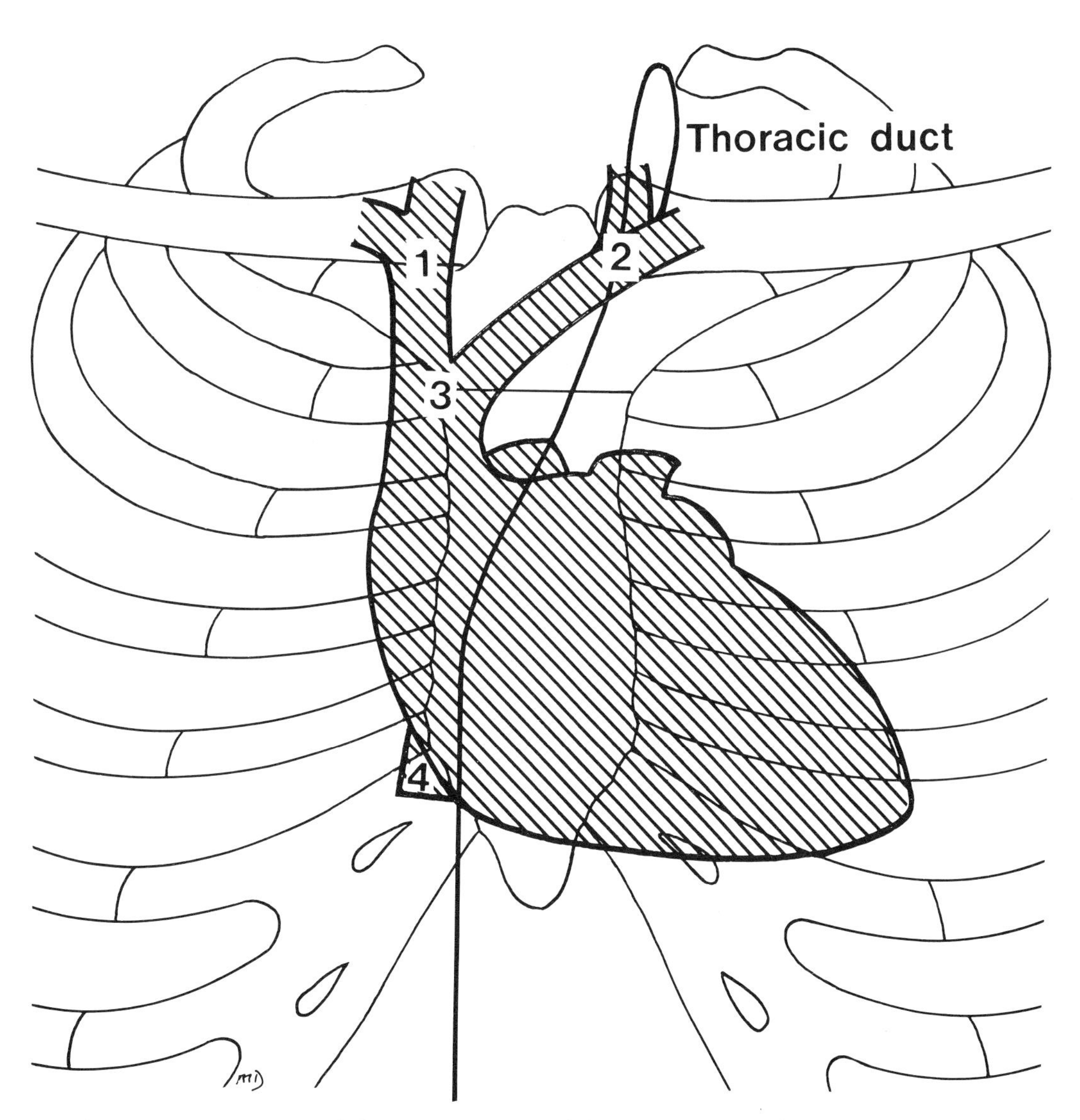

Surface projection of the great veins of the thorax and the thoracic duct

THE CHEST X-RAY

After clinical examination, the most commonly used special investigation of the thorax is the chest x-ray. The usual view is the postero-anterior (PA), in which the subject sits or stands with the anterior surface of his chest against the film, and with the x-ray tube behind his back. The hands are placed on the sides of the iliac crests and the elbows directed forwards, thus swinging the scapulae away from the position in which they cover the lungs.

The film is exposed in suspended deep inspiration, with the breath held after taking a deep inspiration.

The lateral view is now used less routinely. The side of the thorax of which the best image is required is held next to the film and the arms are removed from the field by holding them above the head.

It is essential to view any x-ray systematically. In a chest x-ray the following points should be studied.

The Hard Tissues

The clavicles can be seen across the upper zone of each lung.

The ribs can be seen as they curve around the thorax. The posterior parts are most clearly seen, but the anterior ends can also be identified.

The costal cartilages are radiolucent but may become calcified in later life.

The sternum is largely hidden by the vertebrae in the PA view. In the lateral view the sterno-manubrial joint is an obvious landmark.

The thoracic vertebrae are broader than they are tall. They are further discussed under Radiology of the Back (*see* p. 16).

The vertebral borders and inferior angles of the scapulae can still be seen at the sides of the upper chest wall.

The Soft Tissues

The breasts produce the most important shadows. The normal radiological appearance of the female breast is a homogeneous shadow with a convex lower border similar to that of its partner. Absence of a breast results in increased translucency on that side of the chest. If this is unrecognised there may be confusion with lung pathology. The nipples often produce opacities, 1 cm in diameter, and they can be distinguished from a lung mass as, unlike an intrapulmonary lesion, they change position relative to the diaphragm when the PA film is repeated in expiration.

The muscles of the neck especially the sternomastoids, and the pectoral muscles may produce recognisable symmetrical shadows.

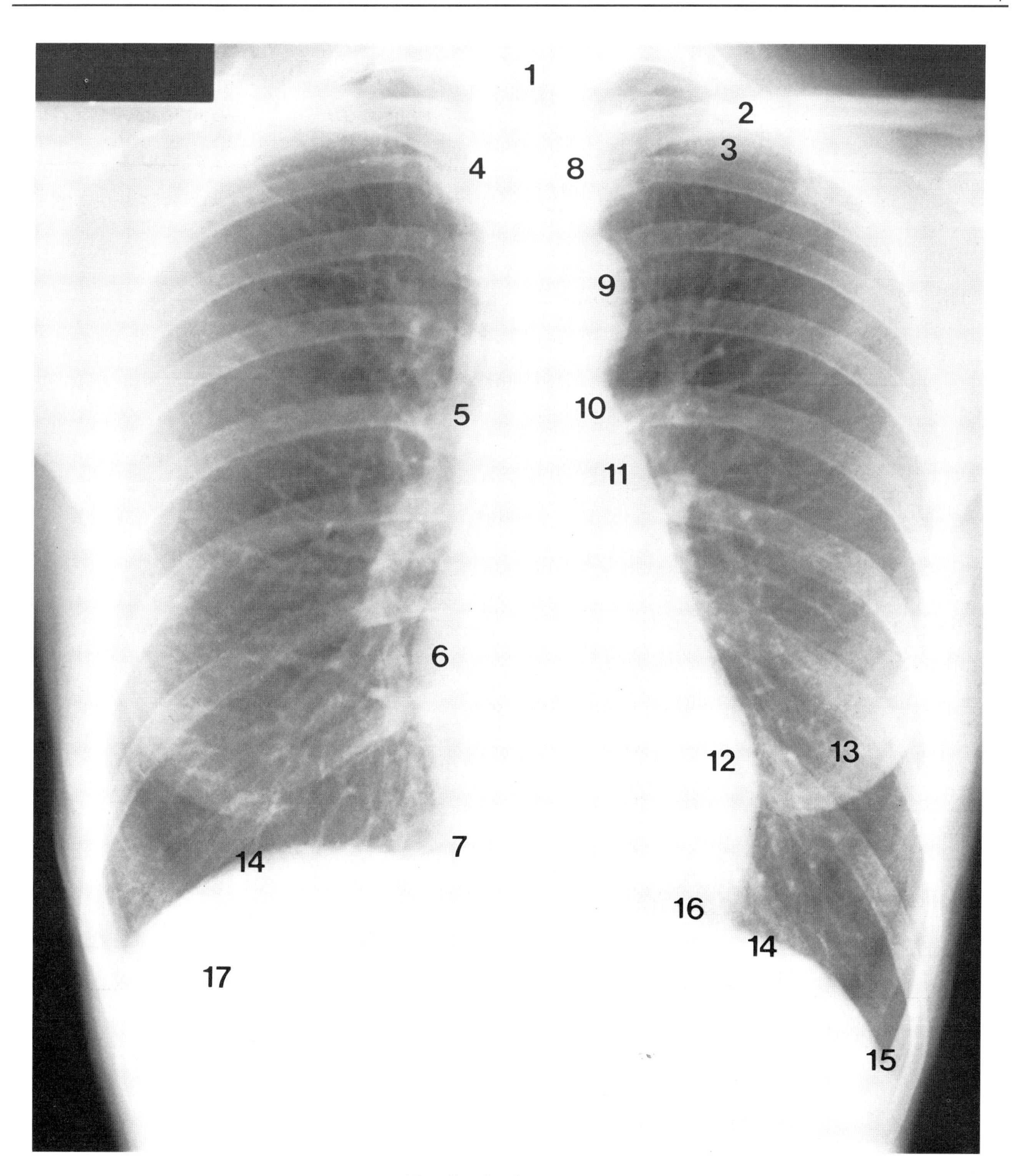

The P – A chest x-ray

1. Trachea 2. Clavicle 3. First costal cartilage 4. Right brachiocephalic vein 5. Superior vena cava 6. Right atrium 7. Inferior vena cava 8. Left subclavian artery 9. Aortic knuckle 10. Pulmonary trunk 11. Left auricle 12. Left ventricle 13. Breast shadow 14. Diaphragm 15. Costophrenic angle 16. Cardiophrenic angle 17. Liver

The Diaphragm

The PA film shows the diaphragm to have two domes, convex upwards, which overlie the viscera of the abdomen. Laterally, the diaphragm meets the chest wall at the acute costophrenic angles. Centrally, the tendinous part of the diaphragm is fused to the pericardium and at each side there is a cardiophrenic angle.

In the suspended deep inspiration of the standard PA film the highest part of the right dome is normally at the same level as the anterior end of the fifth rib, or the body of T10. The left dome lies about 1 cm lower.

The lateral film shows the domes as two convex lines, usually lying one above the other. It may be difficult to decide which line represents which dome, but the fact that the left usually has gas in the fundus of the stomach beneath it may help.

The subdiaphragmatic tissues are easier to recognise on the PA film, where beneath the left dome the gastric air bubble may lie above a recognisable gastric fluid level. Lung markings may be visible through the air bubble. To the left of the gastric bubble, air may also be present in the splenic flexure. The undersurface of the right dome is normally inseparable from the liver shadow.

The Mediastinum

The borders of the mediastinum are made up of the following structures from above down.

On the right:

- the right innominate vein
- the superior vena cava
- the right atrium
- the inferior vena cava.

On the left:

- the left subclavian artery
- the aortic knuckle
- the pulmonary trunk, or left pulmonary artery
- the left auricle
- the left ventricle.

When seen on the lateral film, the heart is shown to lie on the anterior half of the central tendon of the diaphragm. The upper anterior part of the heart is separated from the sternum by a transradiant area, the retrosternal space. In a child, the thymus may produce a shadow in this region, which in PA view may appear as the characteristic 'sail shape'. The retrosternal space is of approximately equal transradiancy to the retrocardiac space which lies behind the lower posterior part of the heart and contains the oesophagus and the other contents of the posterior mediastinum.

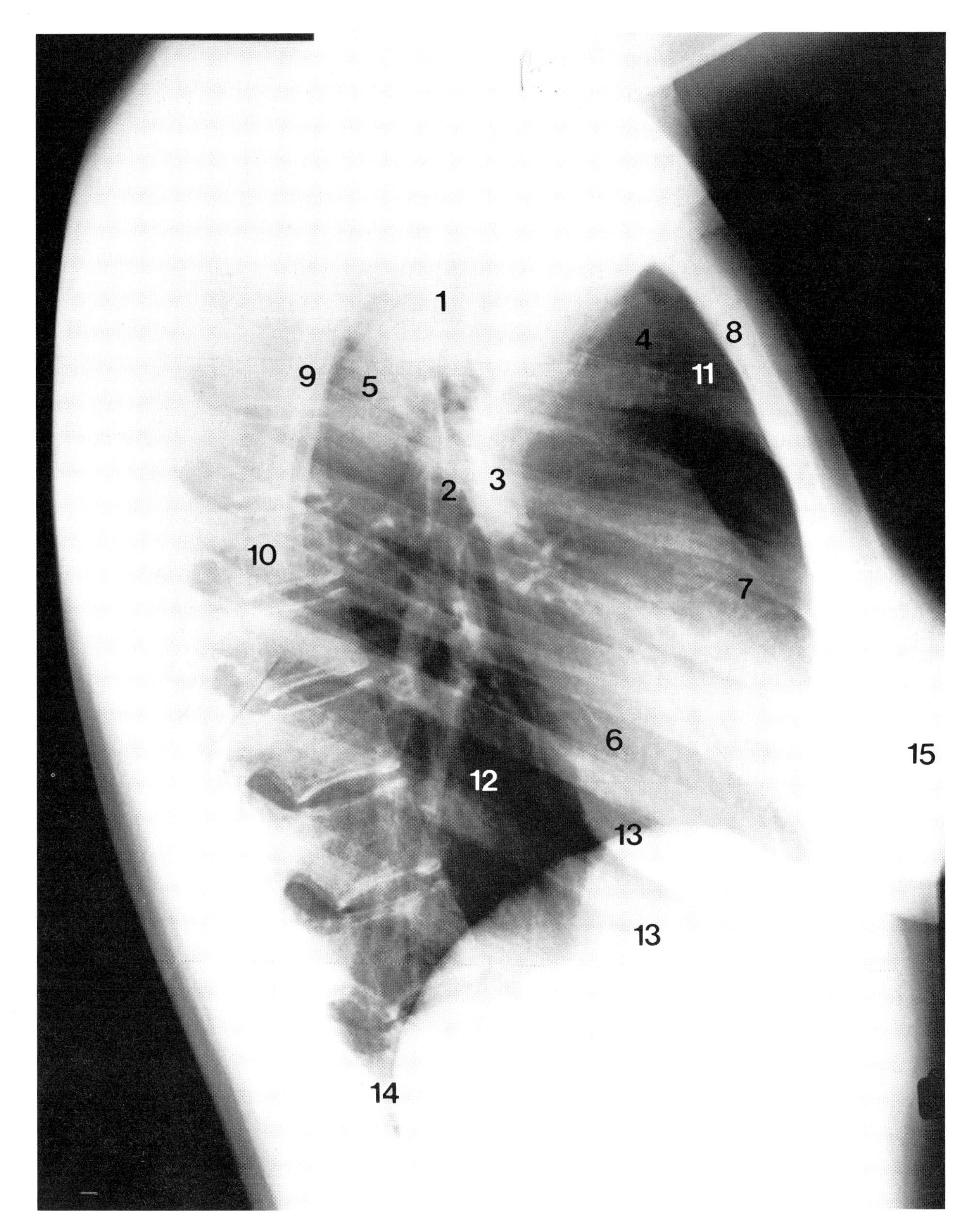

The lateral chest x-ray

1. Trachea 2. Right main bronchus 3. Pulmonary trunk 4. Ascending aorta 5. Descending aorta 6. Oblique fissure. 7. Transverse fissure 8. Body of the sternum 9. Vertebral border of scapula 10. Inferior angle of scapula 11. Retrosternal translucent zone 12. Retrocardiac translucent zone 13. Cupolas of the diaphragm 14. Costodiaphragmatic angle 15. Breast shadow

The Heart

Size. The absolute maximum transverse diameter of the heart varies between 7 cm and 17 cm, although it normally lies between 11 cm and 15 cm. It may be related to the internal transverse diameter of the chest. This relationship is known as the cardiothoracic ratio (CTR) and is usually less than 0.5. Apparent dilatation of the heart may be due to distension of the pericardium by an effusion.

Shape. Changes in heart shape may occur as a result of the enlargement of one or more chambers, or as a result of local aneurysmal dilatation of part of a chamber.

Left atrial enlargement produces a direct change in the left border as the auricular appendage dilates. Posteriorly, left atrial enlargement will indent the oesophagus, a change best seen in the lateral view with the oesophagus outlined by barium. A distended left atrium will also push up the left main bronchus and so widen the angle of bifurcation of the trachea.

Right atrial enlargement produces an increased convexity of the right heart border.

Left ventricular enlargement pushes the apex of the heart laterally and downwards. In the lateral view the lower third of the heart may be shown to displace the oesophagus backwards.

Right ventricular enlargement displaces the apex laterally and upwards.

Cardiac shape and size are also affected by normal respiration and by changes in the cardiac cycle; the latter may become accentuated in arrhythmias.

Calcification. Usually none is seen in the normal radiograph.

Valve calcification can be seen in the lateral film because, in this view, it is not obscured by the sternum. It will lie close to a line connecting the angle between the diaphragm and the sternum to the left main bronchus. The aortic valve is just in front and above the line, while the mitral valve is behind and below.

Coronary artery calcification produces pairs of parallel lines close to the origin of vessels from the aorta.

Pericardial calcification is commonly irregular in distribution.

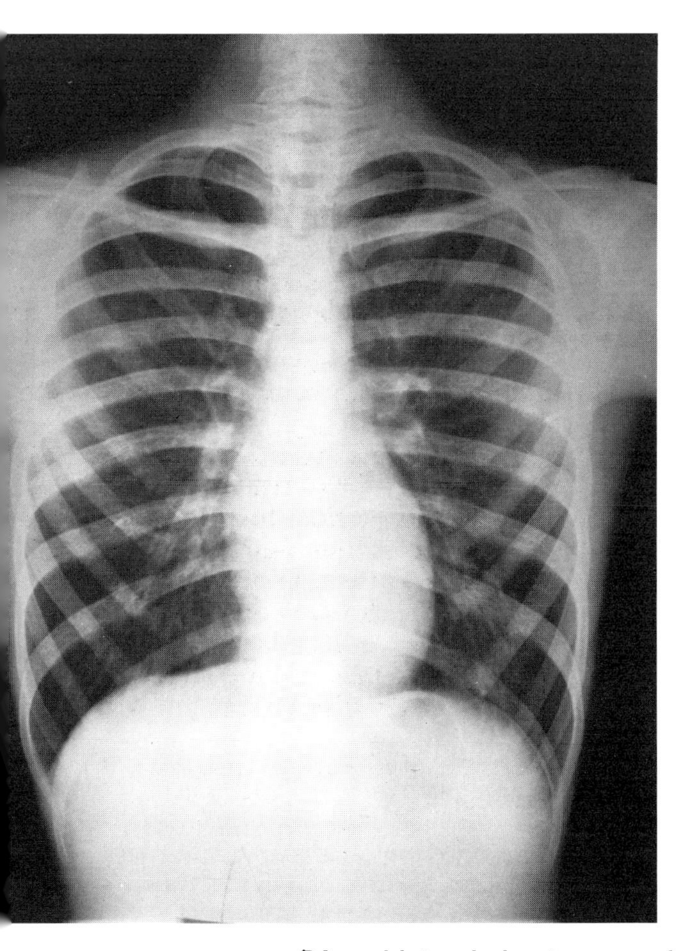

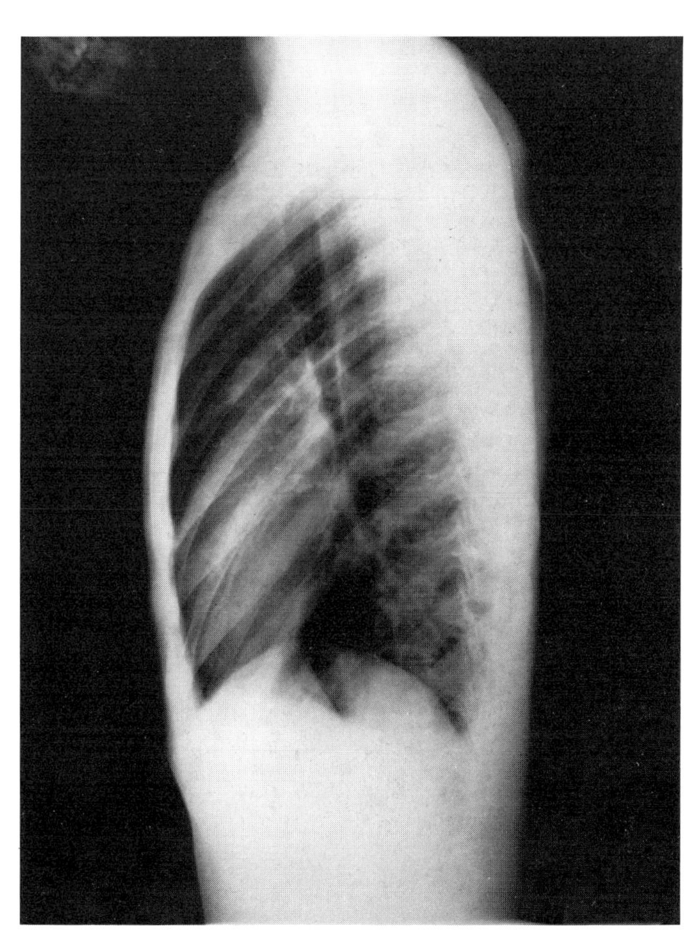

PA and lateral chest x-rays of an asthenic 18-year-old male

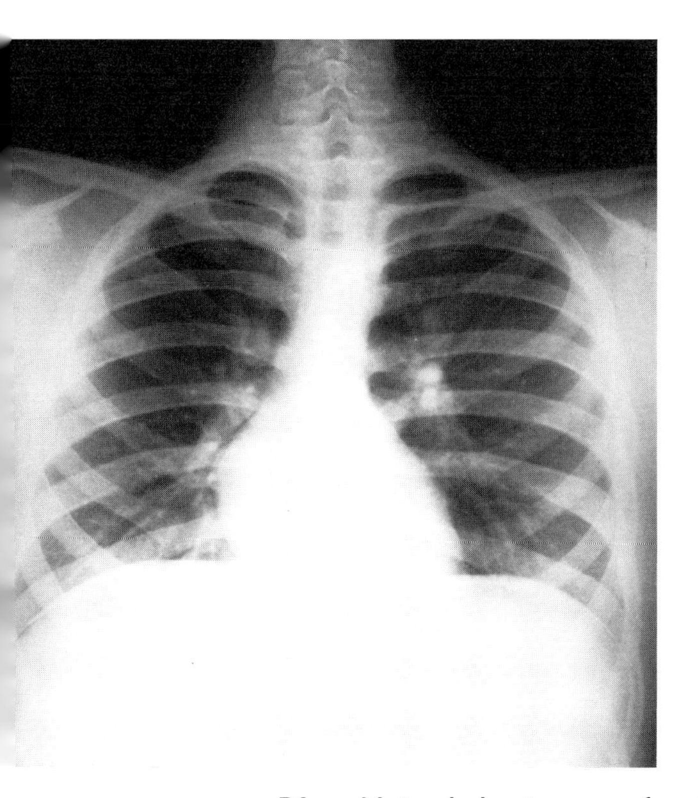

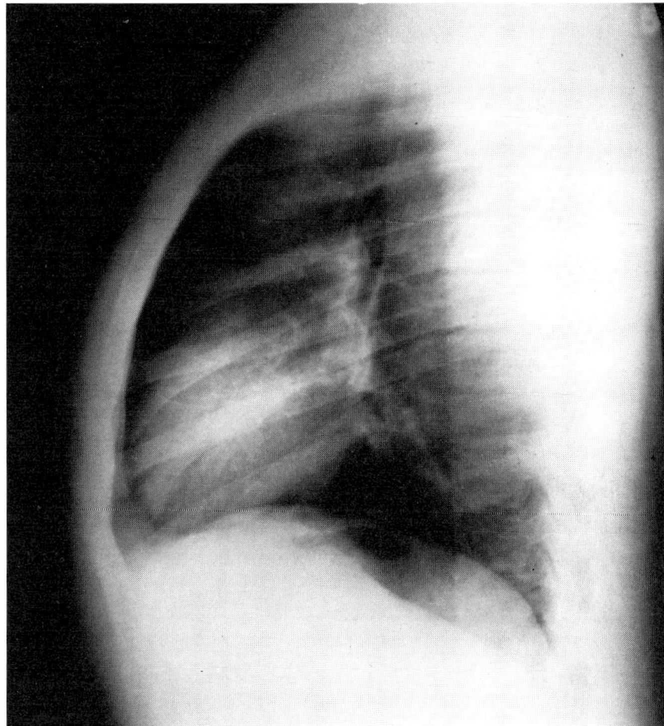

PA and lateral chest x-rays of an hyposthenic 18-year-old male

Special Radiographs and Imaging

Apart from the PA and lateral films, other views and contrast studies may be used in the evaluation of the patient with cardiac disease.

The penetrated PA view may demonstrate intracardiac calcification.

The lateral view, in combination with a barium swallow, will show any part of the heart indenting the oesophagus.

Screening, fluoroscopy allows visualisation of the movements of the heart outline and of intracardiac shadows.

Right heart catheterisation is performed by introducing a catheter into the right atrium by venepuncture. The catheter usually passes freely through the tricuspid valve into the right ventricle, and so into the pulmonary trunk, and often further into the pulmonary vasculature. Combined with the rapid injection of contrast medium through the catheter, a permanent record of the fluoroscopic appearances can be made on cine film. The contrast medium will pass through the pulmonary vasculature and pictures of the left heart can also be taken.

Left heart catheterisation produces better pictures of the left atrium and ventricle. The method relies on identical principles to those used in right heart catheterisation, but the catheter is introduced by puncture of the arterial system. This approach is most commonly used to produce contrast views of the coronary arteries.

Coronary angiography. The coronary arteries fill passively with dye during left heart catheterisation but much clearer views can be obtained by selectively introducing the tip of the catheter into the vessels.

Contrast studies show the right atrium to be a smooth-walled chamber, whilst the right ventricle appears to have an irregular lumen due to the complex arrangement of the trabeculae. The infundibulum leading the blood to the pulmonary valve is smooth walled. The left atrium also has a smooth outline and the cavity of the left ventricle is more regular than its right counterpart. The thickness of the walls of the chambers may be estimated as the distance between the limit of the dye in the chamber and the limit of the cardiac outline.

Radioisotopes are increasingly being used in the examination of the heart. Thallium-201 is taken up by the myocardium in relation to its perfusion and scanning allows an estimation of coronary perfusion to be made. Ventricular function may be estimated by scanning the heart whilst technetium-labelled erythrocytes are injected.

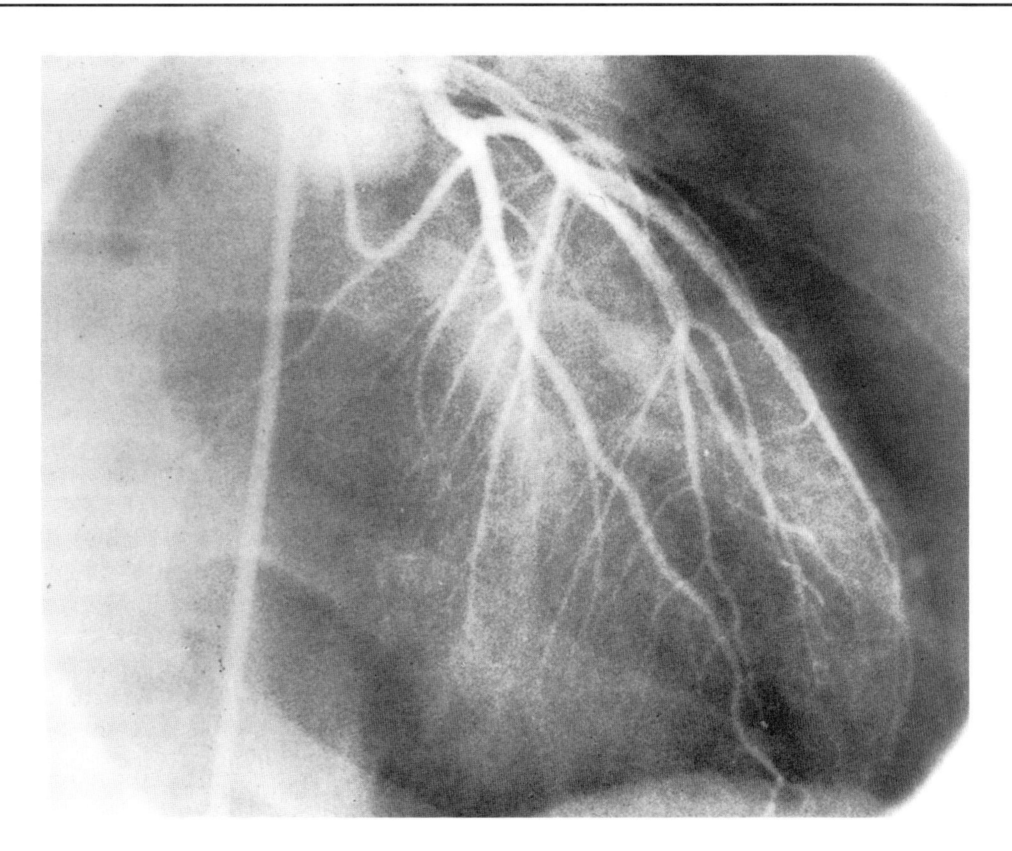

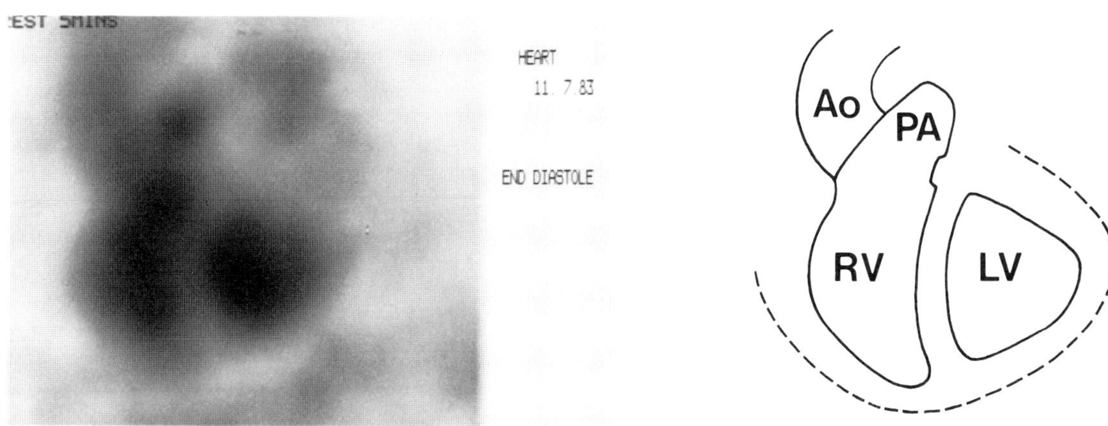

Dye has been injected into the left coronary system

Coronary angiography

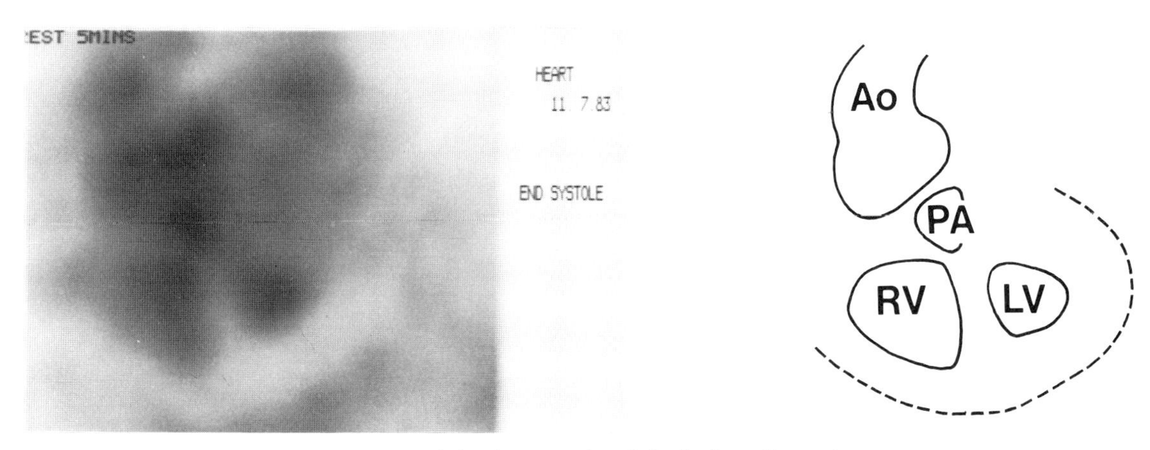

Radioisotope scan of the heart using labelled erythrocytes.
RV – Right ventricle LV – Left ventricle PA – Pulmonary artery
Ao – Aorta

Nuclear angiograph

The Trachea

The radiolucency of the trachea overlies the lower two cervical and upper five thoracic vertebrae. The standard PA view shows only the upper half clearly and special studies are needed to see the lower half and the bifurcation.

The bifurcation is found to have an angle of about 70°. The right main bronchus is 2.5 cm long and lies at an angle of 25° to the vertical. It enters the lung at the level of the body of T5. The left main bronchus is 5 cm long and is at 45° to the vertical. It reaches the level of T6 before entering the lung.

The Hili of the Lungs

The hilar shadows are largely produced by the pulmonary vessels as normal lymph nodes are too small to be recognisable. In a well-penetrated film, air may be seen in the major bronchi but their walls cannot be visualised. The vessels radiating from the hili can be traced to the periphery of the lungs where they are responsible for the lung markings. Details of the arrangement at the hili may be determined by the use of tomography or contrast studies.

Bronchograms are performed by radiography after introducing contrast medium into the respiratory tract. They allow the trachea and bronchi to be seen in outline.

Bronchoscopy. After anaesthetising the pharynx with a local anaesthetic spray, often in combination with a sedative, a bronchoscope, either rigid or flexible, is introduced via the mouth. The instrument is passed down through the pharynx into the larynx where a view of the vocal cords is obtained. In the trachea, the incomplete bands of cartilage can be seen in the walls; they are deficient posteriorly where the wall is flattened. At the lower end of the trachea, the bifurcation is marked by a sharp mucosal fold at the carina, from which the two primary bronchi lead away on either side. In the left main bronchus and the trachea close to its bifurcation, the transmitted pulsation of the aorta may be seen as movements in the walls of the airway. By passing the bronchoscope further, the main bronchi may be inspected.

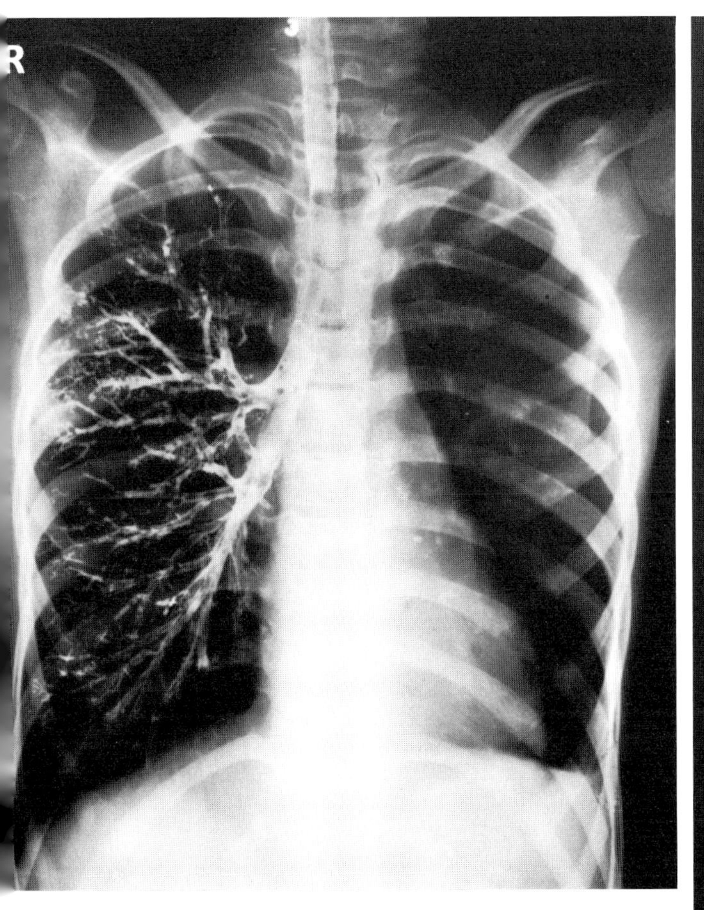

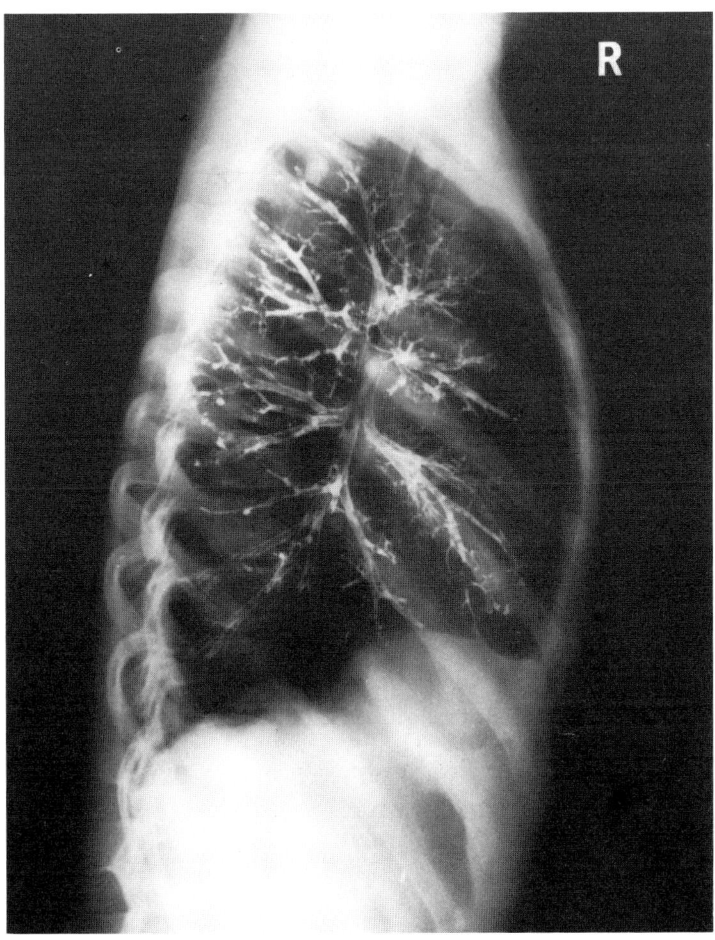

Normal bronchograms of a child demonstrating the bronchi of the right lung PA and lateral views

The Lungs

The lungs are generally markedly radiolucent but the vascular markings and, in appropriate views, the fissures, produce a characteristic pattern. The transradiancy is most marked at the bases of the lungs and is of comparable degree in the two lungs at the same horizontal level.

The vessels are seen as fine white lines on the film, or they may be seen end on as small white circles.

The fissures can only be visualised if they lie tangentially to the path of the x-rays. In the PA film the transverse fissure can be seen from the right hilum to the mid-axillary line in the fifth intercostal space in over 75% of cases. A lateral film is needed to see the oblique fissure. In about 1% of PA films a fine white line runs from the apex of the right lung down towards the hilum. It terminates in a tear-drop shape. This line is produced by the pleura which invests the azygos vein when the vein lies laterally and an azygos lobe is formed medial to the fissure.

Clinically, the lungs are often separated into three zones by imaginary horizontal lines at the level of the anterior ends of the second and fourth ribs. These zones obviously do not correspond to the lobes of the lungs.

The upper zones contain the apices of the upper and lower lobes.

The middle zones contain the remainder of the upper lobes, part of the lower lobes and the upper part of the middle lobe on the right, or the lingula on the left.

The lower zones contain the remaining parts of the middle lobe, or lingula, and the lower lobes.

Radionucleotide lung scanning. A radioisotope is introduced into either the pulmonary vasculature or the respiratory tract and the patient is scanned with a gamma camera.

The scan will indicate the distribution of the isotope.

Intravenous perfusion studies usually use macroaggregates of albumin labelled with the isotope technetium-99m. Injected intravenously it becomes trapped in the pulmonary microvasculature and the scan indicates the pattern of blood flow through the lungs.

Ventilation scans are performed by inhaling a radioactive gaseous isotope, such as xenon-133, and demonstrate air distribution in the lungs.

The two types of study are often used together to demonstrate ventilation-perfusion mismatches.

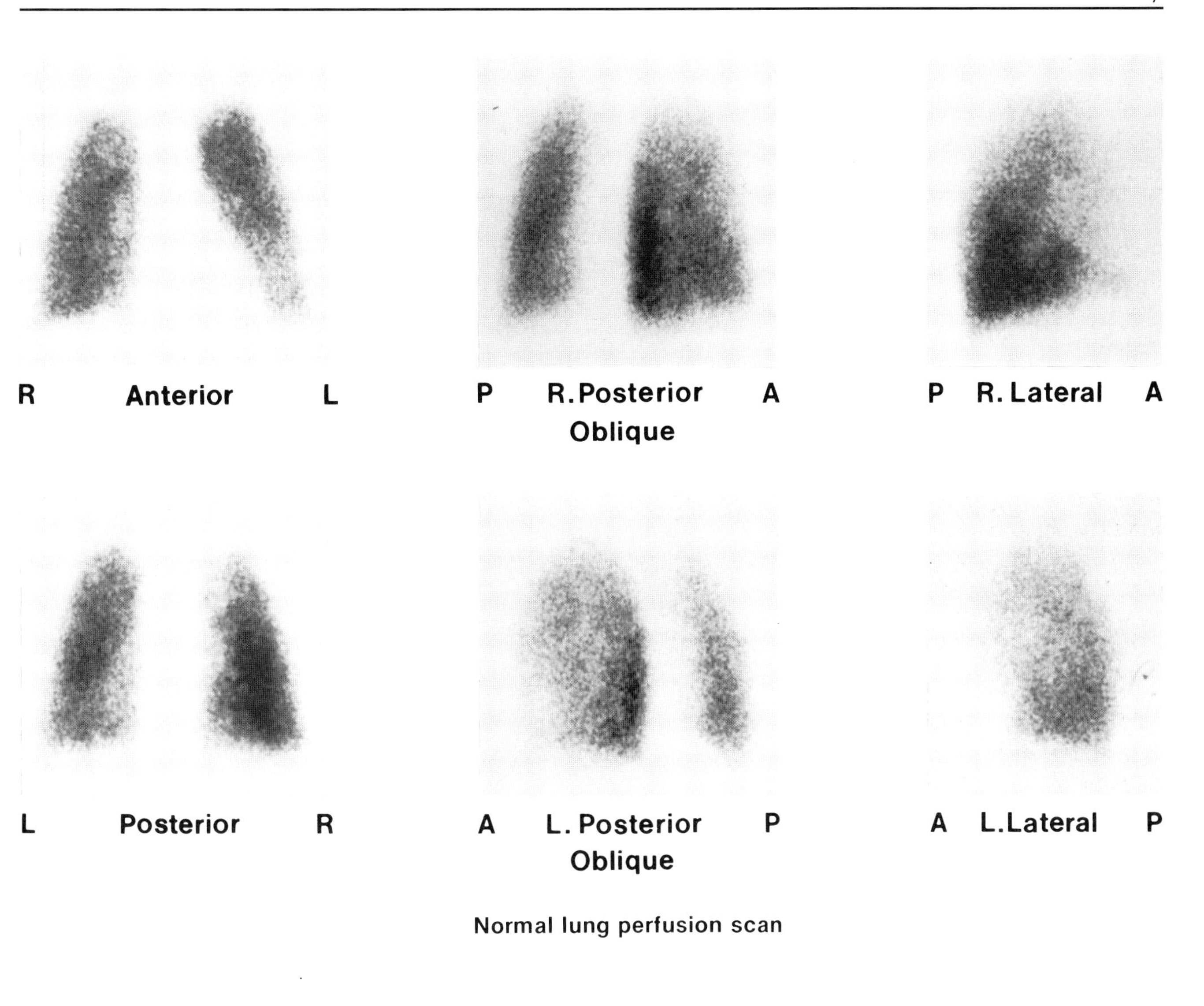

Normal lung perfusion scan

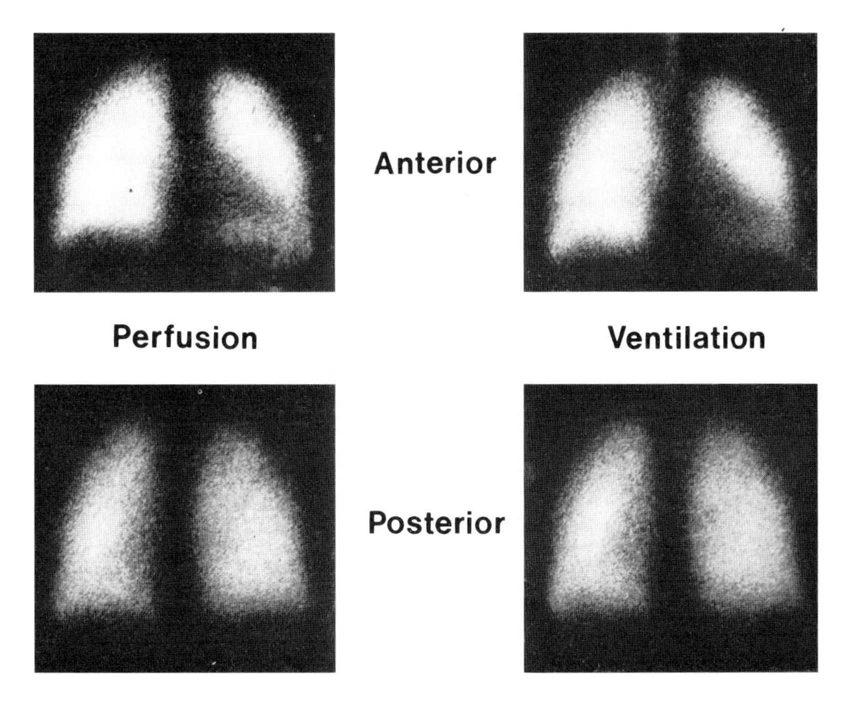

Normal lung ventilation and perfusion scans

7. The Abdomen

THE ANTERIOR ABDOMINAL WALL

The abdomen can be subdivided into regions by a combination of imaginary horizontal planes and vertical lines.

Vertical Lines

The midline [**A**].

The lateral sternal line [**B**] passes along the edge of the sternum at its widest part.

The parasternal line [**C**] lies midway between the lateral sternal and mid-clavicular lines.

The mid-clavicular, or lateral line [**D**] passes through the midpoint of the clavicle which lies about 9 cm from the midline. When extended, the line crosses the costal margin just lateral to the tip of the ninth cartilage, and passes through the femoral point, which lies midway between the anterior superior iliac spine and the pubic symphysis.

The nipple line [**E**] is about 10 cm from the midline.

The anterior axillary line [**F**] passes through the anterior axillary fold produced by the pectoral muscles.

The mid-axillary line lies midway between the anterior and posterior axillary line and, when extended, passes through the tubercles of the iliac crest.

The posterior axillary line passes through the fold produced by the teres major and latissimus dorsi.

The scapular line is drawn vertically through the inferior angle of the bone, with the upper limb in the anatomical position.

The paravertebral line lies along the line of the tips of the transverse processes of the vertebrae.

Horizontal Planes

The xiphisternal plane [**1**] passing through the xiphisternal joint, is at the level of the lower part of the body of T9.

The transpyloric plane [**2**] lies midway between the upper border of the pubic symphysis and the upper border of the manubrium, at the level of the lower part of the body of L1 and the tips of the ninth costal cartilages, approximately one hand's breadth below the xiphisternal joint.

The subcostal plane [**3**] is drawn through the lowest parts of the costal margins (the cartilages of the tenth ribs), at the level of the lower part of the body of L3.

The supracristal plane [**4**] passes through the highest parts of the iliac crests between the levels of the spines of L3 and L4.

The transtubercular plane [**5**] at the level of the spine of L5, is approximately midway between the level of the pubic crest and the transpyloric plane.

The plane of the pubic crest [**6**] is at the level of the lower part of the sacrum.

Usually, the abdomen is divided by the transpyloric and transtubercular planes in combination with the mid-clavicular lines to produce nine regions. Le Clerc's method of subdivision is, however, more clinically useful. The lateral lines are again used but the distance from the xiphisternal joint to the symphysis pubis is divided into thirds.

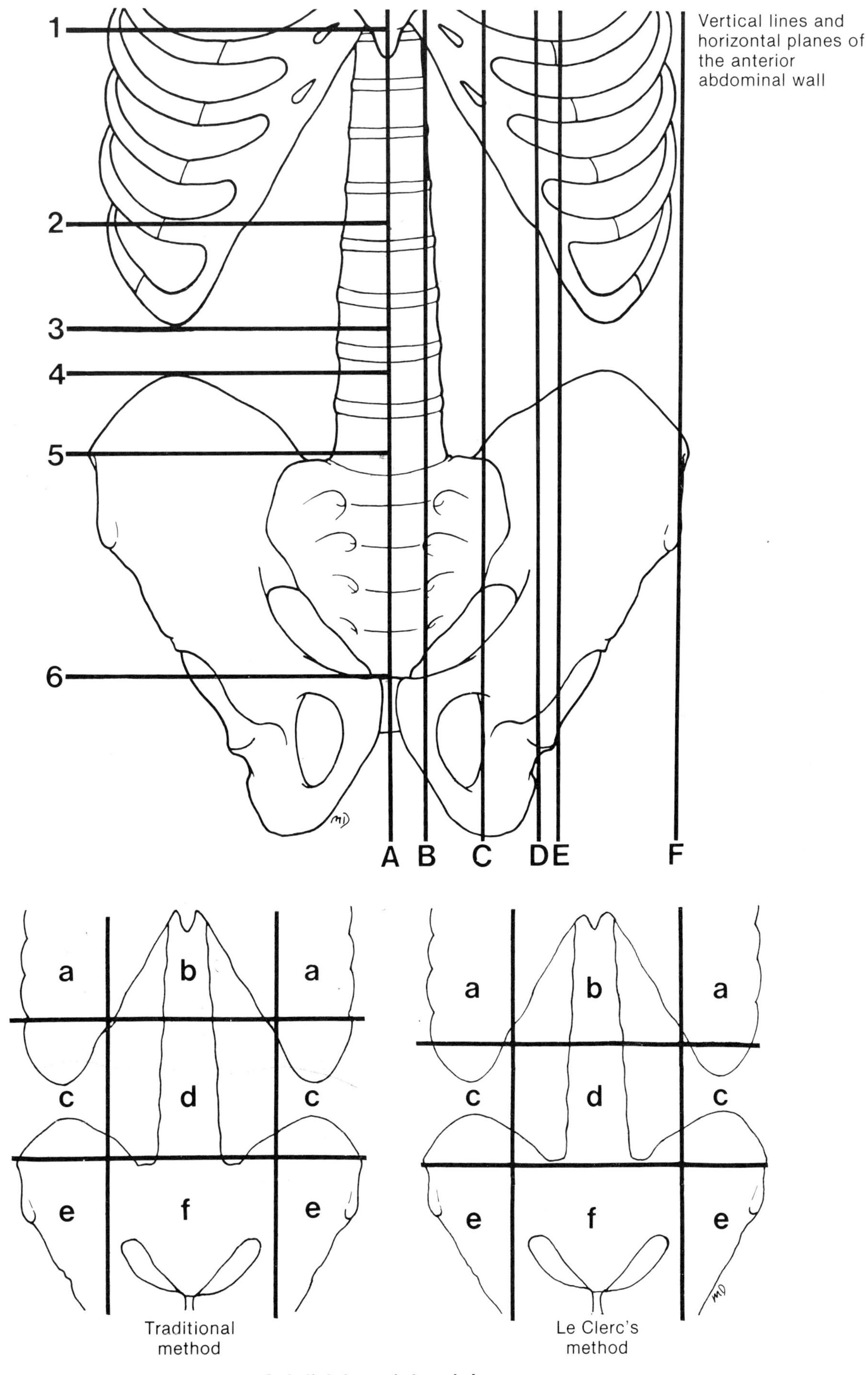

Vertical lines and horizontal planes of the anterior abdominal wall

Subdivision of the abdomen

a – Hypochondrium, b – Epigastric, c – Lumbar, d – Umbilical, e – Iliac, f – Hypogastric

PALPABLE STRUCTURES

The following structures can be felt, and sometimes seen, in the abdomen of a normal individual.

Skeletal Elements

The xiphoid [**1**] lying in the epigastric fossa.

The costal margin [**2**] formed by the cartilages of the seventh to tenth ribs. The eleventh and twelfth ribs and their cartilages can also be felt posteriorly. The tip of the ninth costal cartilage produces a step about halfway along the lower rim of the margin, near the mid-clavicular line.

The iliac crest [**3**] has its highest part posteriorly a little behind its middle. The length of the crest can be felt as far forward as its anterior extremity, the anterior superior iliac spine. The tubercle of the crest produces an obvious protuberance, 5 cm behind the spine and forms the widest part of the pelvis.

The inguinal ligament [**4**] connects the anterior superior iliac spine to the pubic tubercle.

The pubic crest [**5**] extends laterally for 2.5 cm from the midline symphysis, the pubic tubercle, forming its lateral prominence. The tubercle lies in the same vertical plane as the anterior superior iliac spine when standing in the erect position.

The pubic arch [**6**] may be palpated from below between the anterior part of the perineum and the medial side of the thigh. In the male it is more readily felt through the scrotum.

Organs

In a normal individual the following organs may sometimes be palpated.
The lower border of the liver [**A**]
The lower pole of the right kidney [**B**]
The aorta [**C**]
The caecum [**D**]
The vas deferens [**E**]
The sigmoid colon [**F**]

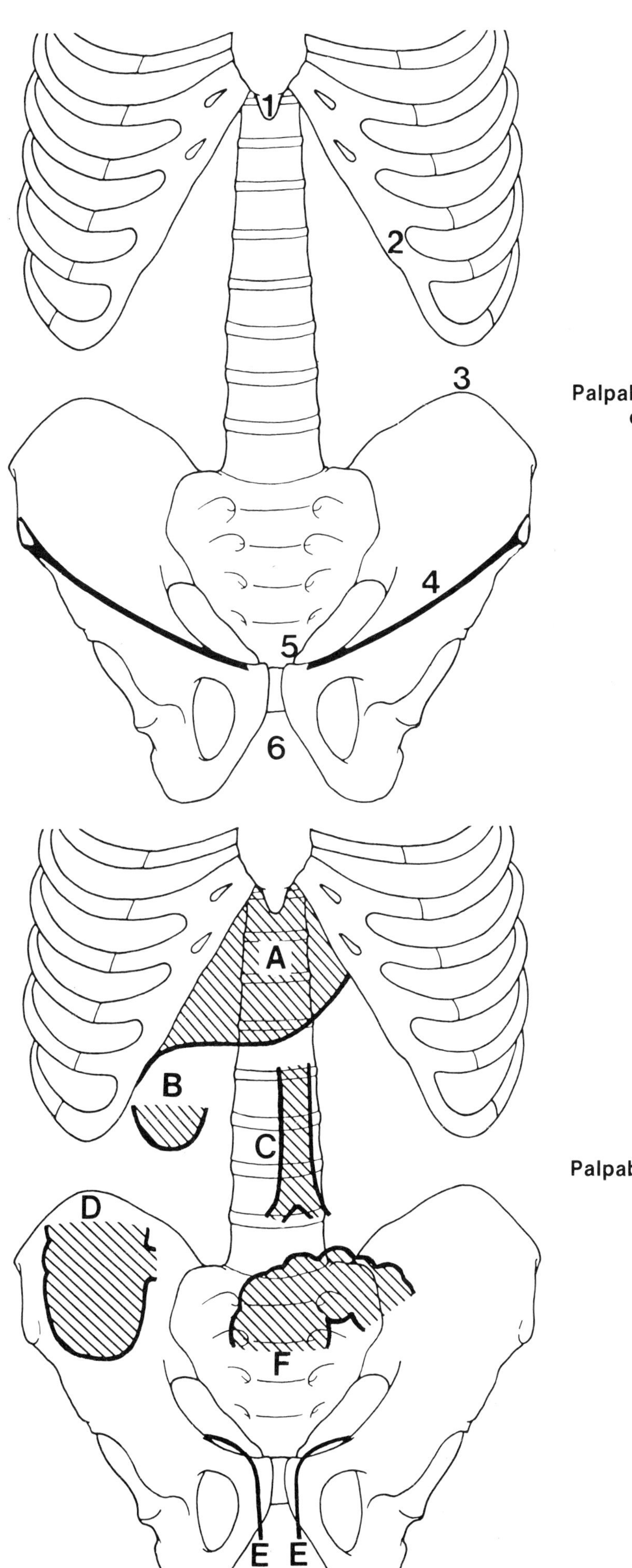

Palpable skeletal elements of the abdomen

Palpable abdominal organs

Muscular and Cutaneous Structures

The umbilicus [1] is an inconstant landmark. In the healthy adult it is approximately at the level of the disc between the third and fourth lumbar vertebrae when recumbent. When standing, in the infant and in the pendulous abdomen, it lies at a lower level.

The rectus abdominis muscle [2] forms a swelling at the side of the midline extending from the pubis to a horizontal line from the xiphoid to the fifth costochondral joint, just below and medial to the nipple. The surface of the muscle may be marked by tendinous intersections, one at the level of the umbilicus, the second at the level of the xiphoid, and the third midway between these two. To demonstrate the muscle, ask the subject to lift his heels 25 cm above the bed whilst keeping the legs straight, or to lift his head against resistance.

The linea alba [3] is overlaid by the median furrow. It is 1.25 cm wide above the umbilicus, but tapers to a narrow band below.

The linea semilunaris [4] produces a shallow curving furrow at the side of the rectus. It runs from the pubic tubercle to the tip of the ninth costal cartilage.

External oblique muscle [5] of which the medial aponeurotic portion and the lateral muscular portions meet at the line between the anterior superior iliac spine and the tip of the ninth costal cartilage. The upper border is marked by a line from the xiphoid horizontally to the fifth rib. The short posterior border runs vertically from the tip of the twelfth rib, at the lateral edge of the sacrospinalis, to the iliac crest. The lower border is the inguinal ligament.

The superficial abdominal reflexes are elicited with the patient supine. The skin of the dermatome tested is stroked lightly towards the midline. A ripple of muscular contraction normally follows the stimulus.

THE NERVES OF THE ABDOMINAL WALL

The iliohypogastric nerve [A] may be marked by a line starting on the back, 5 cm lateral to the first lumbar spine, then running around the trunk, just above the tubercle of the iliac crest to a point 2.5 cm above the pubic tubercle.

The ilioinguinal nerve [B] starts with the iliohypogastric nerve, but passes close to the pubic tubercle to reach the superficial inguinal ring before dividing.

The lateral cutaneous nerve of the thigh [C] is marked on the anterior abdominal wall by a line starting from a point 5 cm lateral to the umbilicus, extending across the inguinal ligament, 1.25 cm from the anterior superior iliac spine, and then a further 10 cm into the thigh. At this point, the gluteal branch is given off and can be represented by a line curving back to the lower part of the buttock. The main body of the nerve is continued to the lateral margin of the patella.

The femoral nerve [D] begins at the same point as the lateral cutaneous nerve of the thigh, but crosses the inguinal ligament 2 cm lateral to the femoral point and then extends for 3 cm into the thigh before dividing.

The obturator nerve [E] begins on the transtubercular plane, 5 cm from the midline, and runs vertically to the level of the anterior superior iliac spines and then curves downwards and medially to a point 2.5 cm lateral to the pubic tubercle.

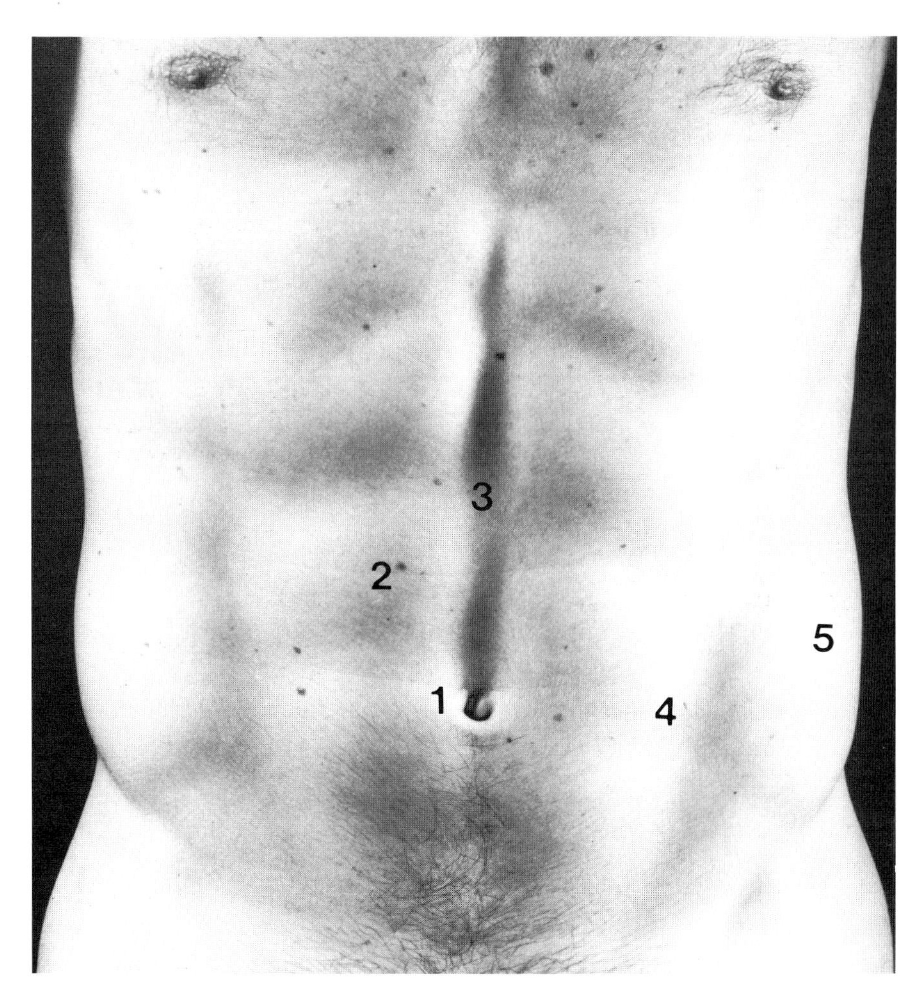

Muscular and cutaneous structures

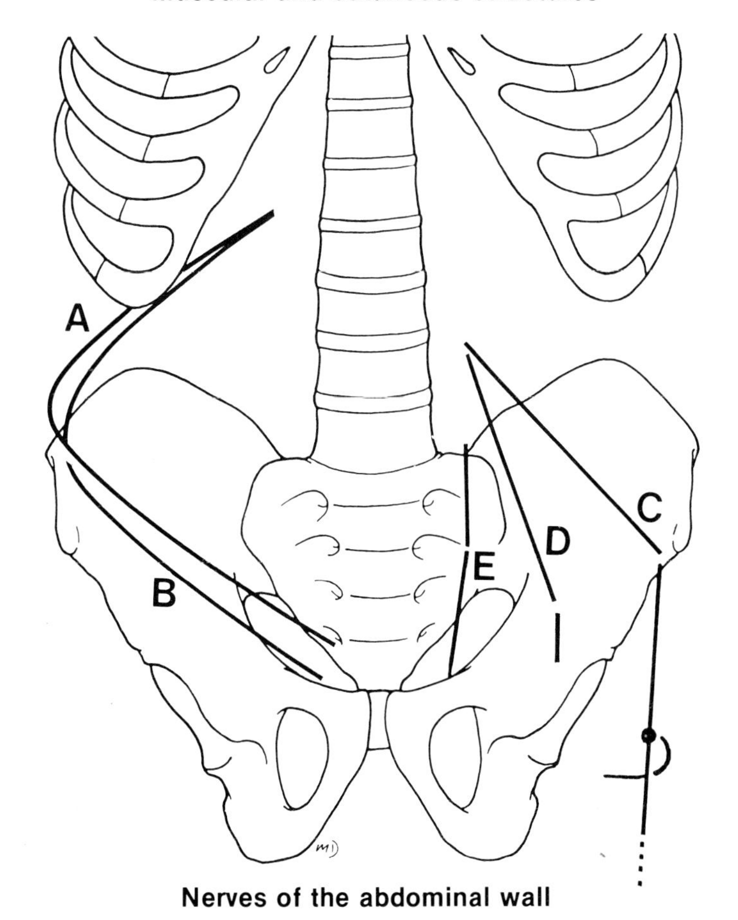

Nerves of the abdominal wall

THE PLAIN ABDOMINAL X-RAY

Both erect and supine views are normally taken of the abdomen. The main purpose of the erect view is the detection of air-fluid levels and free intraperitoneal gas.

Like any other radiograph most information is obtained if the film is studied in a systematic manner.

The bones [**1**]. The bones visible on an x-ray of the abdomen are considered in detail in the other sections of the book.

The stomach [**2**]. The gastric air bubble produces an air-fluid level beneath the left dome of the diaphragm in the erect film; in the supine view the air may be subdivided by gastric rugae to form band-like shadows.

The duodenum [**3**]. An air-fluid level can commonly be seen in the upper part of the duodenum.

The small bowel [**4**]. This lies within the 'picture frame' of the large bowel. Although gas may occupy part of a loop, it is rare for a complete loop to be demonstrated by air. Dilatation of the jejunum by gas reveals the valvulae conniventes which produce complete bands across the bowel giving a 'stack of coins' appearance.

The large bowel [**5**]. Unlike the small bowel, it may normally contain a large amount of gas. Dilatation of the large bowel by gas reveals the haustra which do not cross the full diameter of the lumen; this pattern may be absent distal to the splenic flexure.

The liver [**6**]. This produces a homogeneous shadow beneath the right dome of the diaphragm, tapering to the left. The level of the lower border of the liver may be estimated from the position of gas in the lumen of the hepatic flexure and transverse colon. Enlargement of the liver pushes the diaphragm up, the hepatic flexure, transverse colon and right kidney downwards and the stomach to the left.

The spleen [**7**]. Sometimes the spleen can be seen on the plain x-ray, especially if it is enlarged, when its tip can be seen in the left upper quadrant. Further enlargement displaces the splenic flexure and the left kidney downwards and the stomach to the right.

The urinary tract [**8**]. The appearances of the urinary tract on plain x-ray are considered later in the section on the intravenous urogram.

The psoas muscles [**9**]. These produce soft tissue shadows with well-defined lateral margins extending downwards and laterally.

Calcification. Non-pathological calcification may be seen on the x-ray in the following circumstances:

pelvic vein phleboliths
calcified mesenteric lymph nodes, a result of past tuberculosis
vascular calcification
calcification of the costal cartilages
adrenal calcification.

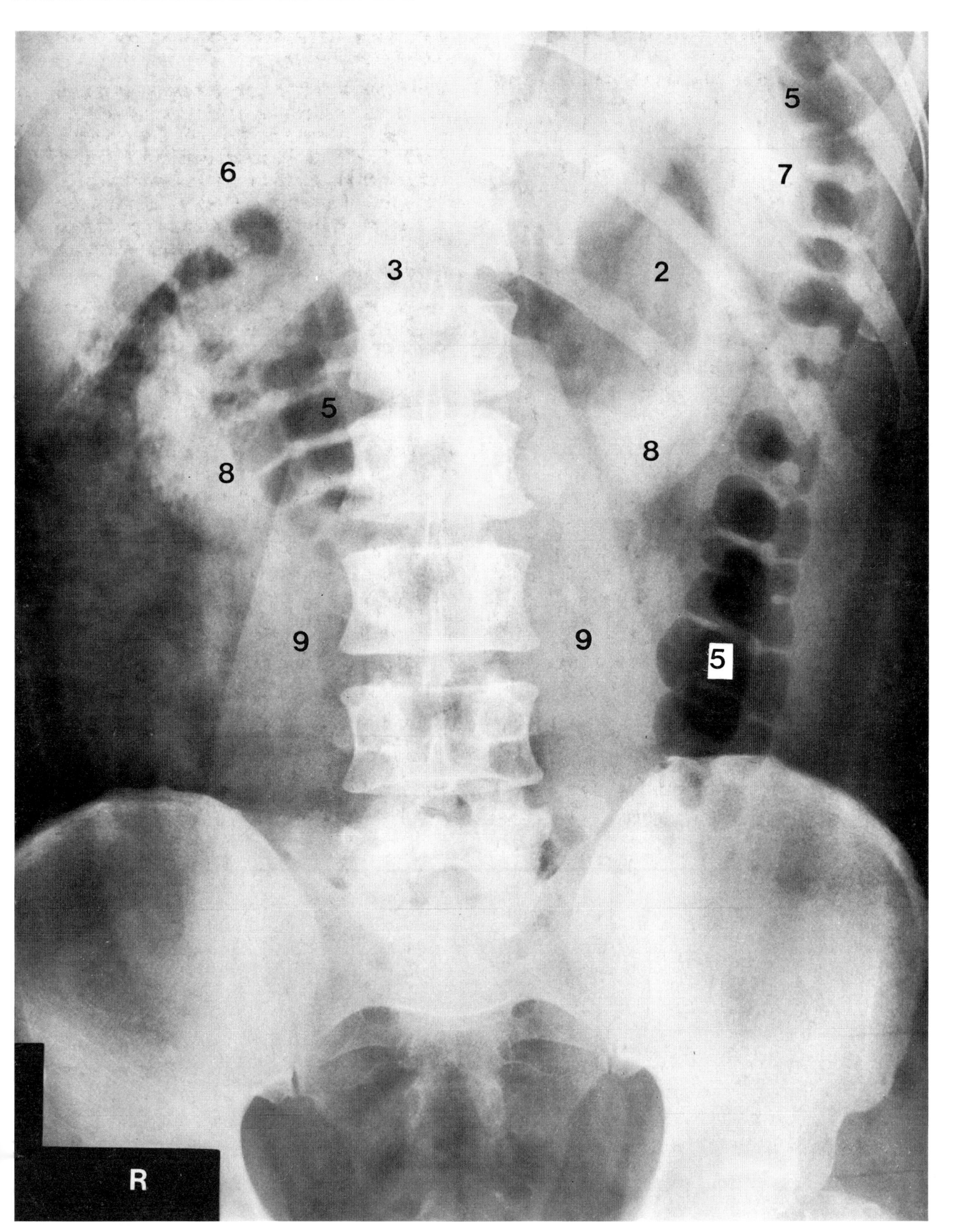

The plain abdominal x-ray

THE INGUINAL REGION

The inguinal region is a confusing area due to similarities in the names of the various constituents and landmarks.

Landmarks

The inguinal groove [**1**]

The femoral or mid-inguinal point [**2**] lies on the inguinal ligament midway between the anterior superior iliac spine and the pubic symphysis. It is 1.25 cm above the femoral pulse and 1.25 cm below the deep inguinal ring. It marks:

the level at which the external iliac artery becomes the femoral artery

the point at which the lateral plane of the abdomen crosses the inguinal ligament

just below the origin of the inferior epigastric and the deep circumflex iliac arteries from the external iliac

just above the origin of the superficial branches of the femoral artery.

The midpoint of the inguinal ligament [**3**] lies on the inguinal ligament midway between the anterior superior iliac spine and the pubic tubercle. It is therefore about 1.2 cm lateral to the mid-inguinal point.

The Inguinal Canal

The inguinal canal is 4 cm long and can be marked as a band 1.5 cm wide between the deep and superficial inguinal rings.

The deep inguinal ring [**A**] like the femoral point lies midway between the anterior superior iliac spine and the pubic symphysis. However, it does not lie on the inguinal ligament but 1.25 cm above it.

The superficial inguinal ring [**B**] lies just above and lateral to the pubic crest and extends laterally no further than the medial third of the inguinal ligament.

The other clinically important structure in this region is the inferior epigastric artery which can be represented as a line drawn from the femoral point to a point 2.5 cm below and lateral to the umbilicus.

It can now be seen that a direct inguinal hernia passing along the inguinal canal can be controlled by pressure at the deep ring, and can be distinguished at operation from an indirect inguinal hernia by the position of the inferior epigastric artery.

THE ILIAC VESSELS

The common and external iliac arteries lie along a line, slightly convex laterally, running from the bifurcation of the aorta at the level of L4, 1.25 cm below and to the left of the umbilicus, to the femoral points.

The common iliac arteries [**1**] are represented by the upper thirds of the lines ending on the intertubercular plane just over 2.5 cm from the midline, where the ureter crosses the pelvic brim in front of the sacroiliac joint.

The external iliac arteries [**2**] are represented by the lower two-thirds of the lines.

The common and external iliac veins [**3**] run along a line drawn from points 1.5 cm medial to the femoral points, up to the point of the formation of the inferior vena cava, at the level of L5, 2.5 cm below the supracristal plane and 2.5 cm to the right of the midline. The common iliac veins lie beneath the upper quarter of these lines.

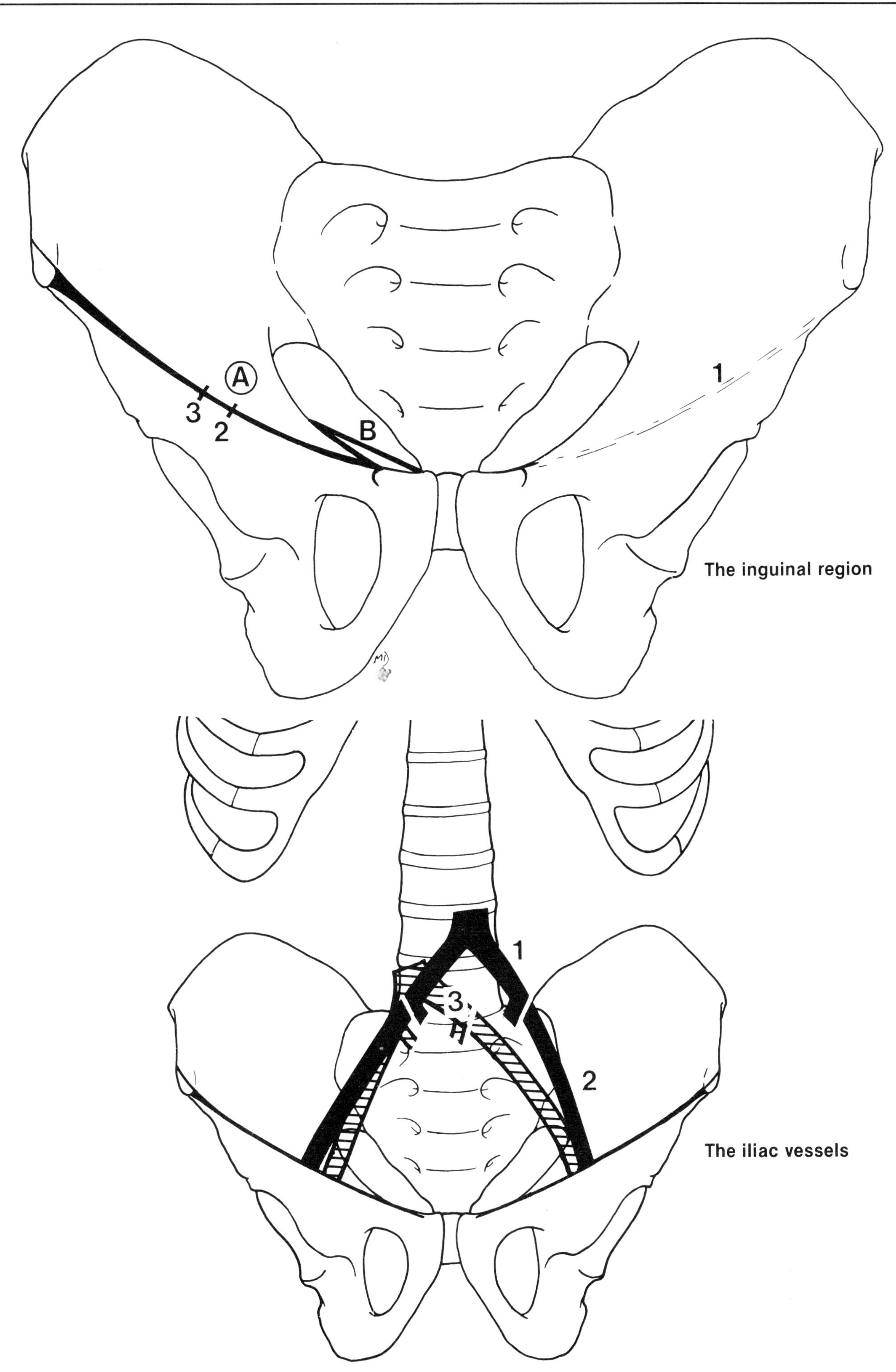

The inguinal region

The iliac vessels

THE OESOPHAGUS

This begins at the lower border of the cricoid cartilage at the level of C6 and runs downwards and slightly to the left to the level of T1. The oesophagus inclines medially to reach the midline at the level of the sternal angle and then continues directly downwards as far as the level of the fifth costal cartilage. Once again it inclines to the left to enter the stomach at the cardiac orifice behind the left seventh costal cartilage 2.5 cm from the midline.

Radiology of the Oesophagus

Although the oesophagus can only be seen in plain x-ray if it is dilated, the film is commonly requested to check for a foreign body. Outlining the oesophagus with barium allows more features to be recognised. Radiographs are taken obliquely so that the vertebrae do not overshadow the oesophagus. The barium swallow shows a smooth outline with three or four long, almost parallel lines, produced by the mucosal folds. The arch of the aorta produces an impression on the left side [**A**]; a smaller impression also on the left may be produced by the left main bronchus [**B**]. The peristaltic waves may be observed if the barium swallow is viewed under fluoroscopy.

Direct visualisation of the lumen of the oesophagus, oesophagoscopy, is most commonly undertaken as part of oesophago-gastro-duodenoscopy.

Radiology of the Gastrointestinal Tract

The amount of information that can be deduced about the tract from the plain x-ray is very limited. The information gleaned may be increased by the use of contrast media. Barium sulphate is the most commonly used medium giving good results as it is markedly radiopaque, it coats the mucosa well and is inert. Problems with barium may arise in the colon where water is reabsorbed and the barium may solidify and impact. Gastrograffin is less radiopaque than barium and is also hypertonic and therefore more irritant if it accidentally enters the lungs.

Because barium is so radiodense, it may obscure the finer details of the mucosa. To some extent this can be overcome by distending the bowel with gas after barium has been introduced. This type of examination is known as a double-contrast study.

Other methods of investigation are being increasingly used in the visualisation of the gastrointestinal tract especially fibreoptic studies, ultrasound, CAT scanning and radioisotope studies.

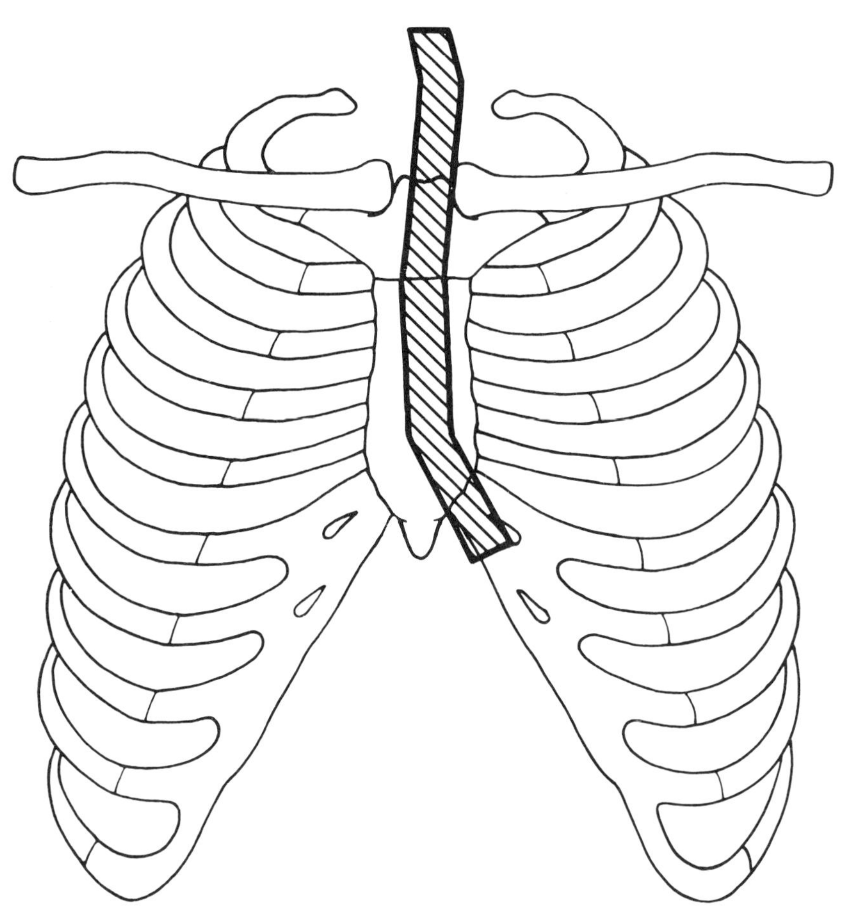

Surface projection of the oesophagus

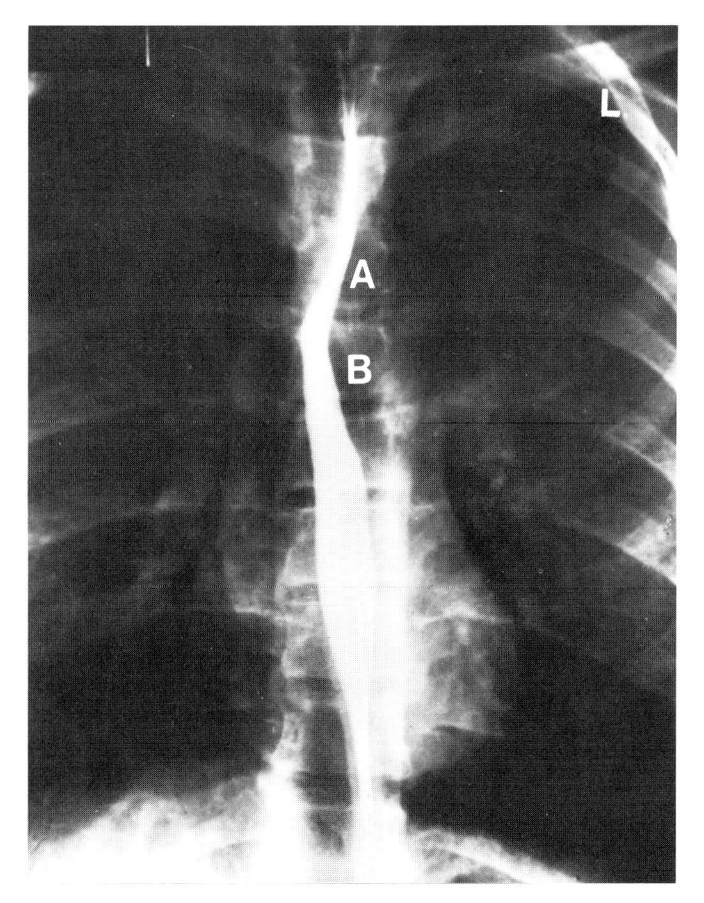

Barium swallow

THE STOMACH

Orifices

The cardiac orifice [**1**] lies behind the left seventh costal cartilage, 2.5 cm from the midline.

The pyloric orifice [**2**] (and prepyloric vein) lies 1.25 cm to the right of the midline on the transpyloric plane.

Contours

The fundus [**A**] can be represented as a semicircle of 7.5 cm diameter starting on the left of the cardiac orifice, the acute angle between the two producing the cardiac notch. The highest point of the fundus reaches the level of the left fifth intercostal space.

The greater curvature [**B**] is continued from the fundus curving gently downwards to the left as far as the tenth left costal cartilage before turning medially to reach the pylorus.

The lesser curvature [**C**] continues from the right side of the cardiac orifice to reach the pyloric orifice. The lowest part of its curving path is marked by a notch, the angular incisure, which lies just to the left of the midline.

Radiology of the Stomach

In a barium meal, the subject swallows 200 ml of contrast medium on an empty stomach, after which films are taken, both erect and supine. Fluoroscopy may also be used to visualise the peristaltic waves. The lesser curve has a smooth edge, which is usually J-shaped, running almost vertically down to the angular incisure then up and to the right. The fundus is outlined when erect by the gastric air bubble filling the concavity of the dome.

The outline of the greater curve is less regular than the lesser curve due to the more prominent mucosal folds. Barium tends to collect in the troughs between the folds, normally three or four in number, extending between the cardiac sphincter and the pylorus. The folds run along a line roughly parallel to the lesser curve and, opposite the incisure, may bulge down and to the left.

In gastroscopy, a fibreoptic tube is introduced into the oesophagus until the distal end arrives at the stomach. The stomach is then distended with air and the tube moved allowing all parts of the lumen to be seen. The mucosa has an orange-red colour and is glistening in appearance. Normally gastroscopy is undertaken as part of oesophago-gastro-duodenoscopy.

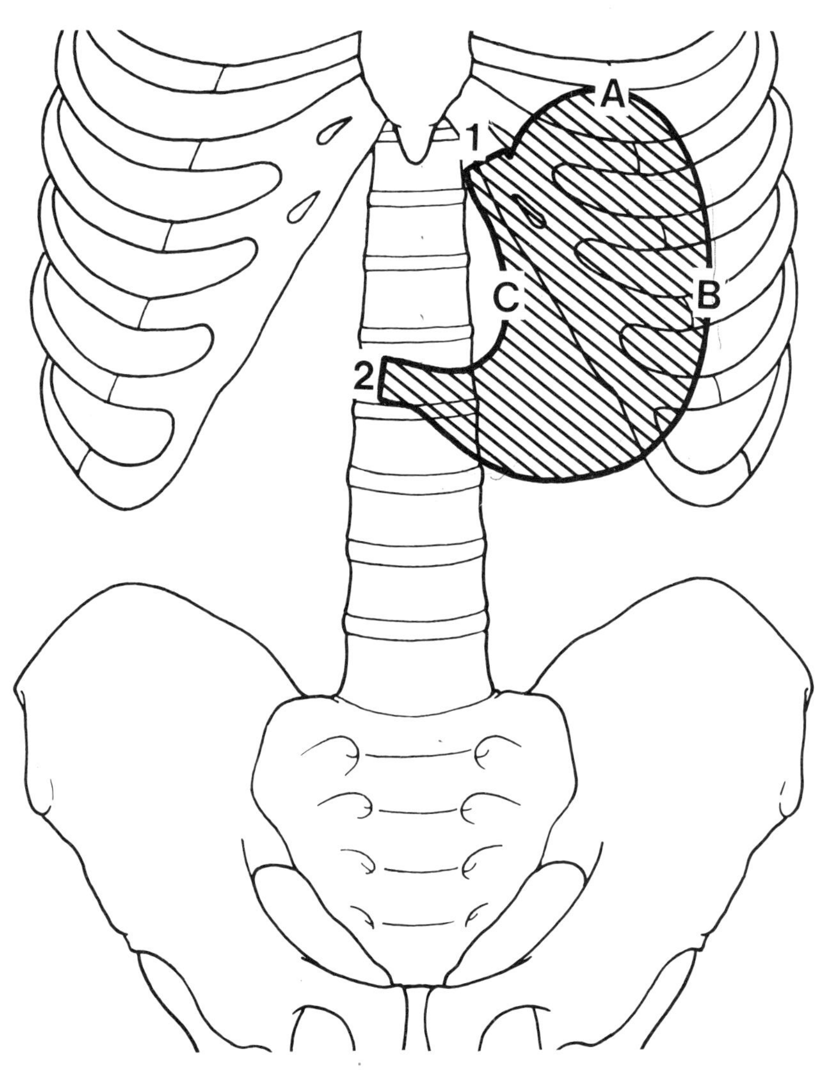

Surface projection of the stomach

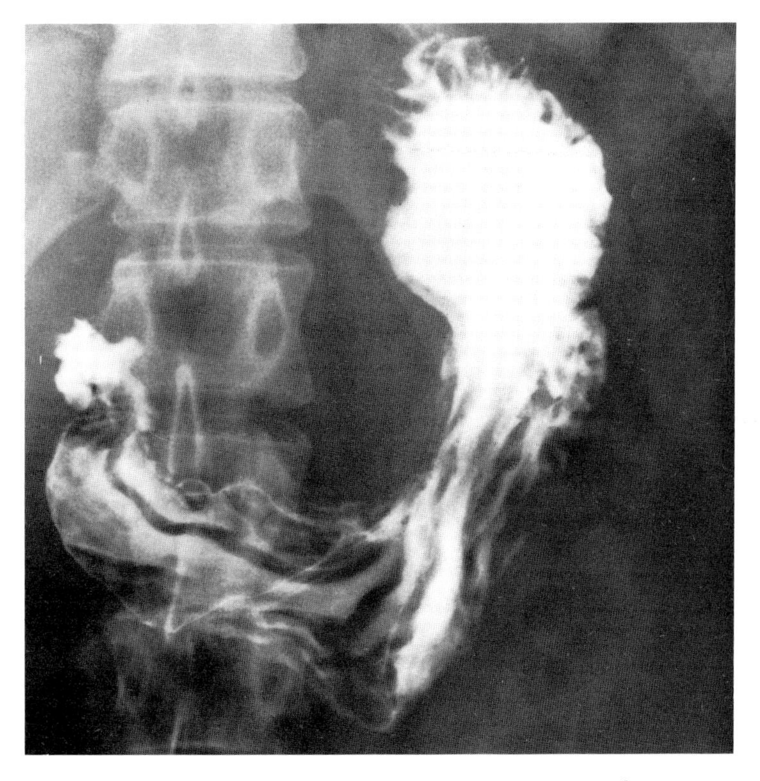

Barium meal demonstrating the stomach

THE DUODENUM

The duodenum is 25 cm long and can be divided into four parts.

The first part [1] is 5 cm long and runs backwards, upwards and to the right. It has a shorter surface projection which can be represented on the body wall by connecting the pylorus to a point about 2.5 cm away just above and medial to the ninth costal cartilage, next to the right side of the middle of the body of L1.

The second part [2] is 7.5 cm long and curves downwards to reach the subcostal plane at the level of L3, 5 cm from the midline. The papilla of Vater, the opening of the common bile and pancreatic ducts, enters halfway along the second part of the duodenum. The accessory pancreatic duct enters about 1.5 cm proximally.

The third part [3] which is 10 cm long, crosses the midline, lying just above the umbilicus, inclining slightly upwards to end just to the left of the midline at the level of the disc between L2 and L3.

The fourth part [4] is 2.5 cm long and ascends to the duodenojejunal flexure, situated 2.5 cm to the left of the midline at the level of the upper part of the body of L2.

The curve between the first and second parts forms the superior duodenal flexure, and that between the second and third parts the inferior duodenal flexure.

Radiology of the Duodenum

A barium meal is squirted through the pylorus into the duodenum by gastric peristaltic waves. The duodenal loop is outlined around the head of the pancreas and may demonstrate abnormalities here. The first part of the duodenum appears radiologically as the duodenal cap or bulb. It is approximately triangular in shape, with the base of the triangle next to the pylorus. Rugae may be seen in its wall running either longitudinally or diagonally. As the barium passes into the second part it becomes more irregular in outline around the transversely running rugae.

Fibreoptic duodenoscopy allows most of the features of the loop to be seen with a forward viewing instrument, but the papilla of Vater is best seen with a lateral viewing scope.

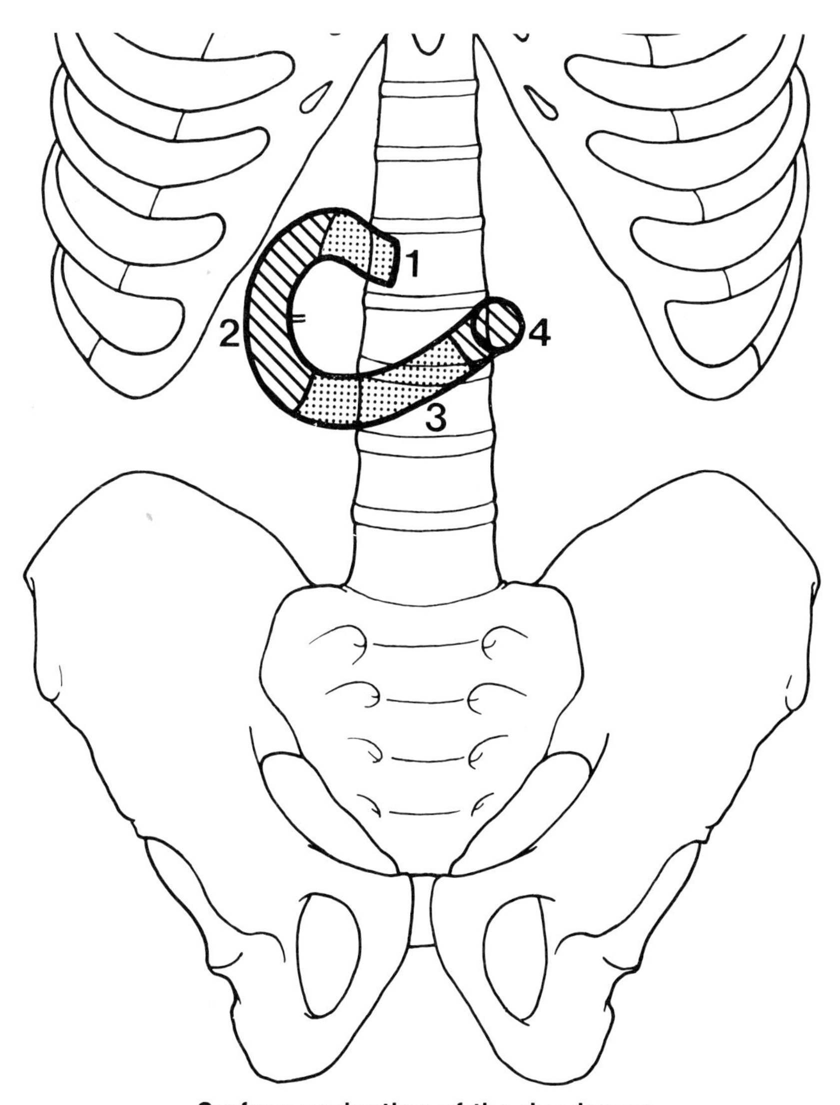

Surface projection of the duodenum

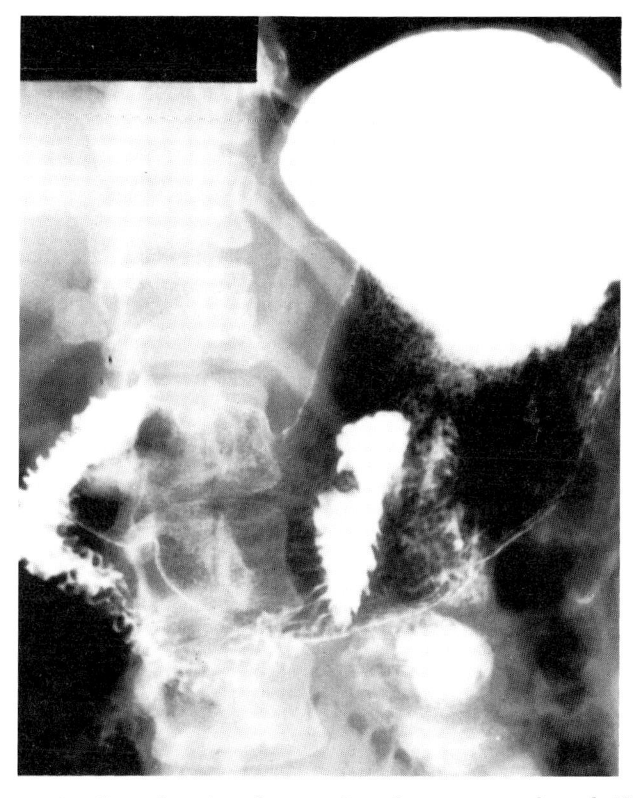

Barium meal demonstrating the duodenum looping across in relation to the stomach

THE SMALL BOWEL AND MESENTERY

The root of the mesentery extends from the duodenojejunal junction, lying 2.5 cm to the left of the midline at the level of the upper part of the body of L2, to the intersection of the transtubercular and right lateral lines lying near the right sacroiliac joint. This line is 15 cm long.

It contains:
the superior mesenteric vessels
the lymphatics draining the small bowel
the autonomic nerves to the bowel.

The small bowel occupies the area beneath the transpyloric plane. The jejunum and ileum may be differentiated by the following features.

	Jejunum	*Ileum*
Length	First 2/5	Last 3/5
Situation	Umbilical region	Hypogastric region and pelvis
Diameter	Two finger breadths	One thumb's breadth
Wall	Thicker, especially proximally	Thinner
Vascularity	Greater, appears redder	Less
Vessels	Fewer arcades	More arcades
Lymphoid tissue	Fewer, smaller follicles	Larger, Peyer's patches
Palpation	'Double thickness'	'Single thickness'
Mesentery	Thinner, less fat	Thicker, more fat

Radiology of the Small Bowel

The traditional method of radiological investigation of the small bowel is to follow the progress of the barium meal through the duodenum and so into the jejunum. Films are taken at regular intervals until the contrast medium reaches the colon. This process usually takes two to three hours but may be even more time consuming in some individuals. The introduction of the contrast medium may be hastened by the small bowel enema, in which the duodenum is intubated and the small bowel flushed with barium.

The diameter of the small bowel, when seen on x-ray, is usually 2 cm, the bore being of roughly the same size throughout except for the terminal ileum which is usually narrower. In the upper parts especially, the outline of the bowel is irregular with a serrated pattern, due to the transversely running mucosal folds, which are about 2 mm wide. The folds are most prominent in the jejunum where, in the contracted bowel, they run diagonally. Distension of the gut shows the mucosal folds as thin lines running directly across the lumen of the bowel, the valvulae conniventes.

Fluoroscopy is often used to assess the function of the ileocaecal valve.

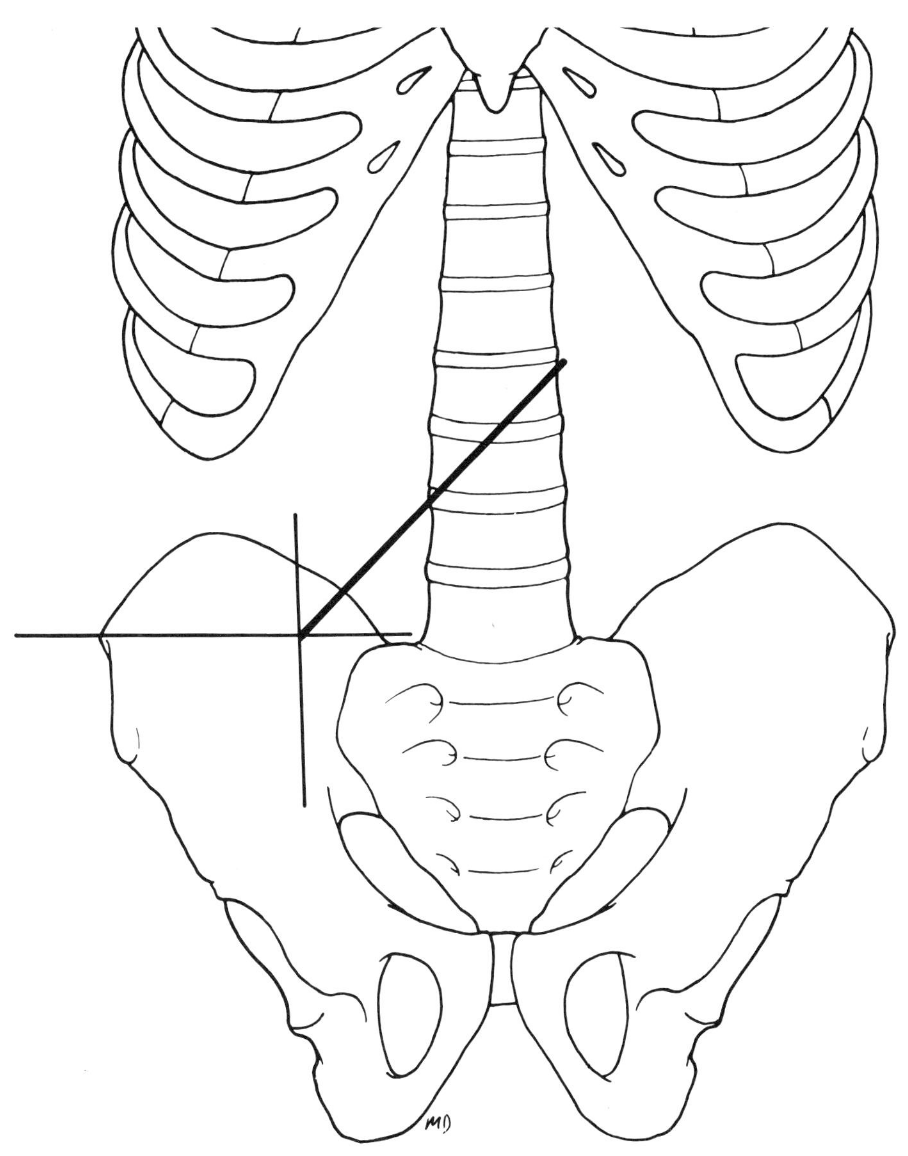

Surface projection of the root of the mesentery

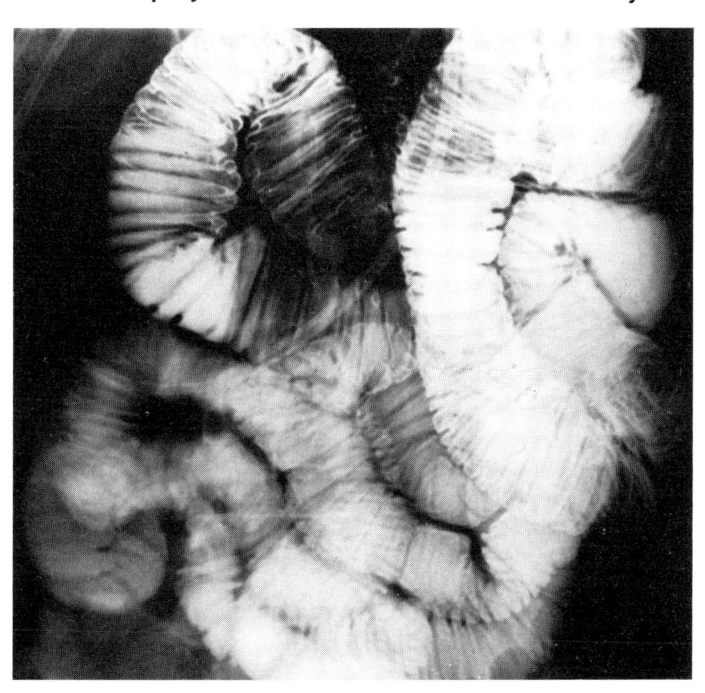

Small bowel enema

CAECUM AND APPENDIX

The ileocaecal valve [**1**]. The valve lies at the point of intersection of the right lateral line and the transtubercular plane near the front of the right sacroiliac joint.

The caecum [**2**]. The caecum lies in the right iliac fossa, usually coming to within 2.5 cm of the inguinal ligament. If distended by faeces or gas it may be palpated through the anterior abdominal wall. It is approximately 7.5 cm in both length and breadth.

The appendix [**3**]. The appendix opens by its base into the caecum, 1.5 cm below the ileocaecal valve. This is marked by McBurney's point which is one-third of the distance along the line connecting the anterior superior iliac spine to the umbilicus.

The epiploic foramen. This is 2 cm wide lying 2.5 cm to the right of the midline and 2.5 cm above the transpyloric plane.

The phrenicocolic ligament. This connects the splenic flexure to the diaphragm in the mid-axillary line at the level of the eleventh rib.

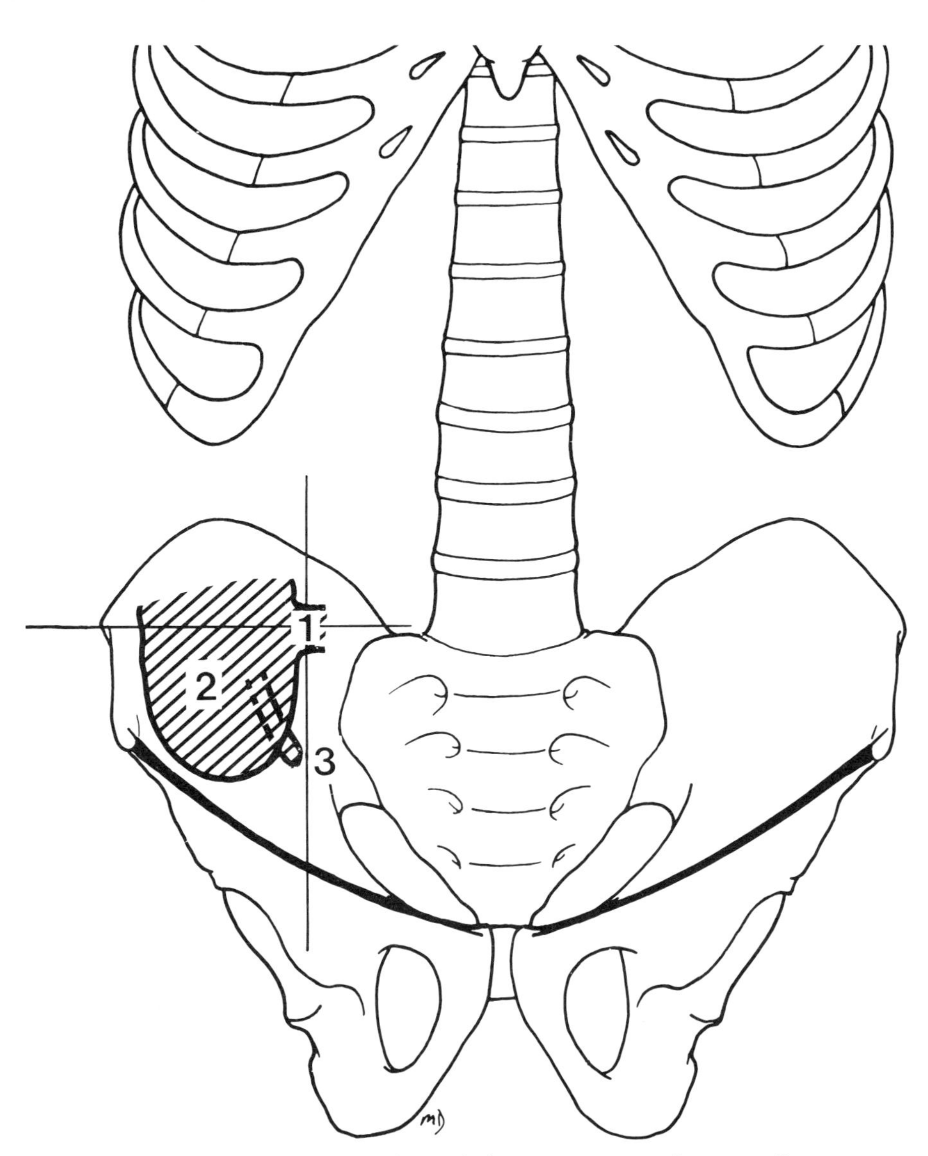

Surface projection of the caecum and appendix

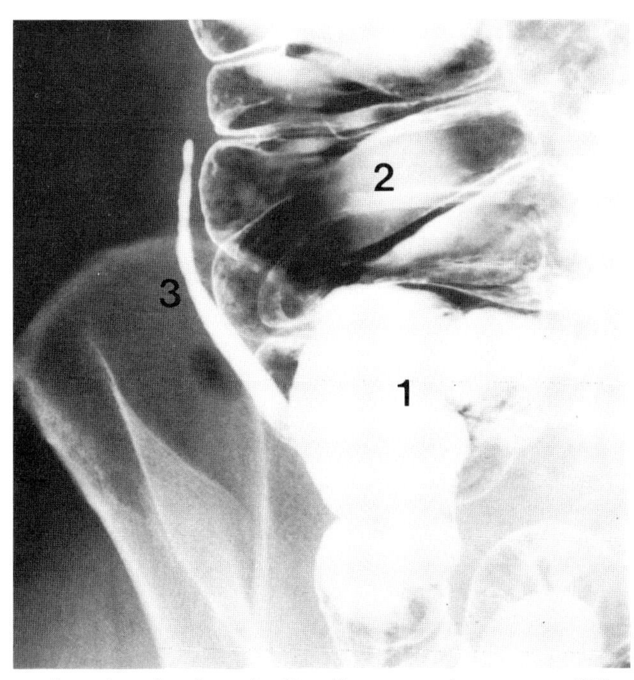

An oblique view showing barium in the ileum and caecum filling the appendix which points rostrally

THE COLON

The ascending colon [**1**]. This is 15 cm long and lies lateral to the right lateral line, ascending from the transtubercular plane as a 5 cm wide band to cross the costal margin and end on the transpyloric plane at the hepatic flexure.

The transverse colon [**2**]. From the hepatic flexure a 5 cm wide band, 50 cm long, is continued across the abdomen, arching around the convex greater curve of the stomach. The lowest point of the curve crosses the umbilicus. The splenic flexure lies under a point 11 cm from the midline just above the transpyloric plane.

The descending colon [**3**]. This can be represented by continuing the band downwards from the splenic flexure for 25 cm. It lies lateral to the left lateral line and descends as far as the inguinal ligament.

The sigmoid colon [**4**]. This is a loop, 40 cm long, whose position and shape are influenced by many factors including the size of its mesocolon, its distension, the state of the rectum and bladder, and the race of the subject. The sigmoid colon starts at the pelvic brim close to the inguinal ligament, and loops across, if unobstructed, into the true pelvis, before being continued as the rectum in front of the third piece of the sacrum at the level of the anterior superior iliac spines.

Radiology of the Large Intestine

A barium enema involves the introduction of barium suspension into the colon via a tube inserted into the rectum. By changing the position of the subject the barium is coaxed into all parts of the colon under the influence of gravity. A double-contrast study is usually carried out and the introduced air may be used to push barium around the colon.

Oblique and supine views are used in order to gain the maximum information around the curvatures of the sigmoid colon, and at the flexures.

The calibre of the colon is shown to decrease distally from the caecum. The ileocaecal valve may project into the caecum causing an apparent filling defect. Sometimes the valve does not prevent the regurgitation of barium into the terminal ileum and, if this occurs, it is important to stop introducing more barium as otherwise the colon will be hidden.

The normal colonic outline is smooth with the exception of the haustrae, these are sacculations that are most marked in the proximal part of the large bowel.

Occasionally the appendix may be filled with barium.

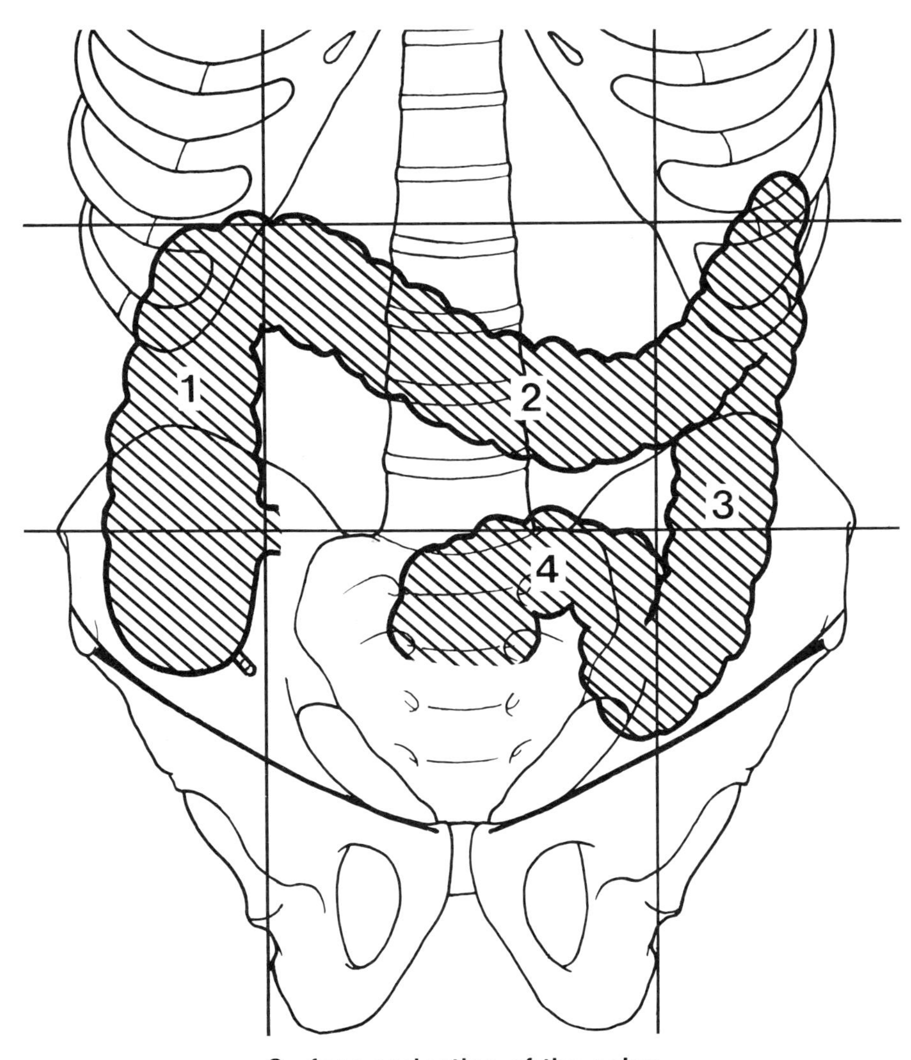

Surface projection of the colon

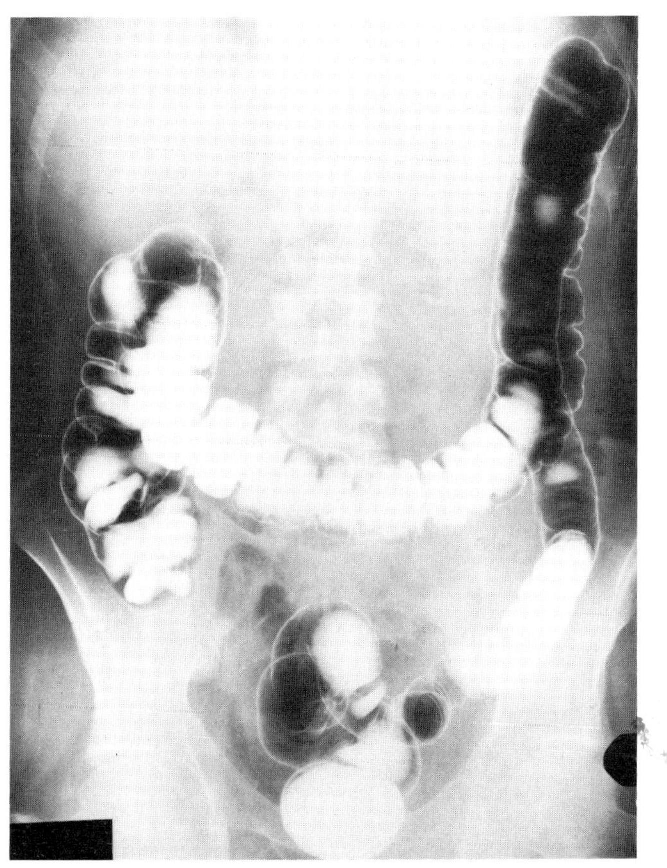

Barium enema

THE RECTUM

The rectum starts in front of the third piece of the sacrum at the level of the anterior superior iliac spines. It ends after 12.5 cm at the anorectal junction, 3.75 cm above the anus, 5 cm from the tip of the coccyx.

Rectal Examination

Position. The patient generally lies in the left lateral position with the neck, hips and knees flexed.

Inspection. The skin around the anal orifice and the orifice itself should be inspected. Normally the skin has folds radiating from the orifice.

Palpation. The tone of the sphincter can be assessed when drawing the buttocks apart and when inserting the right index finger into the bowel. The bowel normally has a smooth wall.

Laterally lie the ischial tuberosities and spines, and the ischiorectal fossae.

Posteriorly the coccyx can be felt and, occasionally, by firm pressure upwards and backwards, the sacral promontory can be reached at the upper limit of the anterior surface of the sacrum.

Anteriorly in the male, the normal prostate is felt as a firm, rubbery structure about 2.5 cm across. The surface is marked by a smooth median sulcus which narrows below. The rectum will slide freely on the surface of the prostate. The seminal vesicles, which lie above the prostate on the base of the bladder, can only be felt if greatly distended. Below the prostate lies the membranous part of the urethra and the bulb of the penis.

In the female, the cervix and part of the body of the uterus may be felt in front of the rectum, and this may be used as a method to assess cervical dilatation during parturition. In a thin woman the ovary may be felt high on the side of the pelvis.

Abnormalities may be detected by rectal examination. These will only be fully appreciated if the examination is performed in a systematic way. The pathological changes can be divided into those:

a. within the lumen
b. in the wall
c. outside the wall which obviously includes both normal and abnormal anatomical relations, e.g. an inflamed appendix, or fluid in the rectouterine or rectovesical pouch of peritoneum. These reach to within 5 cm and 7.5 cm of the anal orifice, respectively.

The length of the average index finger is about 9 cm in the adult male. A mass palpated in the rectum may be pushed down 2 cm or so by asking the patient to bear down. Normally the level of the inferior rectal fold and sometimes the middle rectal fold can be reached.

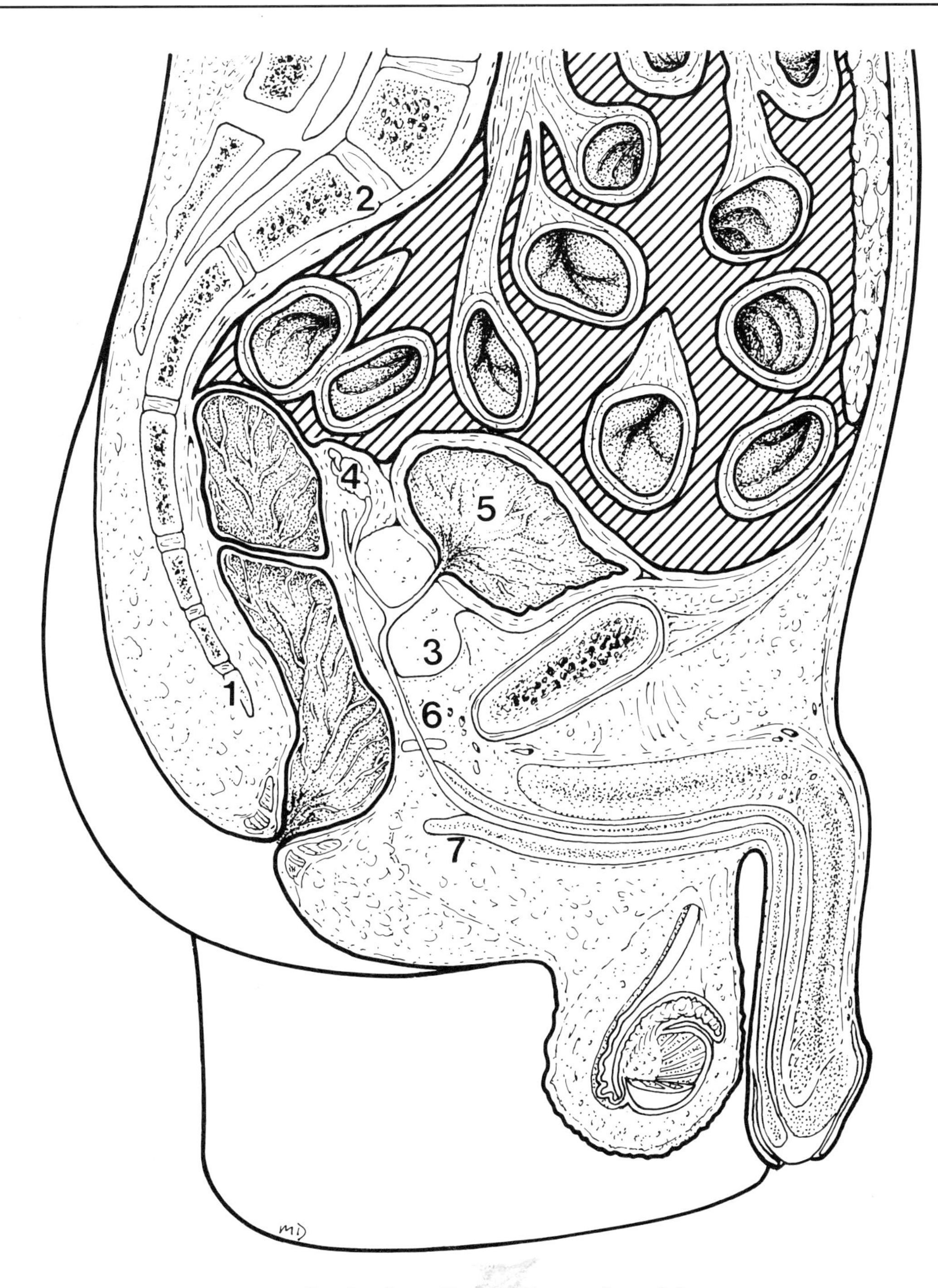

Sagittal section of the male pelvis

1. Coccyx
2. Sacral promontory
3. Prostate
4. Seminal vesicle
5. Bladder
6. Urogenital diaphragm
7. Bulb of the penis

SIGMOIDOSCOPY AND PROCTOSCOPY

Sigmoidoscopy

The most commonly used sigmoidoscopes have a diameter of 1 cm and a length of 25 cm.

With the subject in the left lateral position, the instrument is inserted first in an antero-superior direction and then more posteriorly as the anorectal junction is crossed. The obturator is removed and the light source and bellows for insufflating air are attached. The scope is advanced into the lumen of the rectum using a minimum of air to separate the walls. At 15 cm the rectosigmoid junction is negotiated by angling the instrument to follow the lumen which leads forwards and to the left. The lumen should be inspected systematically while the scope is withdrawn slowly from the fully inserted position.

Proctoscopy

This investigation only allows the visualisation of the lowest 5 cm of the rectum. It is carried out in a similar manner to sigmoidoscopy with the handle of the instrument directed towards the sacrum. The obturator is removed only after the instrument has been fully advanced.

Normal Appearances

The normal rectal mucosa is shining pink and semi-transparent allowing the underlying vessels to be seen. Longitudinal folds are present which disappear when the viscus is distended. Distension, however, only exaggerates the transverse folds. *The upper transverse fold* lies [**1**] at the start of the rectum, *the middle* [**2**] at the upper limit of the rectal ampulla and *the lowest* [**3**] 2.5 cm below the middle fold.

The rectal ampulla [**4**] suddenly narrows at *the anorectal junction* [**5**], at the level of the attachment of the puborectalis fibres of the levator ani. The mucosa here is darker red in colour and more firmly attached to the underlying musculature.

The mucosa of the upper half of the anal canal is plum-coloured due to the internal rectal venous plexus. There are six to ten vertical folds in the mucosa, *the anal columns* [**6**], each containing a terminal branch of the superior rectal vessels. These are most marked in the 3, 7 and 11 o'clock positions. Lower down, the columns are joined by valve-like folds of mucosa, *the anal valves* [**7**], and above each is a recess termed an *anal sinus* [**8**]. The valves lie at the level of the pectinate line, which is opposite the middle of the internal sphincter.

The pecten [**9**] is a 1.5 cm wide transition zone at the start of the lower half of the anal canal. Despite being firmly anchored to the underlying muscle, the submucous veins give the mucosa a shiny blue appearance.

The white line of Hilton [**10**] marks the end of the transitional zone. In the living it is bluish-pink but difficult to see macroscopically. The line marks the gap between the subcutaneous part of the external sphincter and the internal sphincter. The remaining 0.75 cm of the anal canal is formed by *true skin* [**11**] which is white or brown in colour.

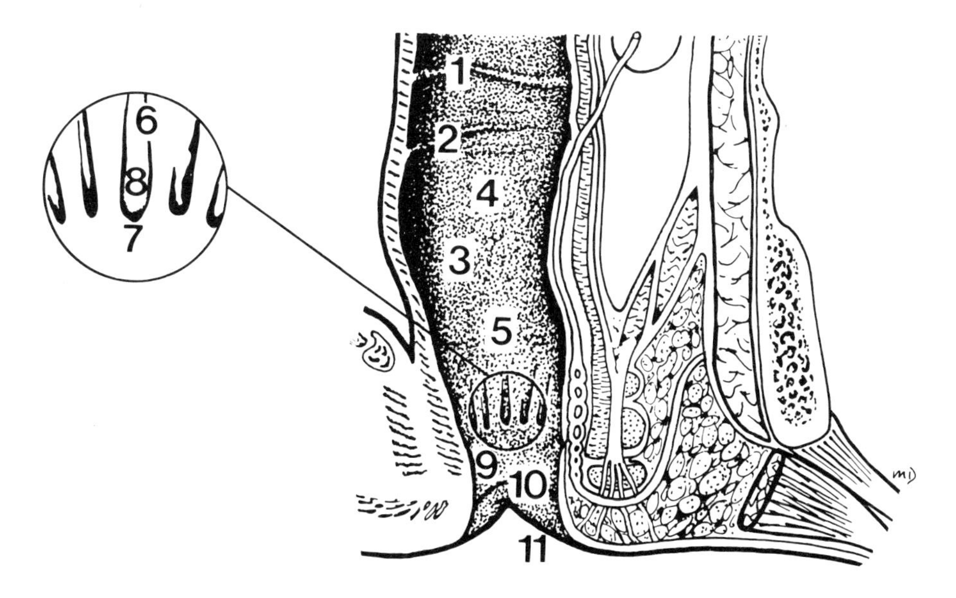

Coronal section of rectum and anal canal

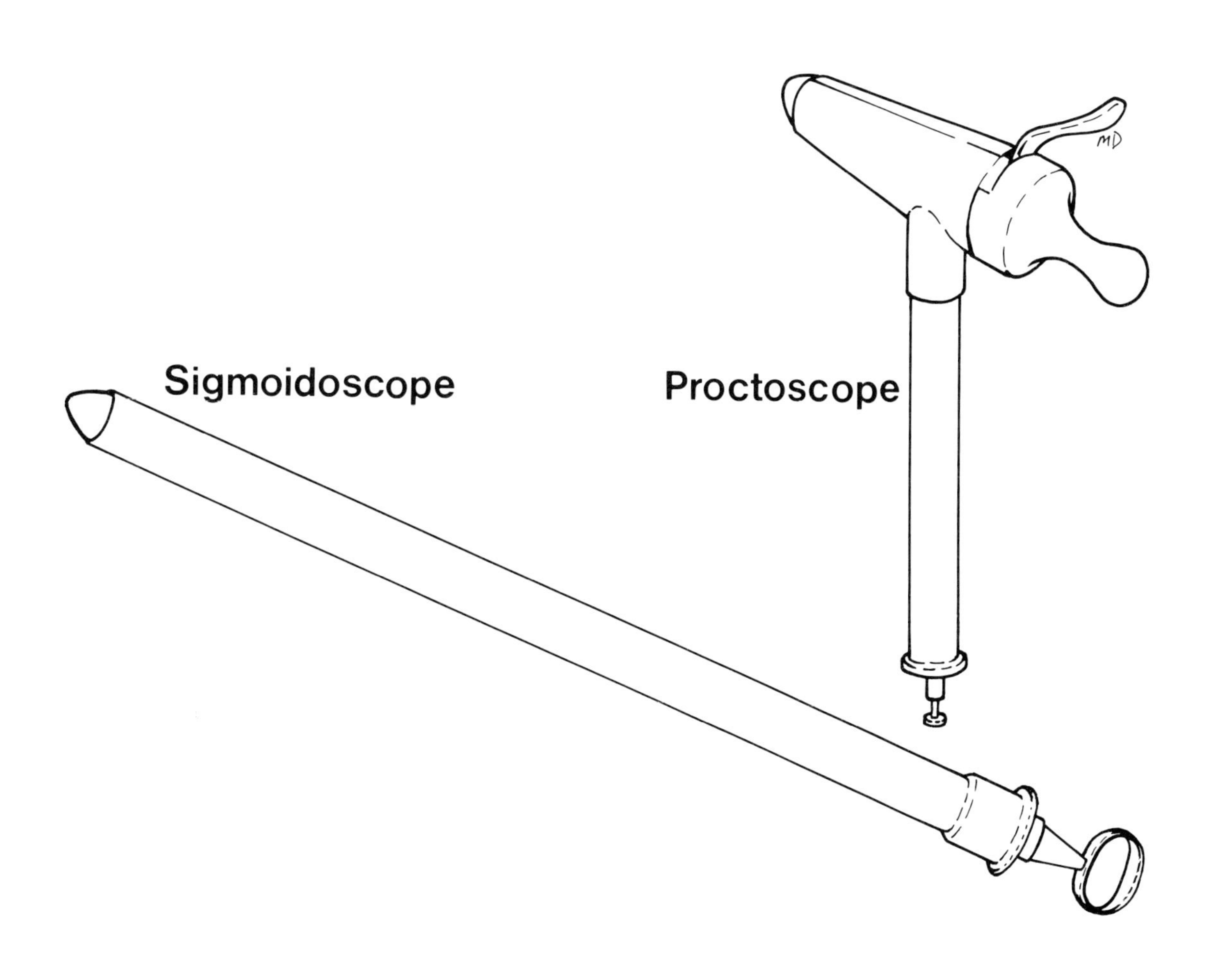

THE ARTERIES OF THE ABDOMEN

The abdominal aorta [**1**] starts when the aorta enters the abdomen at the level of T12, just to the left of the midline, a point about 2 cm above the transpyloric plane. It continues for 10 cm, inclining slightly to the left, to bifurcate at the level of L4, a point most easily represented by measuring 1.25 cm below and to the left of the umbilicus.

The coeliac trunk [**2**] arises from the aorta just after it enters the abdomen, whilst it lies in front of the lowest part of the body of T12.

Branches:

The left gastric artery [**a**] passes up and to the left from the coeliac trunk to lie opposite the seventh costal cartilage, 2.5 cm from the midline, and then curves downwards, convex to the left, to reach the transpyloric plane 2.5 cm from the midline.

The hepatic artery [**b**] runs to the right and slightly downwards from the coeliac trunk for 2.5 cm and then turns sharply to ascend vertically for a further 2.5 cm.

The splenic artery [**c**] is represented by an undulating line from the coeliac axis whose general inclination is to the left and slightly upwards for 8 cm.

The superior mesenteric artery [**3**] leaves the aorta at the level of the transpyloric plane, L1. From this point its course can be represented by extending a line to the intersection of the transtubercular and right lateral lines, and then continuing downwards curving away to the right for a further 7.5 cm.

The inferior mesenteric artery [**4**] begins just to the left of the midline at the level of L3, about 2.5 cm above the bifurcation of the aorta, or 1.25 cm above the umbilicus. It runs to a point 3 cm below the umbilicus, 2.5 cm to the left of the midline.

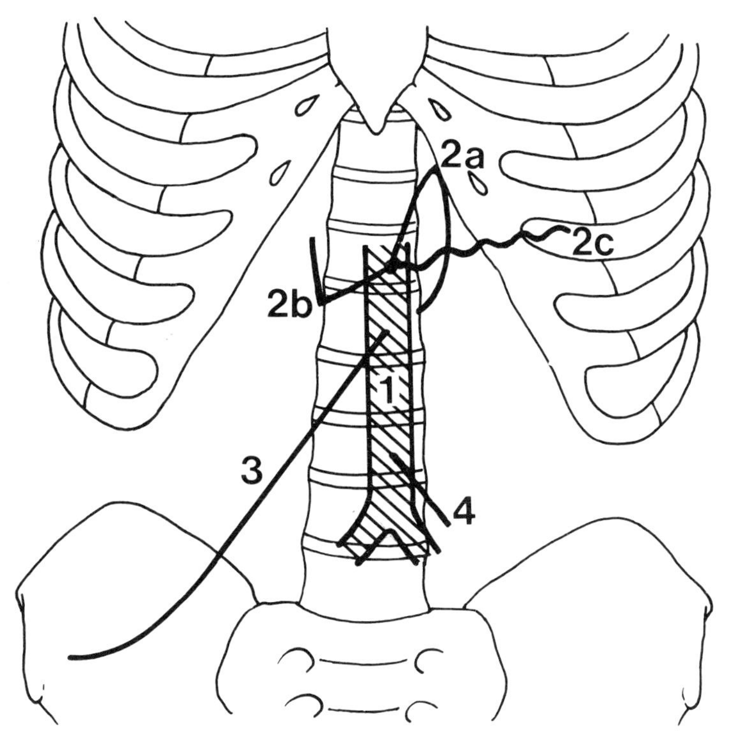

Surface projection of some arteries of the abdomen

Superior mesenteric angiography

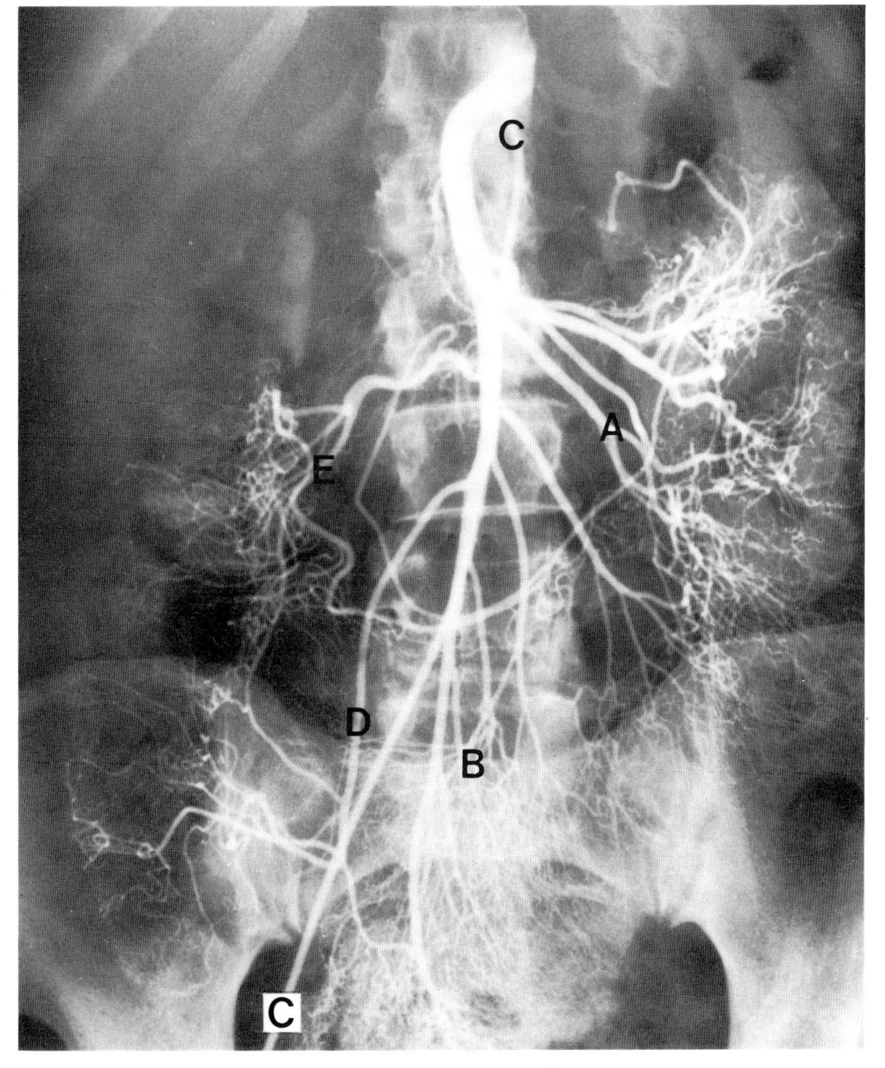

A – Jejunal branches
B – Ileal branches
C – Arterial catheter
D – Ileo-colic artery
E – Anastomosing right and middle colic arteries

The renal arteries [**5**] leave the aorta just below the transpyloric plane, at the level of the upper part of the body of L2, and pass laterally for 3 cm to reach the hilum of the kidney on each side.

The gonadal arteries [**6**] both the testicular and ovarian vessels have a similar origin at the level of the body of L2, and both pass downwards and laterally in the direction of the mid-inguinal points. The testicular arteries reach these points and then continue along the course of the inguinal canals to enter the scrotums. The ovarian arteries, however, only reach halfway to the mid-inguinal points before turning medially to reach the ovaries at the level of the anterior superior iliac spines, just medial to the lateral lines.

THE AUTONOMIC NERVES OF THE ABDOMEN

The coeliac plexus lies at the level of the lower part of the body of T12 and the upper part of the body of L1 around the coeliac trunk and the superior mesenteric artery. The meshwork of fibres forming the plexus connects the coeliac ganglia which lie 2 cm from, and on either side of, the midline.

The abdominal aortic plexus extends downwards from the coeliac plexus as a loose meshwork on the front and sides of the aorta. Inferiorly it continues into the superior hypogastric plexus.

The superior hypogastric plexus lies on the front of the bifurcation of the aorta and between the common iliac arteries at the level of the bodies of the fifth lumbar and first sacral vertebrae, just below the middle of the transtubercular plane.

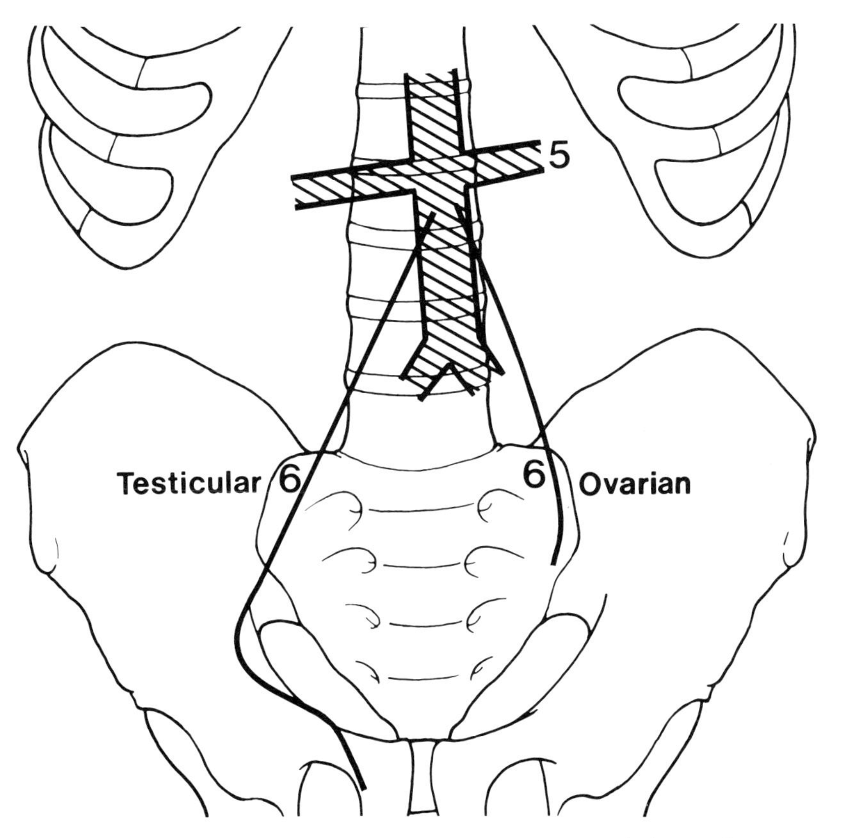

Surface projection of the renal and gonadal arteries

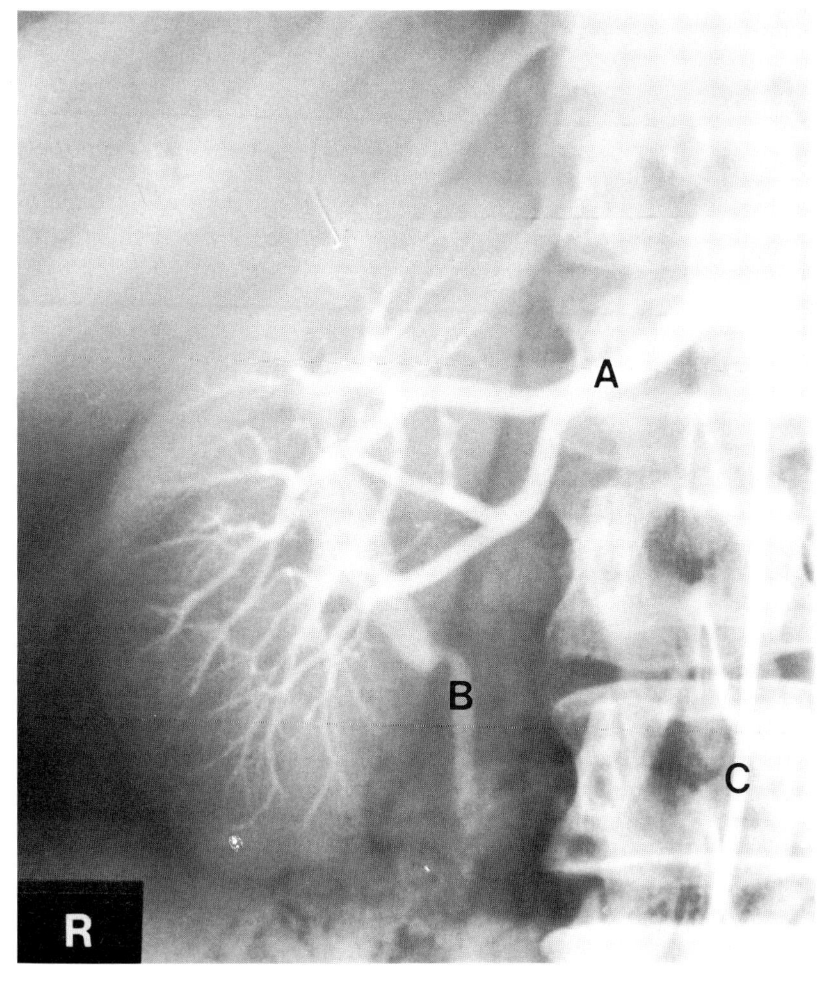

A – Right renal artery
B – Right ureter
C – Arterial catheter

Right renal angiogram

ARTERIES OF THE ANTERIOR ABDOMINAL WALL

The internal thoracic artery divides into its two terminal branches at the level of the sixth intercostal space, 1.25 cm from the lateral border of the sternum.

The musculophrenic artery [1] runs laterally to enter the abdomen behind the eighth costal cartilage and then continues under the cover of the costal margin to reach the tip of the tenth costal cartilage.

The superior epigastric artery [2] descends into the abdomen behind the seventh costal cartilage, and then inclines medially towards the umbilicus to anastomose with the inferior epigastric artery.

The inferior epigastric artery [3] is a branch of the external iliac artery. It can be represented by projecting a line from the mid-inguinal point to 2.5 cm below and lateral to the umbilicus. Hesselbach's triangle, or the inguinal triangle, is bounded by the inferior epigastric artery laterally, the medial half of the inguinal ligament below, and the linea semilunaris medially. A direct inguinal hernia protrudes through this region.

The deep circumflex iliac artery [4] is also a branch of the external iliac artery, and extends laterally from the mid-inguinal point to the anterior superior iliac spine and then along the anterior half of the iliac crest.

The superficial epigastric artery [5] arises from the femoral artery, 1.25 cm below the inguinal ligament, and then runs upwards and slightly medially for 10 cm.

The obliterated umbilical artery [6] produces the lateral umbilical ligament which runs along a line from the umbilicus to a point just lateral to the pubic tubercle.

VEINS OF THE ABDOMEN

The inferior vena cava [**A**]. This can be represented as a 2.5 cm wide band running along a line 2.5 cm to the right of the midline. It begins at the level of the body of L5, 2.5 cm below the supracristal plane, and ascends to leave the abdomen by passing through the diaphragm at the level of T10, directly behind the sternal end of the sixth right costal cartilage.

The renal veins [**B**]. Both lie 2 cm below the transpyloric plane and extend horizontally from the inferior vena cava to reach the hili of the kidneys which lie 5 cm from the midline.

The gonadal veins [**C**]. The testicular veins follow a line drawn from the mid-inguinal points to points lying 2.5 cm below the transpyloric plane, 2.5 cm to the side of the midline. On the right this marks the opening of the vein into the inferior vena cava, on the left the opening into the left renal vein.

The ovarian veins have similar terminations but distally follow a parallel course to the ovarian arteries.

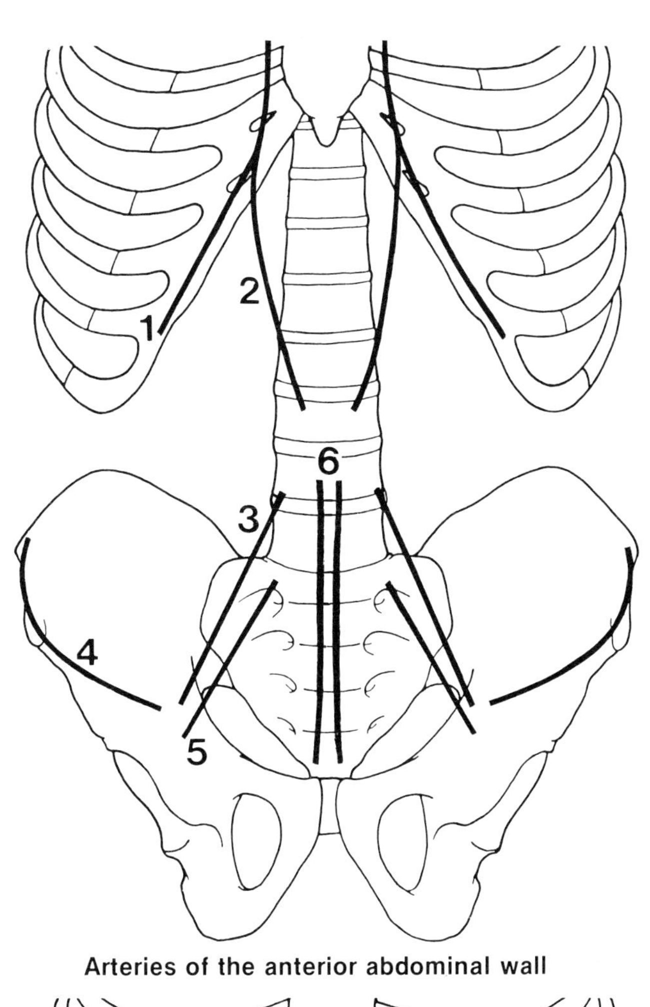

Arteries of the anterior abdominal wall

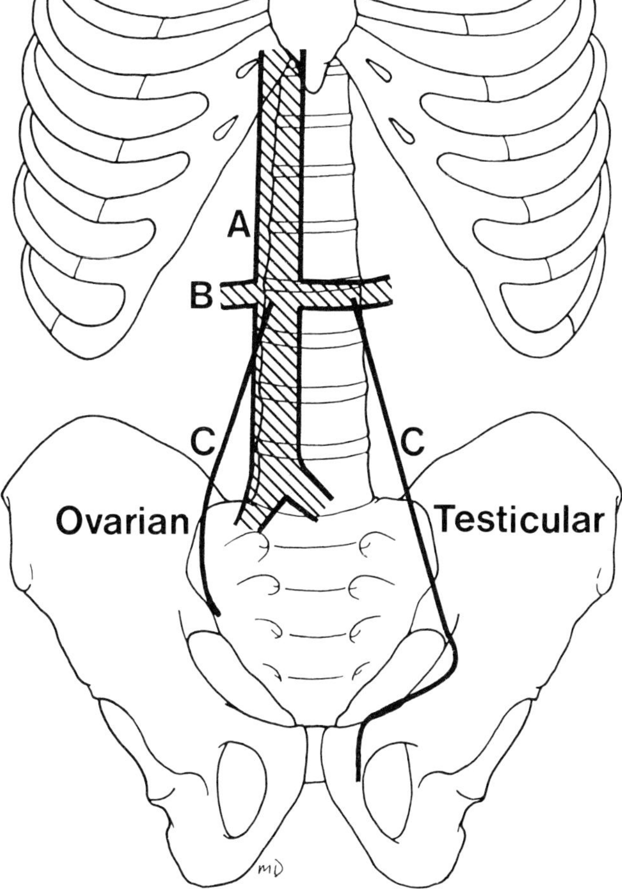

Veins of the abdomen

THE VENOUS DRAINAGE OF THE GUT

The portal vein [**1**] follows a line starting 2.5 cm to the right of the midline just below the transpyloric plane at the level of L2, extending upwards and slightly to the right for 5 cm.

The splenic vein [**2**] is best represented by retracing its course from the origin of the portal vein, along a line running to the left and slightly upwards for 10 cm.

The inferior mesenteric vein [**3**] generally starts 2.5 cm to the left of the midline 3 cm below the umbilicus, entering the splenic vein on the transpyloric plane 2.5 cm to the left of the midline.

The superior mesenteric vein [**4**] begins at the intersection of the intertubercular and right lateral planes and then describes a gentle curve convex to the left to meet the splenic vein and form the portal vein.

Venography

An arterioportogram is performed by injecting contrast medium into the coeliac axis, (or splenic artery or superior mesenteric artery). Films are taken at intervals throughout the investigation, the later films showing the contrast medium in the portal venous system.

In splenoportography, contrast medium is injected into the splenic pulp and drained by the venous system of the spleen into the splenic vein and so into the portal vein which is shown to be 4–8 mm in width.

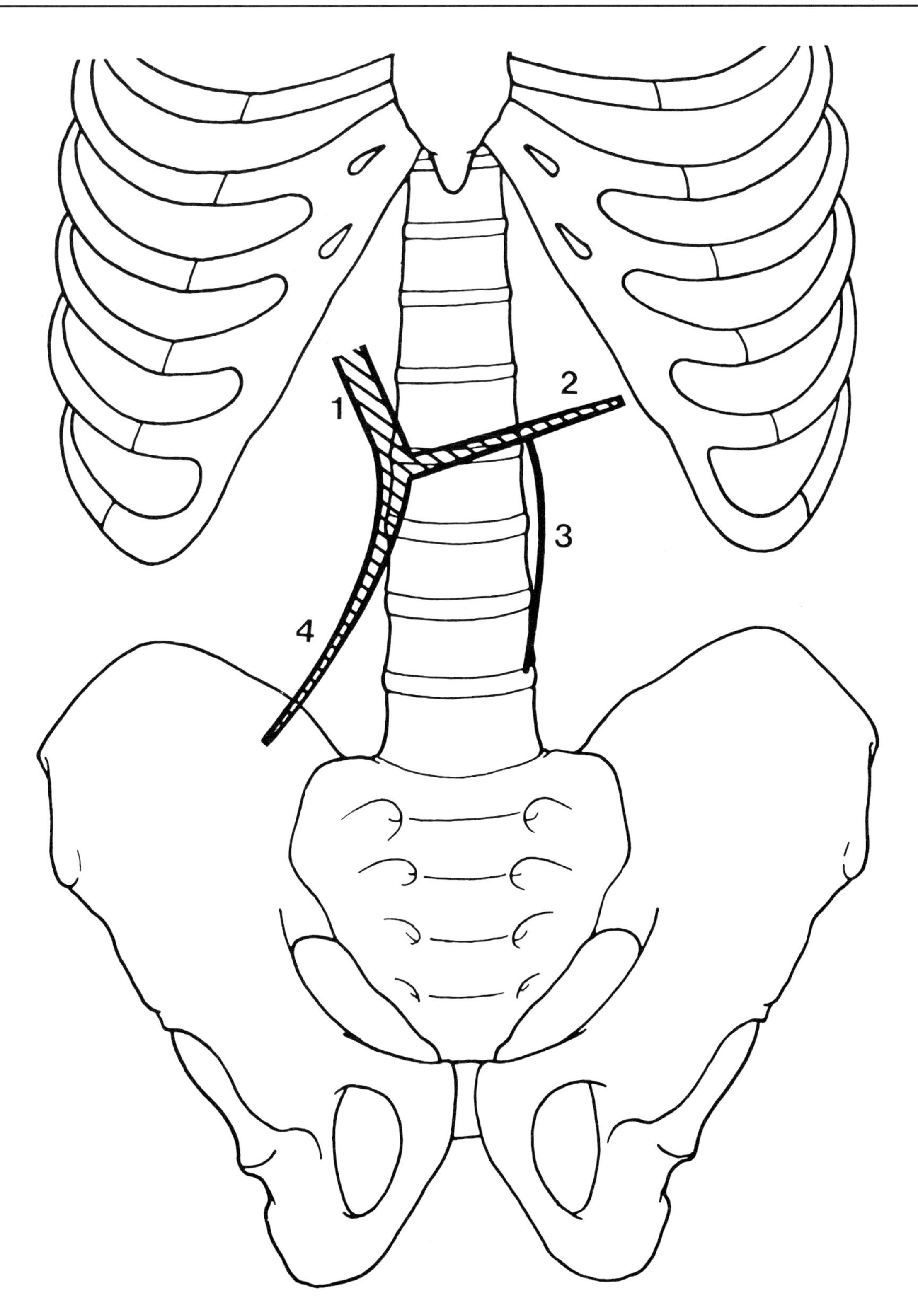

Venous drainage of the gut

THE LIVER

The markings of the anterior surface of the liver are somewhat complex but are very important in clinical practice.

A point marking the right upper limit [**1**] lies in the mid-clavicular line at the upper margin of the fifth rib, just below the nipple.

The upper border [**2**] curves down to cross the midline behind the xiphisternal joint and then runs almost horizontally for over 7.5 cm to the upper left limit.

The upper left limit [**3**] lies in the mid-clavicular line below the nipple on the lower margin of the fifth rib.

The lower border [**4**] descends sharply from this point to cross the left costal margin behind the tip of the eighth cartilage. The border crosses the midline at the level of the transpyloric plane to reach the lower right limit. The lower border is marked by a concavity near the tip of the right ninth costal cartilage. Despite its close relationship with the anterior abdominal wall the lower border cannot be palpated in the majority of subjects, even in women and children in whom it lies at a slightly lower level.

The lower right limit [**5**] lies close to the lowest point of the right costal margin but in the normal subject may be up to 1.2 cm below it.

The right border [**6**] is convex to the right connecting the upper and lower right limits.

The falciform ligament has a free lower border extending from the umbilicus to the lower border of the liver, halfway between the midline and the costal margin close to the transpyloric plane. The parietal attachment of the ligament is marked by a line connecting the umbilicus to the xiphisternal joint.

Imaging of the liver

The plain radiographic appearances of the liver have already been discussed in the section on the plain abdominal x-ray.

CAT scanning is being increasingly used to visualise hepatic architecture. It is normally capable of differentiating between biliary and vascular radicles and, if problems of this nature do exist, then a contrast medium can be introduced intravenously to highlight the vascular elements.

Ultrasonic scanning still remains the commonest method used to image the liver. It has disadvantages in that it can only detect pathologically enlarged intrahepatic ducts, and in 10% of patients the examination is unsatisfactory, either due to bowel gas or obesity. The common bile duct and portal vein have a similar diameter when seen on the scan, the vein lying immediately posterior to the duct; this produces a view similar to that looking down the barrels of a gun, the 'shot gun sign of Weill'.

Radionucleotide examinations involve the intravenous injection of radioactive technetium-labelled iminodiacetic acid derivatives. These compounds are taken up by the Kupffer cells and handled like bilirubin. Scanning the upper abdomen shows whether all or part of the liver is participating in the excretion of these compounds.

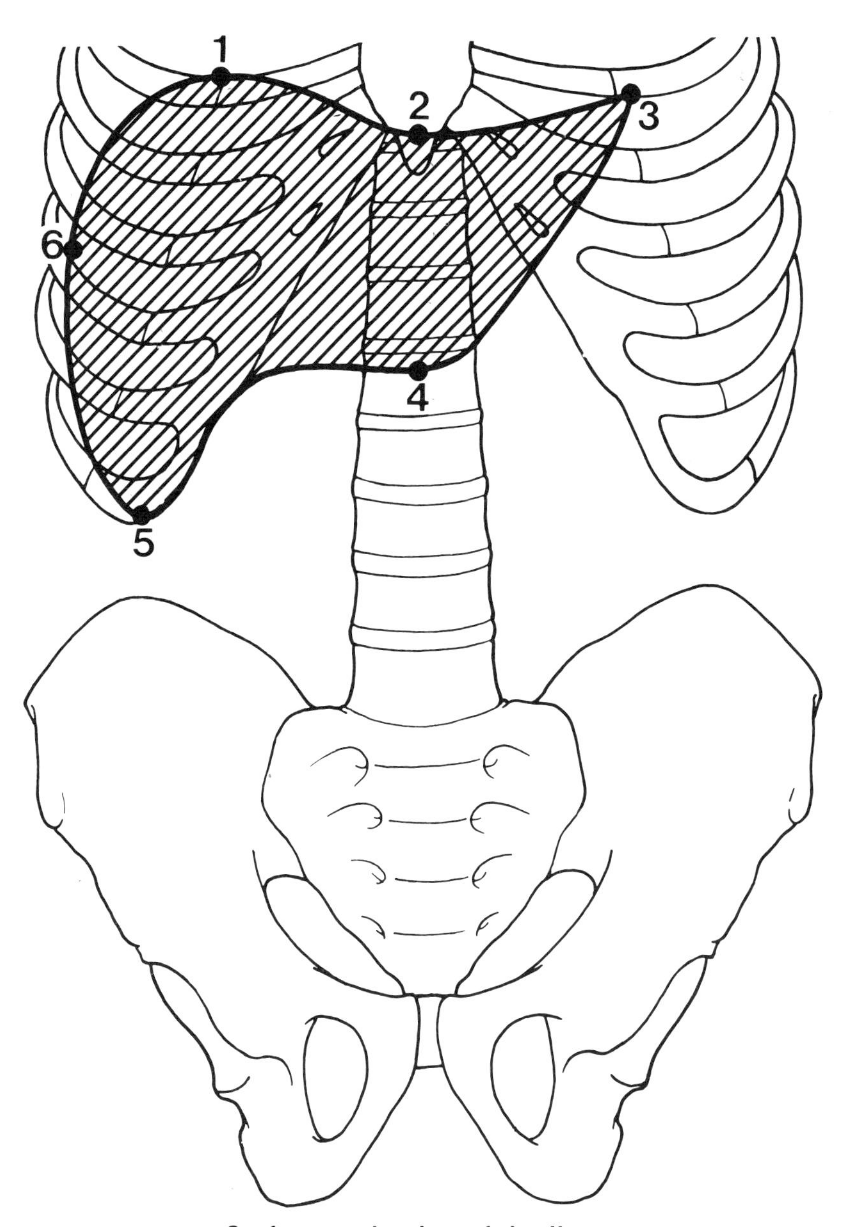

Surface projection of the liver

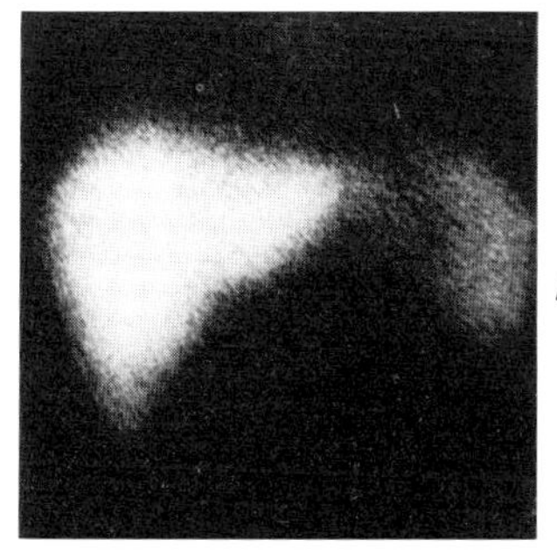

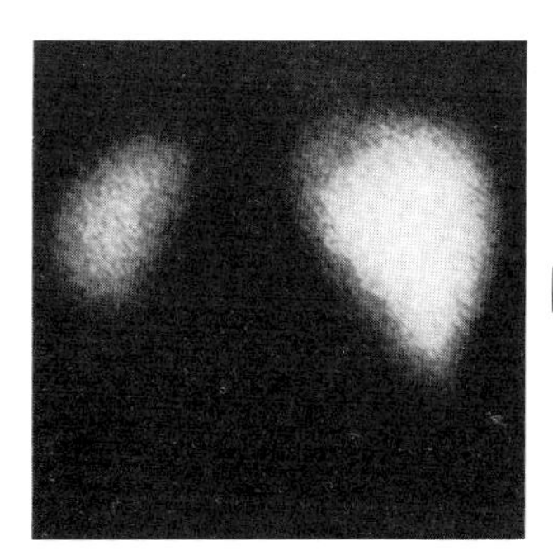

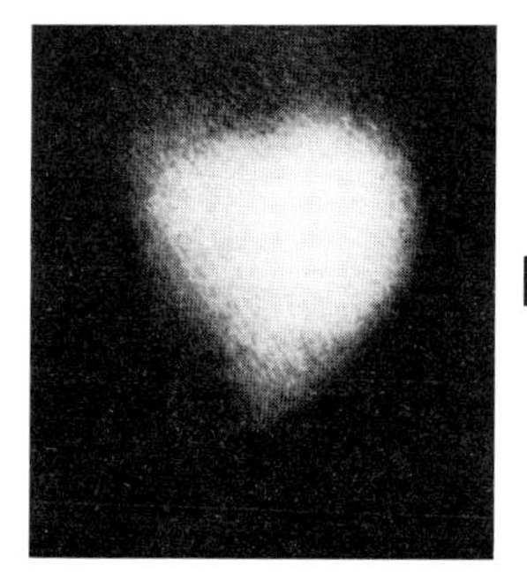

Isotope imaging of the liver and spleen

THE BILIARY SYSTEM

The fundus of the gall bladder [**1**] projects below the lower border of the liver to contact the abdominal wall where the right lateral border of the rectus abdominis crosses the ninth costal cartilage, i.e. the transpyloric plane.

The body of the gall bladder [**2**] lies beneath a line drawn for 5 cm upwards and to the left from the fundus.

The bile duct [**3**] runs vertically downwards to reach the transpyloric plane 5 cm to the right of the midline, and then for a further 2.5 cm downwards and to the right, to enter the ampulla of Vater halfway along the second part of the duodenum.

Radiology of the Hepatobiliary System

The plain film may reveal a limited amount of information about the hepatobiliary system. Apart from assessing the size of the liver as previously discussed, the film may reveal calcification in the liver, pancreas and biliary system. Gall bladder enlargement may be obvious as a pear-shaped opacity in the right hypochondrium, between the twelfth rib and the upper lumbar vertebrae. A lateral film shows that an enlarged gall bladder lies in the anterior third of the abdomen.

Many of the more complicated methods of investigation of the hepatobiliary system are designed to determine the site of pathology in cases of jaundice. Jaundice, an elevated bilirubin level, results from an abnormality in the excretion of this product of red blood cell breakdown. The problem may be within the hepatobiliary system:

1. within the hepatocyte
2. involving the intrahepatic ducts
3. involving the extrahepatic ducts (this results in proximal accumulation of bile in the intrahepatic ducts which will then dilate).

The oral cholecystogram is performed as follows. On the first day a plain film is taken of the right hypochondrium and then an iodine-based contrast medium is swallowed, but no food is consumed overnight. The following morning radiographs are taken of the region. A fatty meal is then taken which stimulates the gall bladder to contract and the films are repeated. The examination is unsuccessful if there is obstruction to bilirubin excretion producing a serum bilirubin in excess of 40 μmol/l.

The oral cholecystogram shows the gall bladder to have a smooth outline unless it contains gall stones. The normal cystic duct has a diameter of less than 1 cm. The examination also demonstrates the common bile duct.

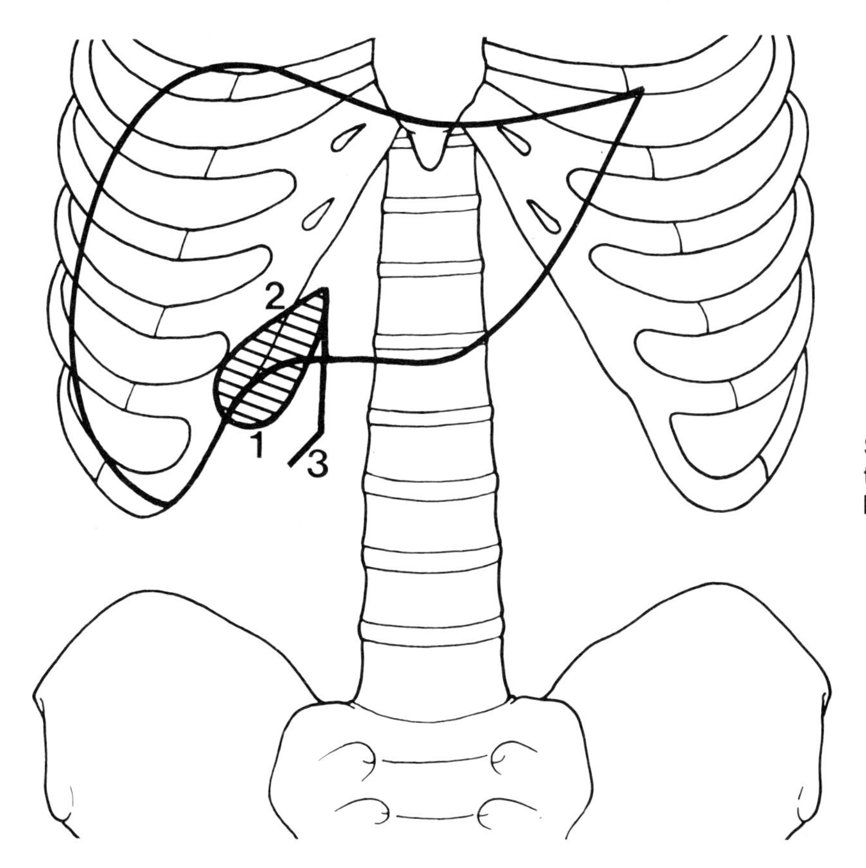

Surface projection of the gall bladder and bile duct

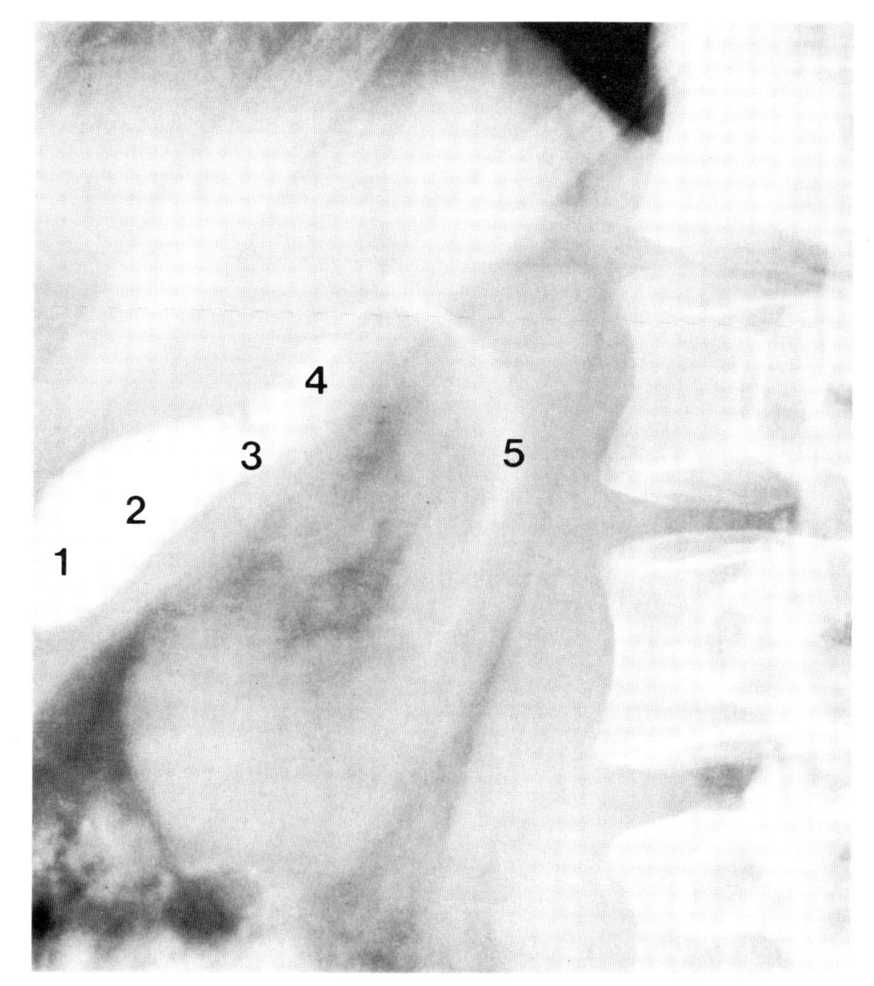

Oral cholecystogram

1. Fundus } of gall
2. Body } bladder
3. Neck
4. Spiral valves of Heister in cystic duct
5. Common bile duct

Ultrasound is useful in distinguishing intra- and extrahepatic duct obstruction as this method will identify swollen intrahepatic ducts. Gall stones may also be detected by ultrasound as the stones reflect sound waves.

Intravenous cholangiography relies on an intravenously introduced contrast medium with an iodine base, which is excreted by the hepatobiliary system. The contrast medium reaches the extrahepatic ducts in normal circumstances in such a concentration that the ducts and gall bladder can be seen without further concentration of the contrast medium by the gall bladder. The image of the gall bladder is not as clear as in the oral cholecystogram.

Percutaneous transhepatic cholangiography. As its name suggests this procedure depends on contrast medium being introduced through the skin via a very fine needle directly into the bile ducts. It is normally performed in jaundiced subjects to show the site of obstruction in an extrahepatic duct.

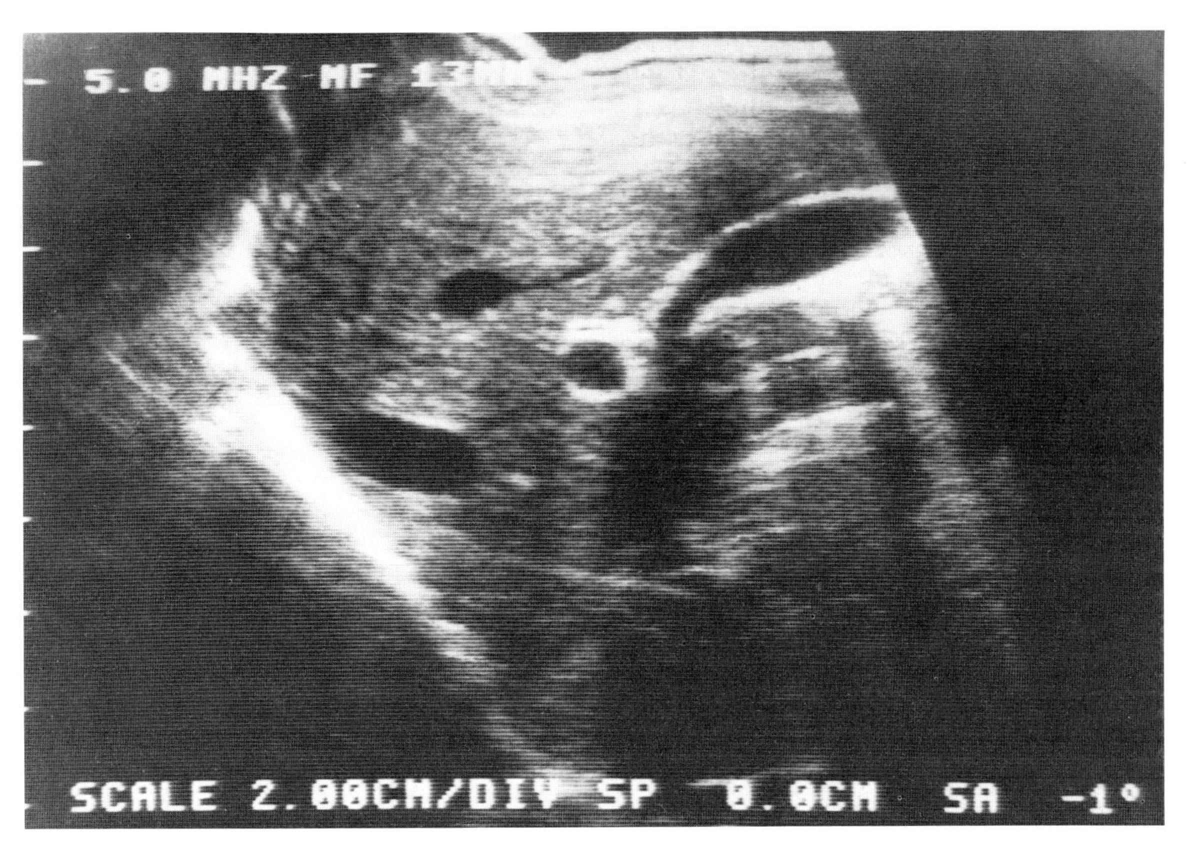

Ultrasonic scan of the biliary system

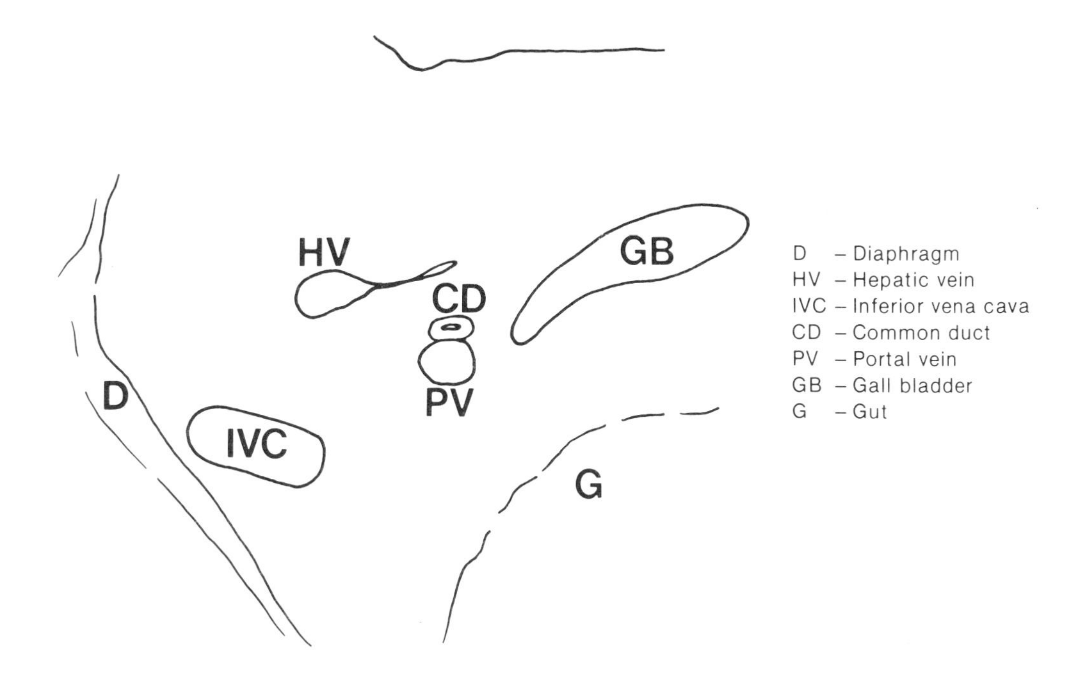

Endoscopic cannulation of the common bile duct. Under endoscopic control a catheter is inserted through the papilla of Vater into the common bile duct. Contrast medium is injected and obstruction of the extrahepatic ducts can be identified. The pancreatic duct system is also seen in most cases.

Operative cholangiogram. During an operation on the hepatobiliary system contrast medium is injected directly into the gall bladder, cystic duct or common bile duct. This is most commonly used during the operation of cholecystectomy (removal of the gall bladder) for gall stones, to check whether any stones have passed down the ducts. They are likely to cause problems if they are not removed during the operation.

T-tube cholangiography. If the common bile duct is opened at operation then the cross piece of a hollow T-shaped tube may be inserted into the duct. Bile is allowed to drain freely through the tube and can pass either along the cross of the T to the duodenum or down the stem of the T to an external collecting system. About ten days after the operation, contrast medium is injected through the stem of the T to demonstrate both any remaining stones and the free passage of the contrast into the duodenum. If no stones remain and there is no obstruction to flow then the tube is removed.

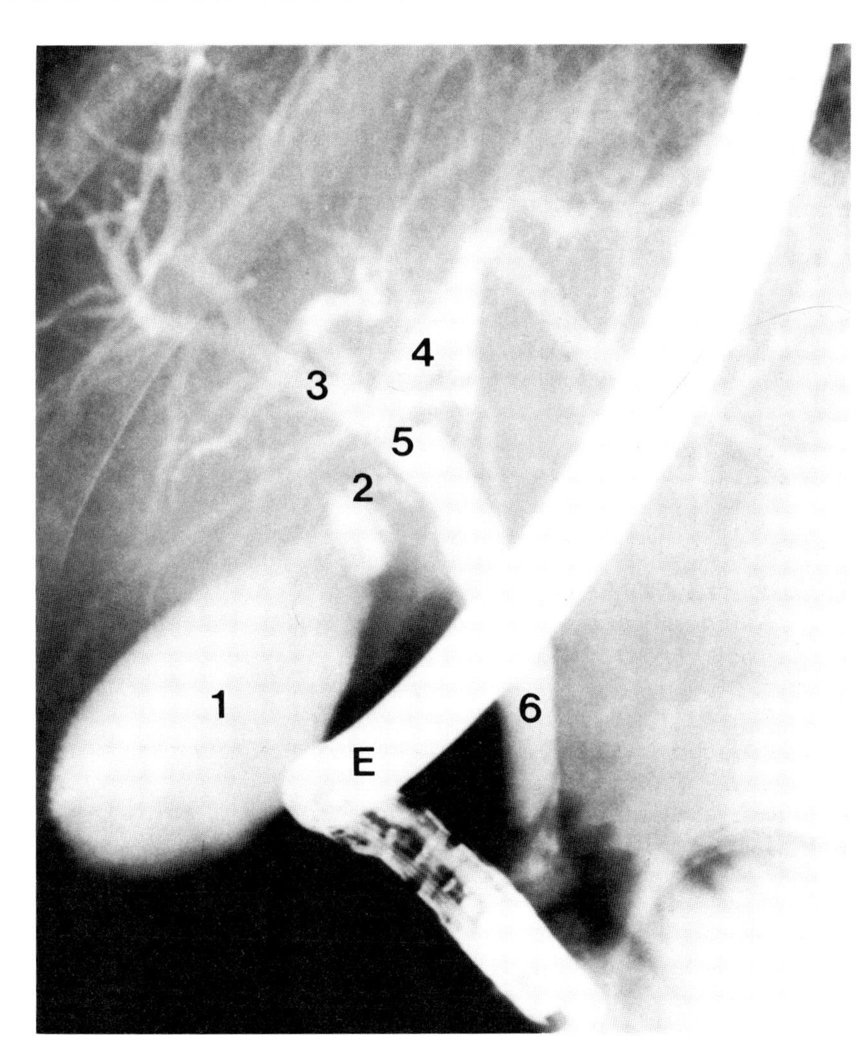

Endoscopic cannulation of the common bile duct

1. Gall bladder
2. Spiral valves
3. Right hepatic duct
4. Left hepatic duct
5. Common hepatic duct
6. Common bile duct
E – Side viewing endoscope cannulating common-bile duct

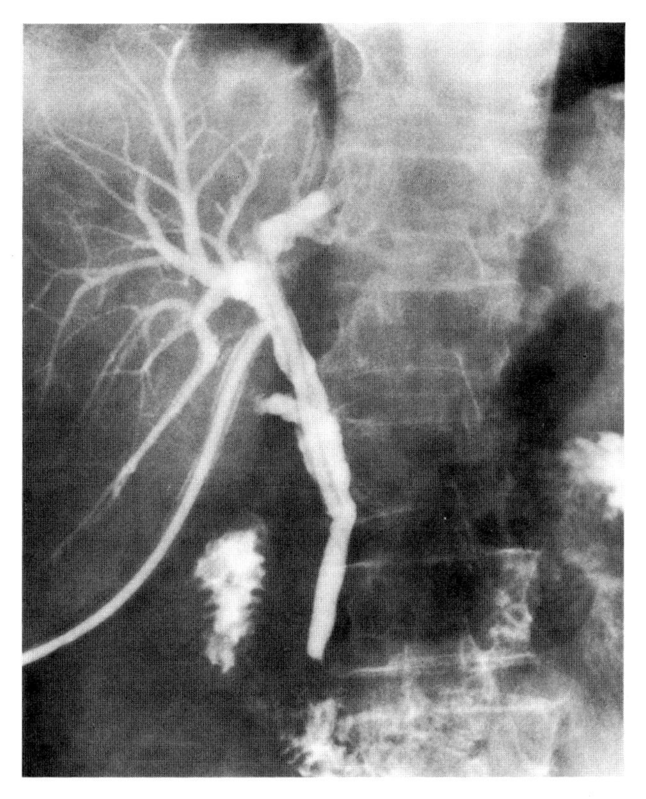

T-tube cholangiogram

THE SPLEEN

The postero-medial end of the spleen lies 5 cm to the left of the midline at the level of the space between the spines of T10 and T11. The axis of the spleen extends antero-laterally along the line of the tenth rib and normally reaches no further forward than the mid-axillary line, so that the normal organ lies beneath ribs 9, 10 and 11.

Imaging the spleen

The spleen is a difficult organ to view by conventional methods. On a plain radiograph, it may be seen as an opacity of similar radiodensity to the kidney and liver in the upper left part of the abdomen, to the left of the gas in both the splenic flexure and the stomach.

LYMPH NODES OF THE ABDOMEN

Following the general rule, the deep lymphatics of the abdomen run along the arteries. The superficial lymphatics are important clinically and their position should be known.

The superficial inguinal nodes lie in three groups. The lateral and medial groups lie in the superficial fascia just below the inguinal ligament. The vertical group lies around the terminal part of the long saphenous vein (see page 170).

The pubic nodes are few in number and lie in front of the pubic arch.

The inferior epigastric glands follow the vessels of that name.

THE PANCREAS

The best way to mark the pancreas is to start at its neck. This lies on the transpyloric plane 1·2 cm to the right of the midline, behind the pylorus. The body extends from the neck along a line running upwards and to the left for 10 cm to reach the hilum of the spleen. The head of the pancreas lies downwards and to the right of the neck, within the confine of the duodenal loop.

Imaging the pancreas

The pancreas is a difficult organ to visualise. On the plain film, pancreatic calcification may be seen or enlargement deduced from the displacement of surrounding structures. Information about the head of the pancreas may come from a barium meal which outlines the surrounding duodenal loop. Successful ultrasound examination is difficult because of the gas in the overlying gut. CAT scanning is useful, but not widely available. Endoscopic retrograde cannulation of the pancreatic duct may be performed as described for the cannulation of the common bile duct. These two investigations together comprise endoscopic retrograde cholangiopancreatography (ERCP).

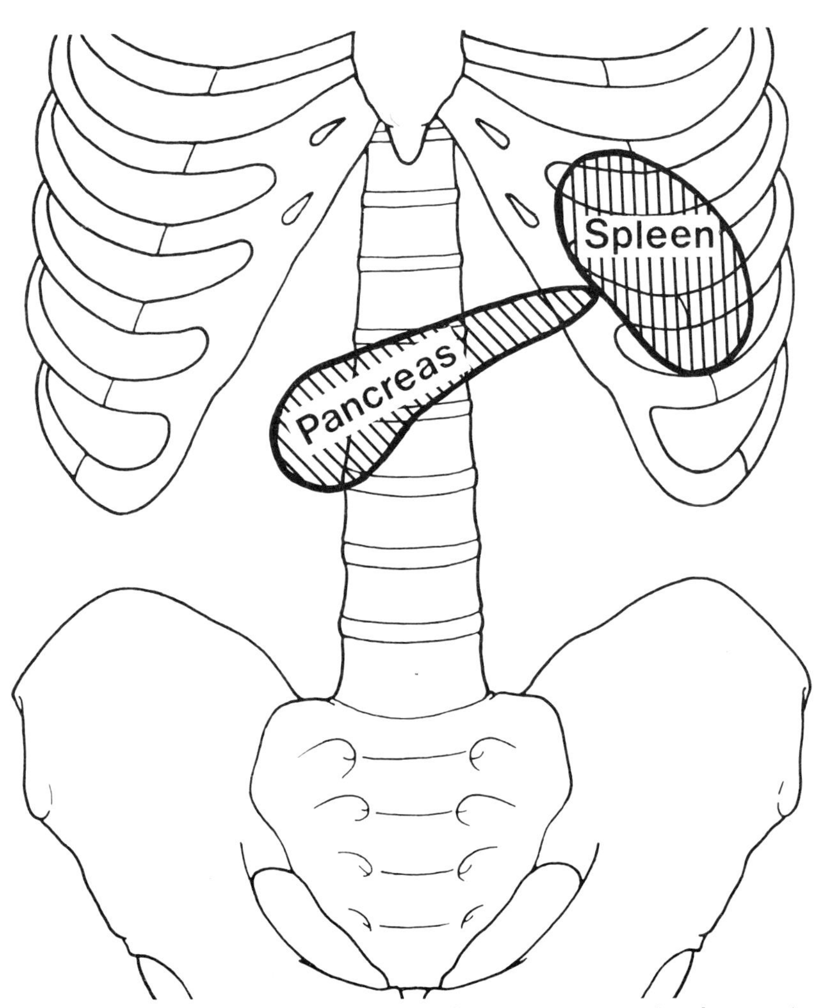

Anterior surface projection of the pancreas and spleen

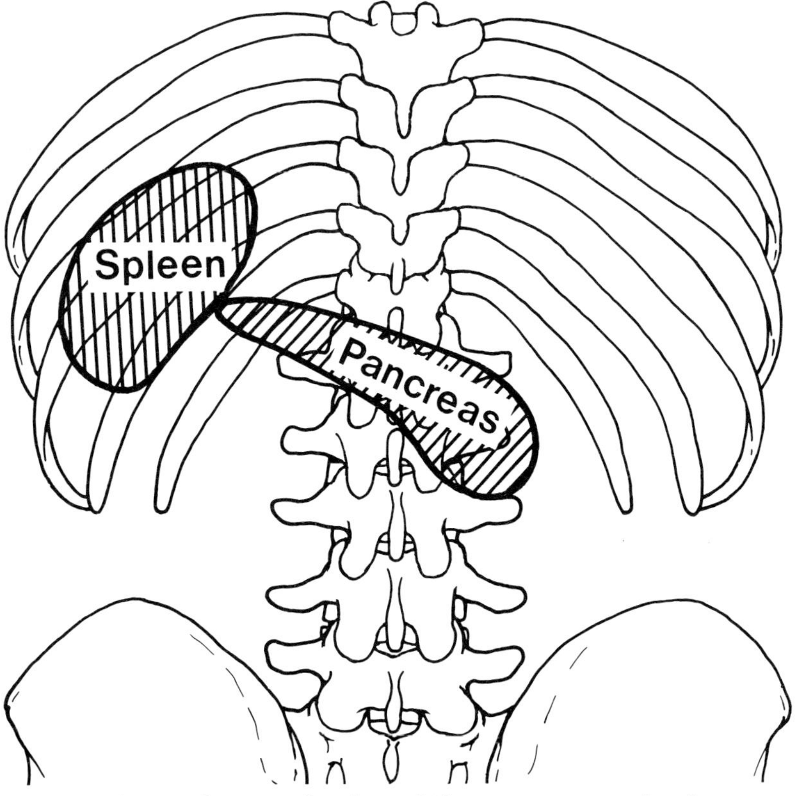

Posterior surface projection of the pancreas and spleen

THE ADRENAL GLANDS

The centres of both adrenal glands lie symmetrically 3 cm above the transpyloric plane, 3 cm from the midline.

Radiology

The normal adrenal glands cannot be seen on plain x-ray, but their position may occasionally be indicated by calcification. They can be visualised by injecting gas in front of the sacrum; this tracks up in the retroperitoneal tissues to surround the adrenals and kidneys. The adrenals appear as triangular shadows 2 cm wide resting on the upper poles of the kidneys.

THE KIDNEYS

The hila of the kidneys lie approximately on the transpyloric plane, the right just below and the left just above, both 5 cm from the midline. These points lie just medial to the tips of the ninth costal cartilages. The lower poles of the kidneys lie 7.5 cm from the midline at a level 2.5 cm above the supracristal plane. The upper poles lie only 2.5 cm from the midline at a level midway between the xiphisternal joint and the transpyloric plane. The length of each kidney is 11 cm; the true width of 6 cm is only represented as 4.5 cm on the surface due to the oblique position of the organs on the sides of the lumbar vertebrae.

Posteriorly, each kidney can be projected as the same obliquely set bean-shaped outline around the hilum, which lies at the level of the spine of L1, 5 cm from the midline. The upper poles reach the level of the spine of T11 and so lie deep to the twelfth rib. The lower poles lie at the level of the spine of L3.

THE URETERS

The ureters begin at points on the transpyloric plane, 5 cm from the midline and run downwards and slightly medially to enter the bladder at points marked superficially by the pubic tubercles.

They are of approximately 3 mm in diameter but they are constricted at three sites:

1. At their junction with the renal pelvis.
2. Where they enter the bony pelvis, 2.5 cm from the midline on the transtubercular plane.
3. At their entry into the bladder.

THE BLADDER

The neck of the bladder lies deep to the lower part of the pubic symphysis. The openings of the ureters lie behind the pubic tubercles. As the bladder is distended, the peritoneum is separated from the suprapubic region so that the bladder lies directly beneath the anterior abdominal wall. When the desire to micturate is perceived in the normal individual, approximately 300 ml of urine are contained in the organ and the bladder is in contact with the abdominal wall for 5 cm above the pubic symphysis.

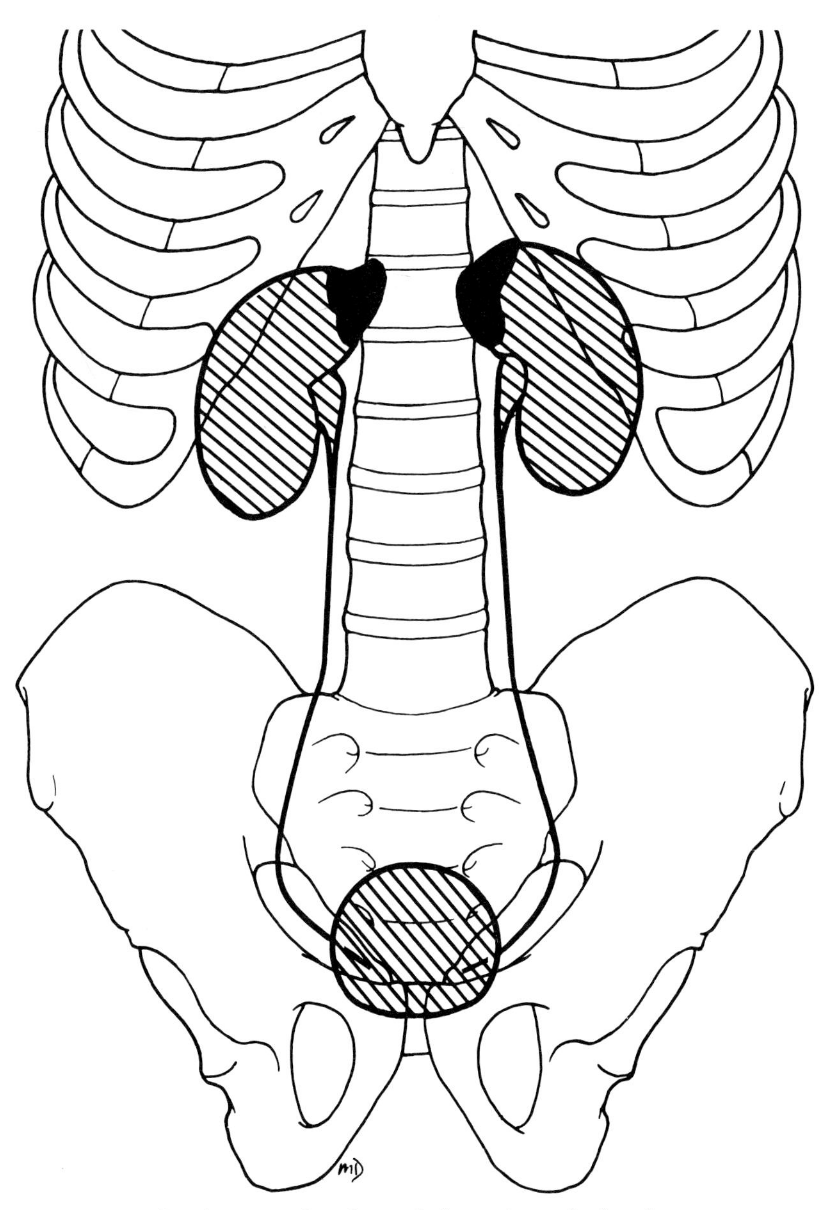

Surface projection of the adrenal glands and the urinary tract

Imaging the urinary tract

The straight radiograph of the abdomen reveals a few details of the architecture of the urinary tract. It is usually possible to see the renal outlines surrounding the homogeneous shadows of renal tissue; they are roughly oval with a concavity on their medial borders. Occasionally, there is evidence of developmental fetal lobulation. On the plain film it is important to identify all calcifications as they are obscured by contrast medium if they lie within the urinary tract.

The intravenous urogram depends upon the ability of the kidney to concentrate and excrete organically-bound iodine compounds. Iodine-based contrast medium is injected intravenously and is filtered by the glomeruli into the urine. A film exposed one minute after the injection of the medium shows the dye within the renal parenchyma, and is especially useful in demonstrating the renal outlines. This image is known as a *nephrogram.*

On the nephrogram the following points should be noted:

1. That there are two kidneys in their normal positions.
2. The renal outlines are normal.
3. The renal lengths are between 10 and 16 cm. There is a diurnal variation of 1 to 2 cm. Normally there is under 1.5 cm difference between the two sides. The absolute size is related to age, sex and body size.
4. The width of the renal parenchyma should be uniform and symmetrical, measuring 1.5 to 2 cm except at the upper and lower poles where it is usually 1 cm thicker. It is estimated by drawing an imaginary line connecting the pyramids, roughly parallel to the renal outline.

A pyelogram is the image of the collecting system which is seen on later films. Whereas the nephrogram depends upon the amount of contrast reaching the kidneys, the pyelogram depends upon the ability of the kidney to concentrate urine. The density of the pyelogram can be increased by restricting the fluid given to the subject prior to the examination.

The pyelogram should be examined to check that the calyces are roughly evenly scattered and have a normal appearance. They should appear concave when they lie in the plane of the film, and round with a dense rim if seen end on. The pelvis of the kidney is very variable in its anatomy. It may lie either within or outside the renal parenchyma. Laterally, the pelvis usually divides into upper, middle and lower major calyces, which connect in turn with the 6–12 minor calyces. The size and shape of the pelvis is as variable as its relationship to the renal parenchyma. Commonly, the superior border is convex and the inferior concave, (but it too may be convex). The pelviureteric junction is another region with variable normal anatomy, but usually the pelvis funnels to join the ureter at an obtuse angle.

After the five-minute film has been taken pads are applied to the abdomen. These produce a denser view of the collecting system as they delay emptying by compressing the ureters as they cross the brim of the pelvis.

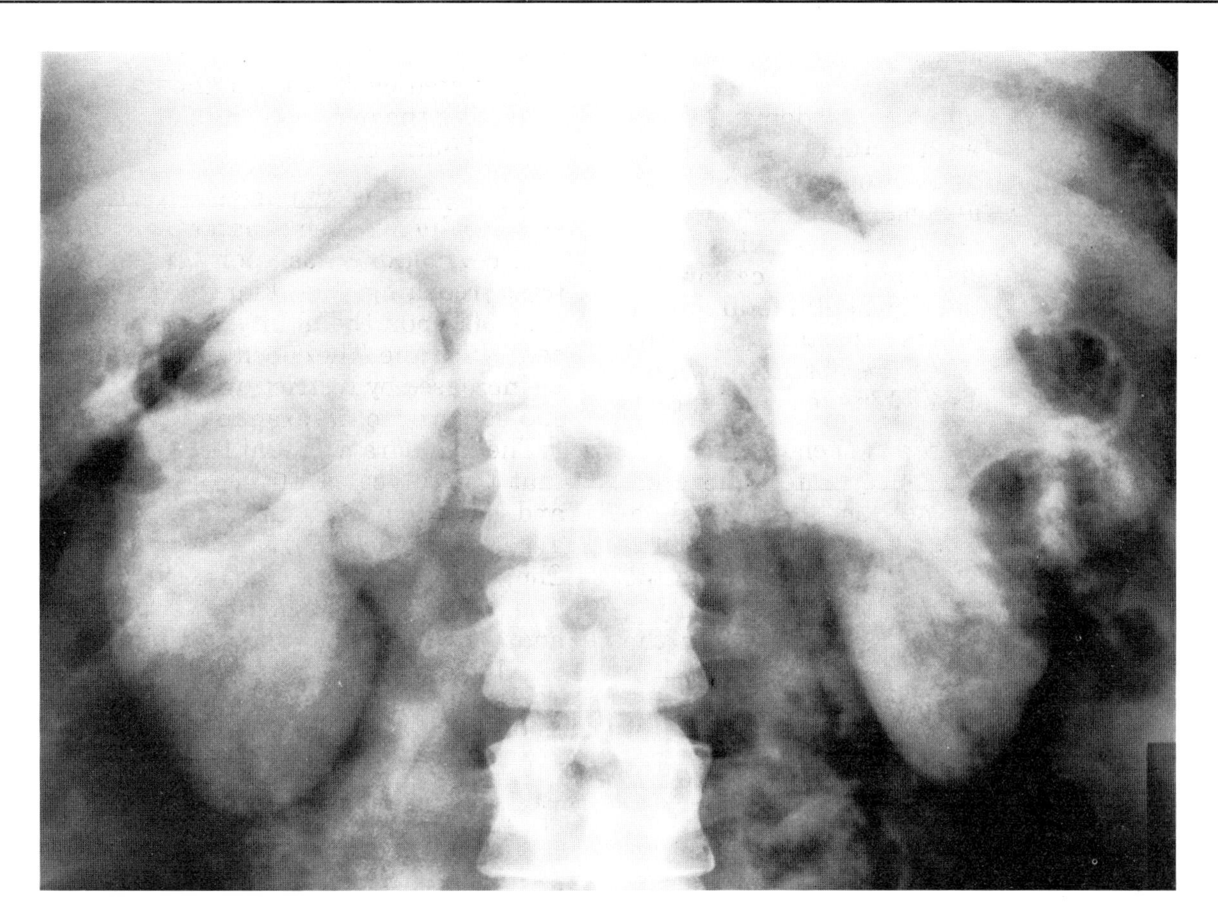

Nephrogram

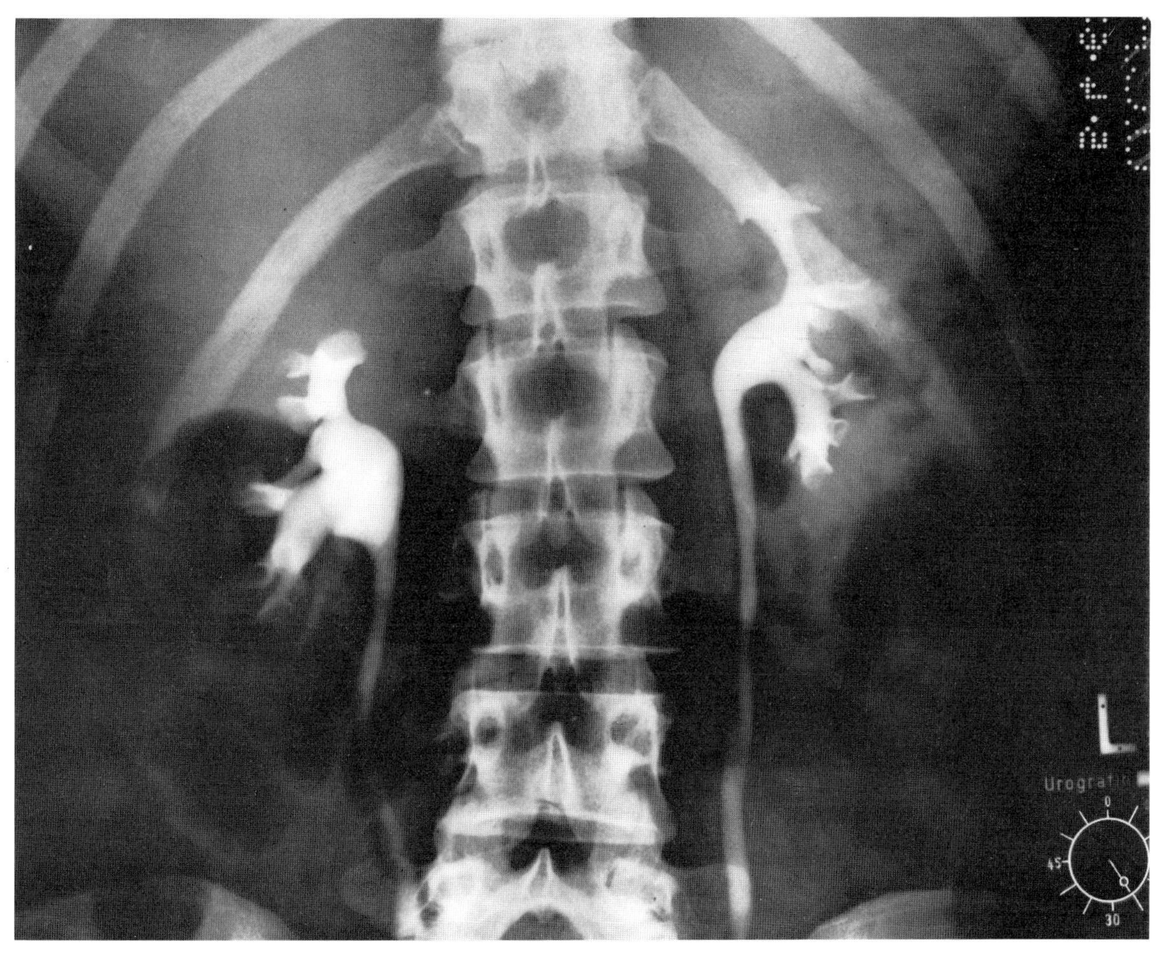

Pyelogram

Immediately after the compression is released, the rush of contrast medium into the ureters gives the best chance to see them throughout their length. At other times part of their lumens is usually obliterated by peristaltic waves. Their normal diameter is less than 7 mm. They pass down close to a line joining the tips of the transverse processes of the lumbar vertebrae, across the sacroiliac joints, and then run about 1 cm medial and parallel to the bony pelvic brim until they reach the ischial spines where they turn medially to enter the bladder.

When full of urine, the bladder may be seen on the plain film as a faint shadow above the pubic symphysis. When full of contrast medium, it has a smooth lumen, with a convex upper border but this may be indented by the uterus, sigmoid colon or ileum. The circular outline has a lower border which lies close to the pubic symphysis and may be indented by the levator ani. After micturition the bladder is emptied leaving only a small amount of residual fluid trapped by the folds of mucous membrane.

A retrograde pyelogram is more dangerous than a retrograde ureterogram as the catheter is passed further into the ureter and a larger amount of contrast medium is injected. In both examinations a cystoscope is introduced into the bladder and a catheter is passed under direct vision into each ureteric orifice. Whilst contrast medium is injected its passage is watched by fluoroscopy. These methods reveal details of the collecting system but not of the renal parenchyma.

Percutaneous antegrade pyelography involves first localising the kidney with ultrasound or radiology in combination with a large dose of contrast medium. A fine needle is introduced through the skin, usually of the back, into the collecting system and contrast medium injected. The needle may have a Teflon sheath which may be left in place to drain the urine through a percutaneous nephrostomy in cases of outflow obstruction.

Micturating cysto-urethrography is undertaken by first catheterising the bladder and draining the urine. The bladder is then filled with contrast medium through the catheter. Whilst the subject micturates the bladder and urethra are viewed under fluoroscopy to check the mechanical aspects of micturition.

Urethrography may be undertaken either as an integral part of micturating cystourethrography, or by injecting contrast medium through a cannula in the urethral meatus.

Radiological studies of the renal vasculature are performed by injecting contrast medium into the renal arteries by a catheter introduced into the aorta through the femoral artery (see page 79). The renal veins may be seen in late films but better images are produced by selectively threading a catheter into the renal veins via the inferior vena cava.

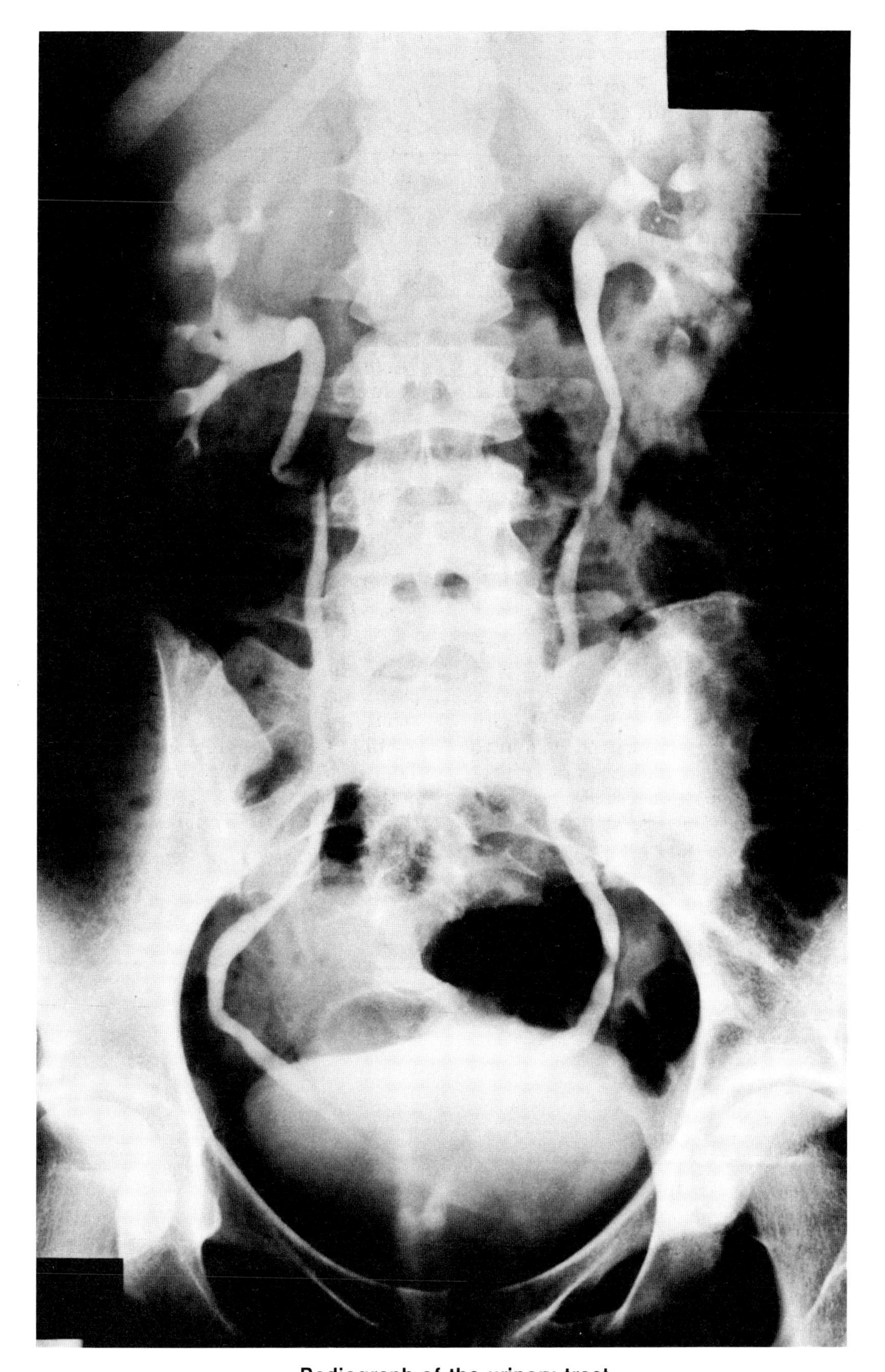

Radiograph of the urinary tract

Dye can be seen passing from the collecting systems of the kidneys, entering the ureters and so reaching the bladder. In this film jets of urine squirted from the ureters are inter-lacing in the cavity of the bladder.

Ultrasound can only demonstrate a limited amount of information about the normal urinary tract. The collecting system appears as an area of high level echoes within the renal substance. It is not possible to see the normal ureter consistently.

Radioisotope investigations yield details about renal structure in relation to function which are not obtainable in such detail by the other methods of investigation.

Static functional imaging uses DMSA (dimercaptosuccinic acid) labelled with a technetium isotope. After injection, this compound is slowly taken up by the kidneys and after about three hours one-half of the injected dose can be found within the renal parenchyma. Scanning at this time allows an assessment of the regional function of the parenchyma.

Dynamic functional imaging uses DTPA (diethylenetriamine pentacetic acid) similarly labelled with technetium. This is only filtered by the glomerulus. After injection of a bolus, a sequence of images is obtained. The image produced at 30 seconds indicates the differential regional perfusion of the kidneys, whilst that at two minutes indicates the regional glomerular function. Later images show the collecting system and may indicate obstruction to the flow of urine.

Cystoscopy

The visualisation of the interior of the bladder and the urethra is made possible by the introduction of the cystoscope, a rigid mirror and telescope combination. On entering the bladder the slit-like internal urethral orifices can be seen connected by the inter-ureteric ridge. This ridge forms the posterior border of the trigone, an equilateral triangle. The mucosa has a shining appearance through which the vessels can be seen, the arteries are paler than the veins, but both are most numerous in the trigone, extending laterally to encircle the ureteric orifices.

The interior of the urethra may also be inspected using the cystoscope. The most important clinical anatomy is that of the male prostatic urethra because of the more frequent use of transurethral prostatectomy. On the posterior wall of this part of the urethra is a median longitudinal ridge, the urethral crest, which is surmounted by a prominence, the colliculus seminalis, (or verumontanum). At the apex of the colliculus is the opening of the prostatic utricle, with the small openings of the ejaculatory ducts on each side. The shallow grooves on either side of the urethral crest are the prostatic sinuses into which the prostatic ducts open.

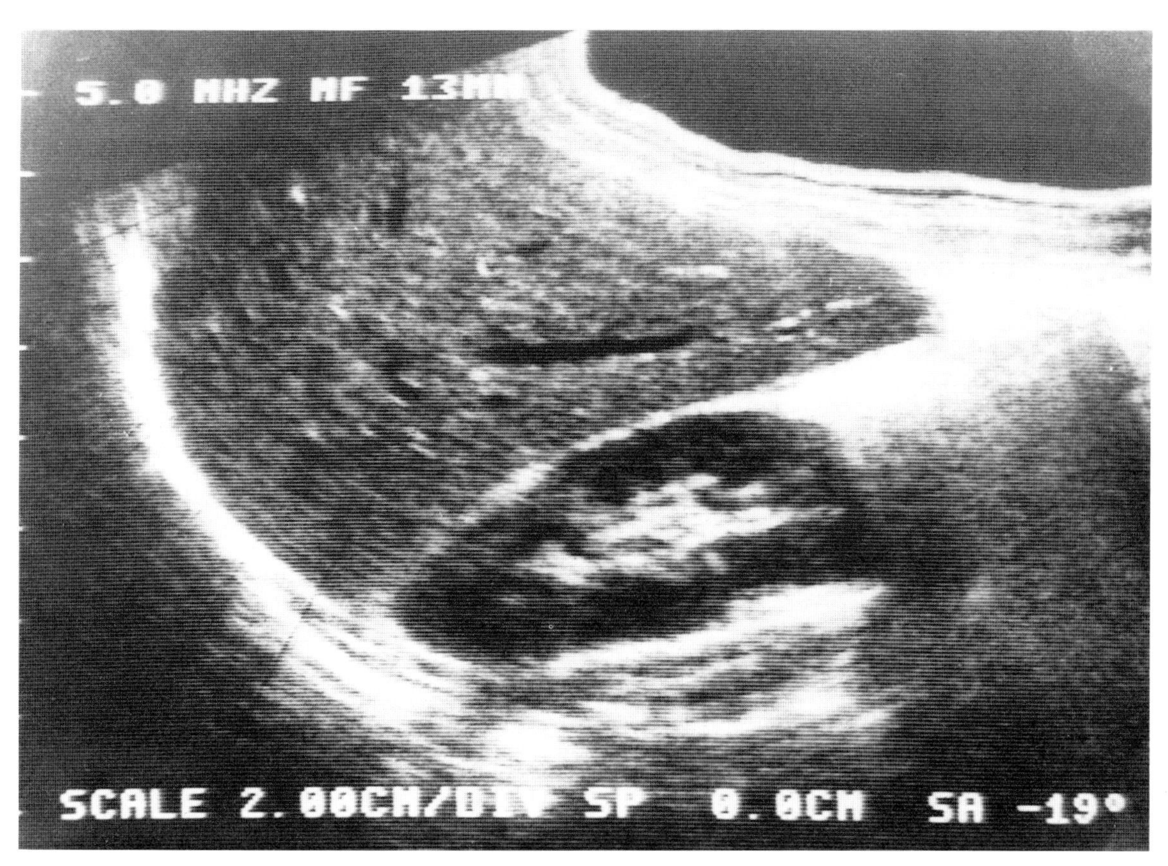

Ultrasonic scan of the right kidney

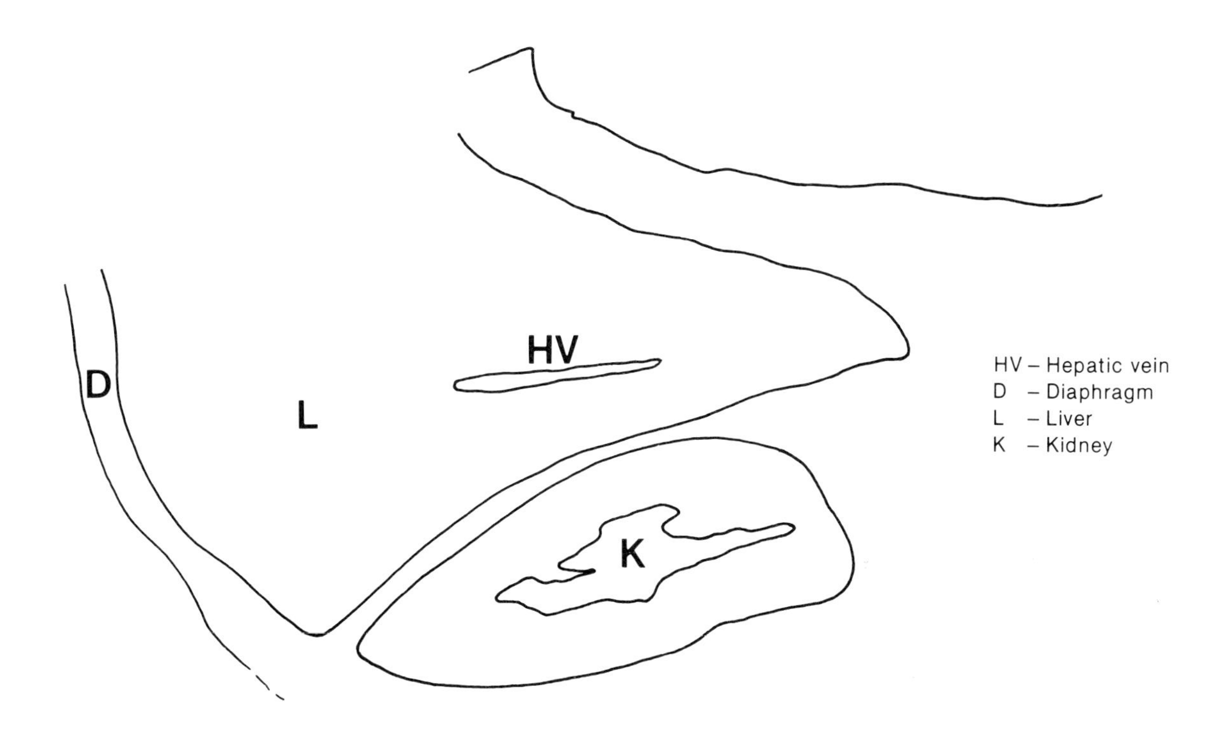

THE PERINEUM

The perineum is the diamond-shaped area bounded laterally by the ischial tuberosities, anteriorly by the pubic symphysis and posteriorly by the coccyx.

The anal triangle is the part which lies behind a line joining the ischial tuberosities. The anus forms an antero-posterior slit in the midline with radially arranged corrugations in the surrounding skin which lead into the anal canal.

The urogenital triangle lies in front of the anal triangle. Its appearance obviously differs between men and women.

The Male Urogenital Triangle

The sides of the triangle [**1**] are marked laterally by the palpable conjoined rami.

The bulb of the penis [**2**] occupies the midline where it can be felt as a smooth firm structure separated from the rami by a small gap, with the crura of the corpora cavernosa lying more laterally as they diverge to attach to the medial sides of the pubic arch.

The scrotum [**3**] is further to the anterior and has a midline raphe. Within the scrotal sacs the two testes are palpable, each approximately 5 × 2.5 × 2.5 cm with a volume of 10–15 g. The size may be measured by comparison with a series of egg-shaped models of known volume, (an orchidometer).

The ellipsoid-shaped testes lie obliquely with the upper ends pointing forwards and medially. They are firm to palpation with smooth convex surfaces except along the lateral part of the posterior border where the epididymis is attached.

The body of the penis [**4**] is covered by thin, hairless skin which extends over the glans penis as the prepuce or foreskin.

The Female Urogenital Triangle

The mons pubis [**A**] is the swelling covering the front of the female pubic symphysis which becomes covered with pubic hair.

The labia majora [**B**] are two antero-posterior directed folds which extend backward from the mons and join at the posterior commisure. The gap between the posterior commisure and the anus overlies the perineal body and is 2.5 cm long. The region between the labia majora is known as the pudendal cleft.

The labia minora [**C**] are folds of skin, each about 4 cm long, running antero-posteriorly in the pudendal cleft. Anteriorly, each bifurcates to join its partner on each side of the clitoris. The most anterior union forms the hood-like prepuce of the clitoris, while behind the clitoris, the frenulum of the clitoris is formed by the union. The vestibule is the region enclosed by the labia minora.

The clitoris [**D**] is a minature homologue of the penis, but it is not traversed by the urethra.

The urethra [**E**] opens to the perineum, 2.5 cm behind the clitoris. It produces a saggital slit with slightly elevated edges.

The vaginal orifice [**F**] is a midline aperture in the posterior part of the vestibule. The hymen is a variable fold of mucous membrane partially obstructing the vaginal orifice. When torn, for example at the first coitus, its remnants produce the small carunculae hymenales.

Immediately lateral to the hymen in the vestibule are the orifices of the greater vestibular glands.

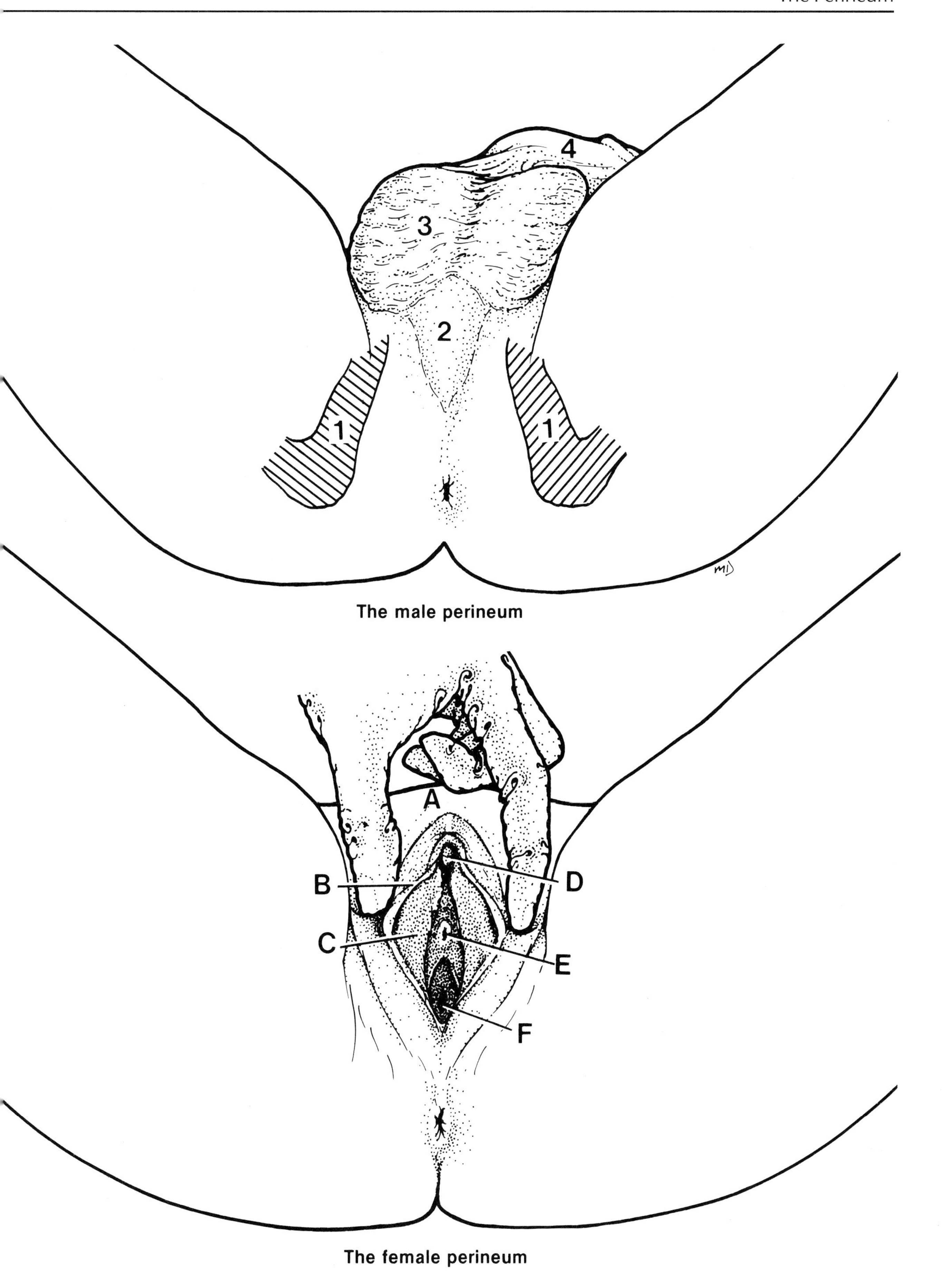

The male perineum

The female perineum

VAGINAL EXAMINATION

Position. The woman generally lies supine with the hips and knees partially flexed, the feet together and the knees apart. Alternatively, the left lateral position may be used.

Inspection. Observe the perineum carefully and ask the patient to strain down to reveal gross prolapse. Part the labia minora and inspect the hymen; if this is intact it is unusual to proceed further.

Speculum examination. After separating the labia minora with the left hand the speculum is gently inserted with the broadest part of the blade running antero-posteriorly. In cases of prolapse a Sim's speculum is generally used with the patient in the left lateral position. Otherwise, with the patient supine, a bivalve speculum is inserted. The blades of the speculum are parted and the cervix and external os can be seen. The vaginal walls can be visualised by rotating the speculum.

Digital examination. After the speculum is withdrawn one or two fingers of the right hand are gently introduced into the vagina while the left hand still parts the labia minora. The cavity of the vagina points upwards and posteriorly. The anterior wall is about 7.5 cm long and the lining mucosa is marked by transverse rugae. The cervix projecting into the apex of the cavity is surrounded by the four fornices, anterior, posterior and right and left lateral.

A bimanual examination is performed by moving the left hand on to the abdominal wall just below the umbilicus, the fingers are then extended down the abdominal wall to reach into the pelvis. The fundus of the uterus can then be caught between the left hand and the fingers of the right hand in the anterior fornix. When assessing the uterus the following factors should be considered:

- position in relation to the cervix
- size: normally 7.5 cm long, 5 cm wide, 2.5 cm deep
- shape: usually described as pear-like
- consistency: firm
- mobility: normally mobile in all directions
- tenderness: tender if squeezed during bimanual examination
- abnormal attachments

In some subjects the ovaries can be felt bimanually. They are situated at the same level as, but 5 cm medial to the anterior superior iliac spines. If palpated they are very tender. Normal uterine tubes cannot be felt.

Anterior to the fingers in the vagina, the bladder and symphysis pubis can be felt.

Laterally is the ischiorectal fossa. Here the perineal muscles may be assessed, normal ovaries and abnormal tubes felt and theoretically a stone palpated in the ureter as it lies in direct relation to the lateral fornix.

Behind the vagina lies the recto-uterine pouch and behind that the rectum.

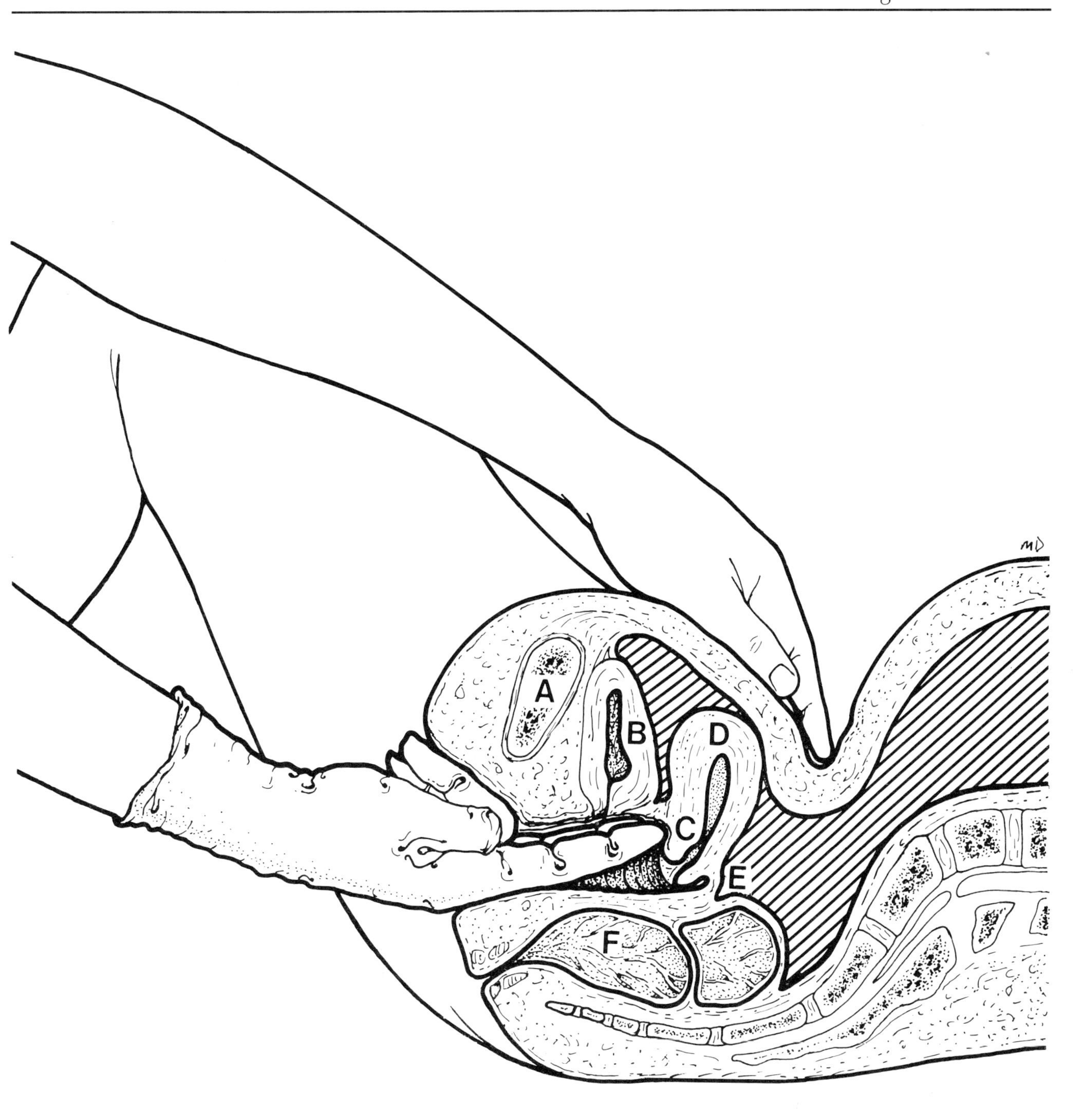

Sagittal section of the female pelvis during vaginal examination

A – Symphysis pubis
B – Bladder
C – Cervix
D – Fundus of uterus
E – Recto-uterine pouch
F – Rectum

RADIOLOGY OF THE PELVIS

The normal film of the pelvis is an antero-posterior view taken with the subject lying supine.

Lying anteriorly the size of the pubis is increased in this view. The pubic symphysis appears as a 5 mm gap between the bones.

The obturator foramina are extremely variable in radiological appearance and the ischial tuberosities also have an irregular appearance. Both these structures should be roughly symmetrical.

The iliac crests have rough outlines. The alae are covered by the partially gas-filled bowel which suggests apparent demineralisation in some areas of the bone.

Assessment of the pelvic brim is important prior to child bearing and for this purpose the lateral view is usually employed. This aspect is considered further under the section on imaging in pregnancy.

RADIOLOGY OF THE FEMALE REPRODUCTIVE TRACT

A hysterosalpingogram is performed by introducing contrast medium into the cavity of the uterus by a catheter inserted into the cervix. The interior of the normal uterus has a smooth, inverted triangular shape. The contrast medium spills from the upper angles of the uterine cavity to outline the Fallopian tubes taking a convoluted path towards the side walls of the pelvis. The internal diameter of the tubes is usually less than 1 mm, but in the lateral third they widen to about 4 mm. Dye should spill freely from the tubes into the pelvis.

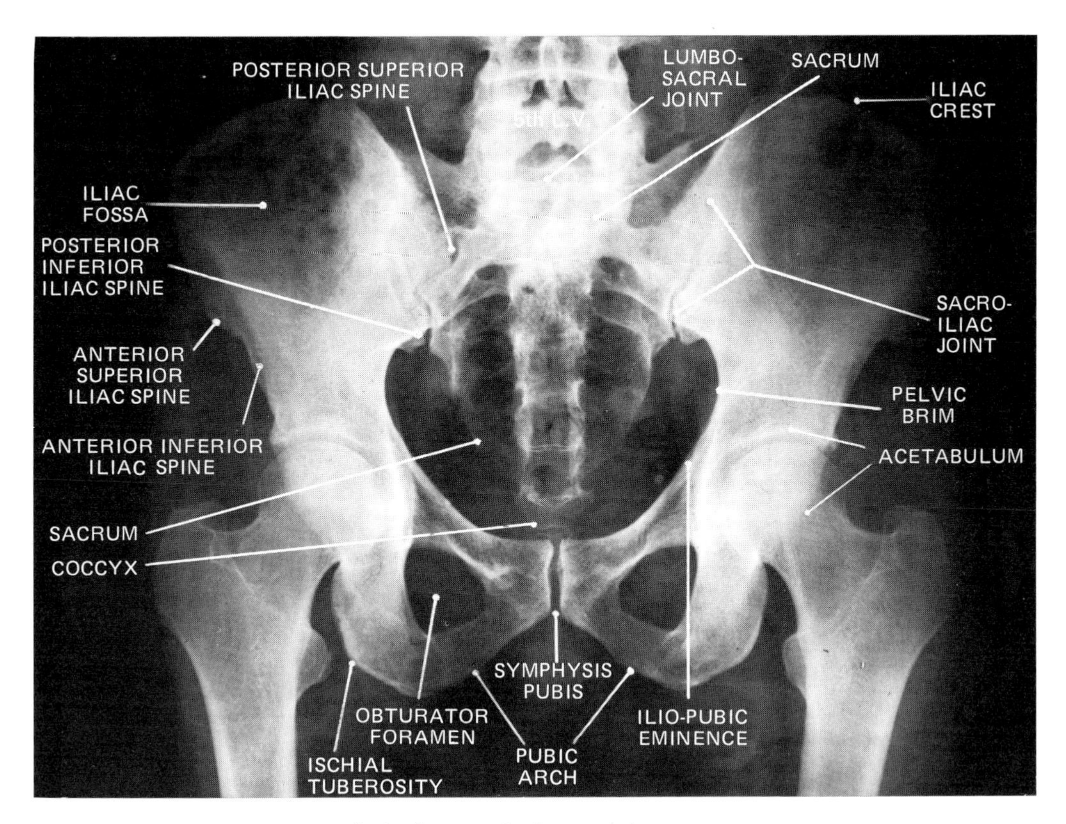

Anterior-posterior pelvic x-ray

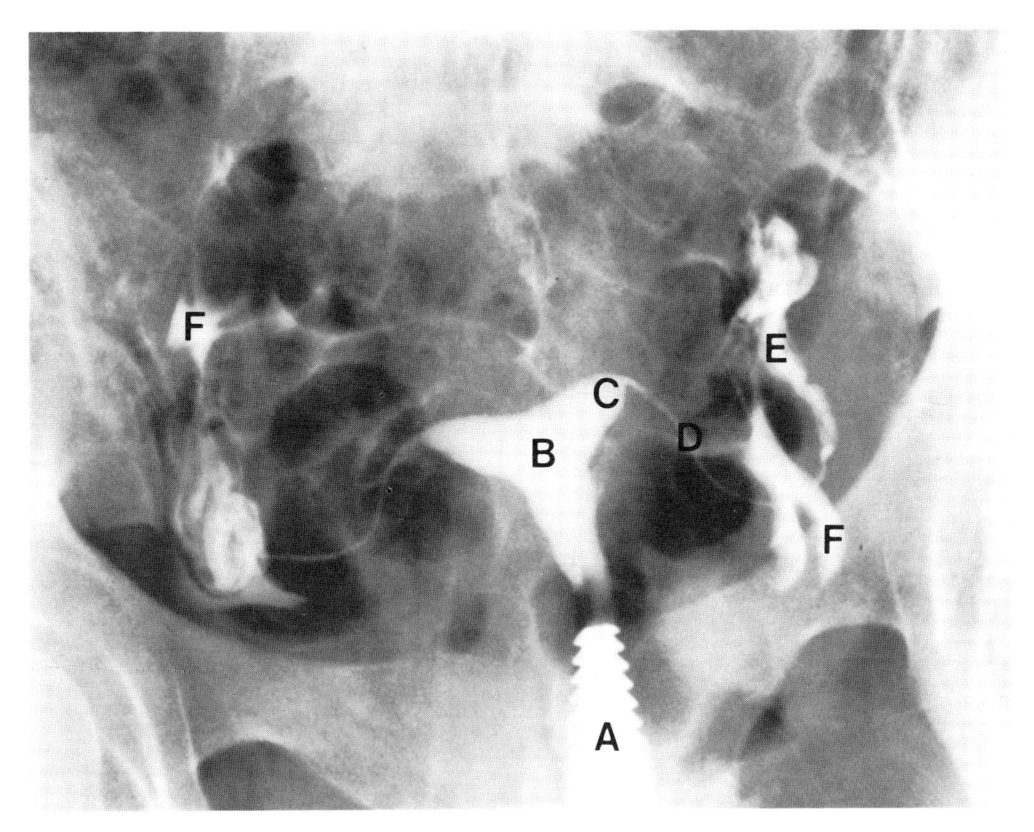

Hysterosalpingogram

A – Catheter. B – Body of uterus. C – Uterine cornu. D – Proximal uterine tubes. E – Contrast medium surrounding fimbriae. F – Peritoneal spill of contrast.

PREGNANCY

The 40 weeks of pregnancy produce physiological changes in the anatomy of the pelvis. During pregnancy vaginal examination should be directed towards assessing the size of the pelvis to establish whether or not there will be obstruction to normal labour. Particular reference should be paid to the following points:

1. The promontory of the sacrum, which may be occasionally reached by firm pressure upwards and backwards in the posterior fornix, but is not normally felt.

2. The curve of the sacrum, which should be smooth.

3. The coccyx which should not be angulated forwards.

4. The ischial spines lie on the same imaginary plane as the cervix. Check that they are not angulated into the cavity of the pelvis.

5. The sacrospinous ligament should be over two finger breadths in length.

6. The subpubic arch usually has an angle over 90°.

7. The intertuberous diameter should be at least four knuckle breadths.

Imaging in Obstetrics

Radiology during pregnancy is falling into disuse because of the dangers to the unborn child. The main application now is that of pelvimetry to determine the relative dimensions of the head of the fetus and the pelvis of the mother. Measurements are made on the supine-posterior view and also on the more useful erect-lateral. Corrections are made for distortion on the film by the inclusion of a graduated metal rule but can take no account of transradiant soft tissues. Radiographs also have a more limited application in the diagnosis of skeletal deformities in the fetus.

Ultrasound has largely replaced radiology and no harmful effects on the fetus have yet been shown. It is possible to recognise a gestational sac from the fifth week of pregnancy, but the developing fetus within the sac cannot usually be separated until the seventh week. The longest demonstrable length of echoes within the sac is taken to indicate the crown-rump length and is used to date the fetus between the seventh and twelfth weeks. Beyond the age of 12 weeks the age of the fetus is estimated from the biparietal diameter, when the scan cuts across the falx and lateral ventricles. The placenta can be scanned from the ninth week when it appears as a region of variable thickness over part of the uterine cavity.

Lateral pelvic x-ray

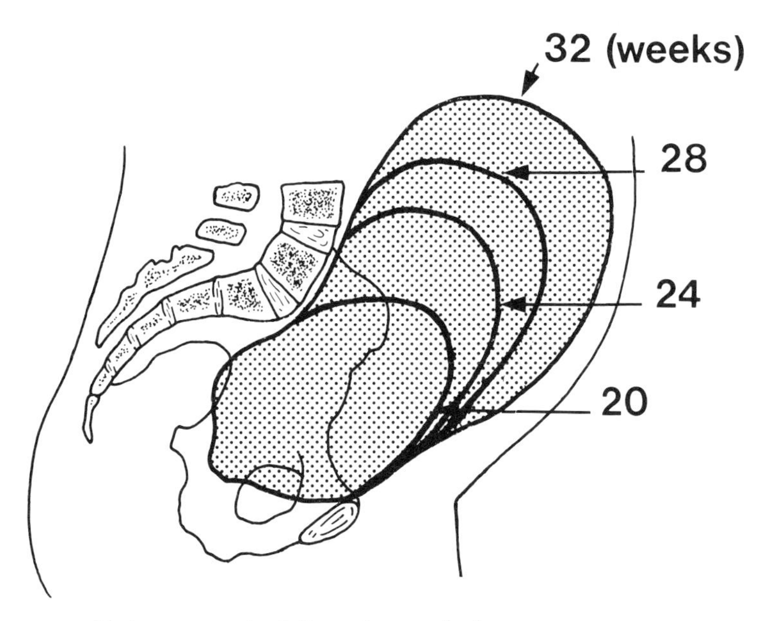

Enlargement of the uterus during pregnancy

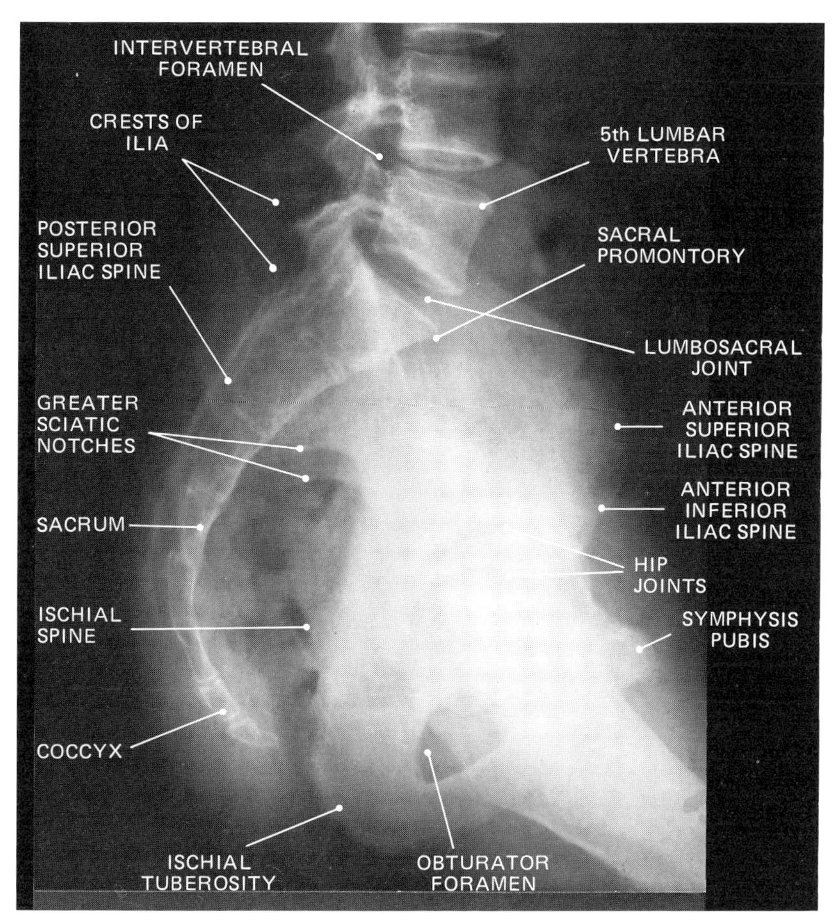

PELVIS LATERAL

Symptoms and Signs of Pregnancy

The symptoms and signs of pregnancy are a combination of anatomical and physiological changes. The average duration of pregnancy is usually 266 days from conception, or 280 days from the last menstrual period. This can be divided into three equal trimesters.

First Trimester

Amenorrhoea. The cessation of the normal menstrual pattern. In a few women a degree of bleeding continues at the times the normal menstrual periods would have occured—'the placental sign'—thought to be due to a partial failure to suppress menstruation.

Bladder irritability

Breast changes. The breasts enlarge and become tender from the sixth week of pregnancy. At 12 weeks the nipple and areola increase in size and permanently darken in colour. Montgomery's tubercles appear and it is possible to express a clear secretion from the breasts. By 20 weeks the secondary areola is becoming pigmented. This will disappear after pregnancy.

Changes in the vagina. Oestrogens cause a bluish discolouration on the vulva in the fourth week (Jacquemier's sign). At the tenth week pulsations can be felt in the lateral fornices (Oslander's sign). Vulval varicosities (Kluge's sign) may appear in the tenth week.

Uterus. The softened lower part of the uterus may be compressed during bimanual examination from the sixth to the tenth weeks (Hegar's sign).

Temperature. The normal elevation of the second half of the menstrual cycle persists with a temperature of 37.2 to 37.8°C.

Pigmentation. Apart from the nipples and areola, the face (cholasma) and the abdomen (the midline linea nigra and the diffuse striae gravidarum) may bear witness to the pregnancy.

Second and Third Trimesters

The uterus enlarges as the infant develops.
At 7 weeks it is the size of an egg.
At 10 weeks the size of an orange.
At 12 weeks as large as a grapefruit and palpable in the abdomen just above the pubic symphysis.
At 16 weeks it is halfway to the umbilicus.
At 20 weeks it has reached the umbilicus.
At 28 weeks it is halfway to the xiphisternum from the umbilicus.
At 36 weeks it reaches the xiphisternum, but as the fetal head descends into the pelvis its upper limit drops to its level at 32 weeks, three-quarters of the way from the umbilicus to the xiphisternum.

Between 16 and 20 weeks fetal movements are perceived by the mother as 'quickening'. After 30 weeks the irregular uterine contractions of Braxton Hicks first occur.

Internal ballottement of the fetus is possible from the fourteenth week and individual fetal parts can be felt from 24 weeks onwards. The fetal heart beat can be detected from the twelfth week with the aid of Doppler ultrasound and from week 20 it can be heard with a traditional fetal stethoscope.

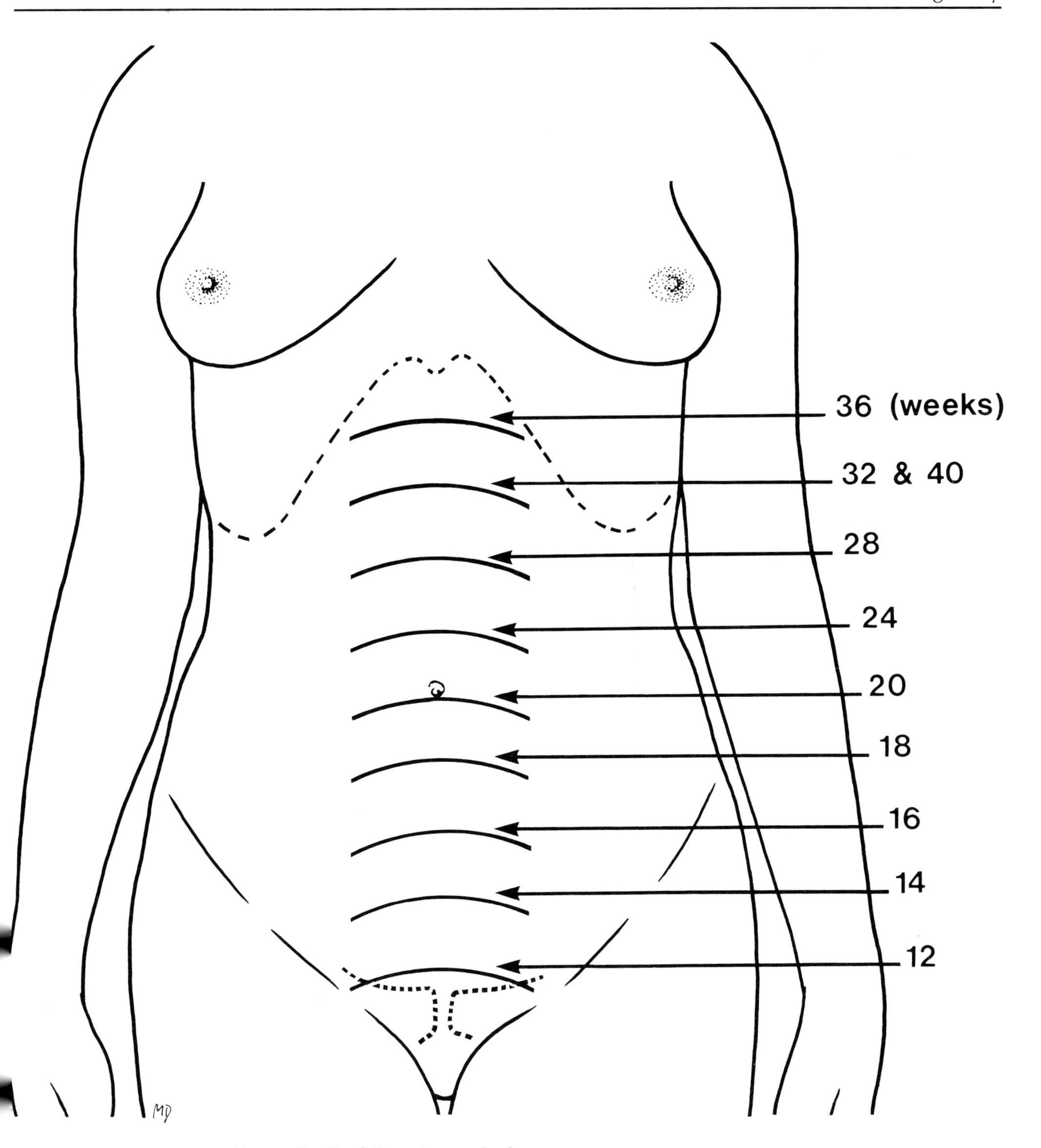

Upper limit of the uterus during pregnancy

8. The Upper Limb

BONY LANDMARKS

The Clavicle

The clavicle is usually visible throughout its whole length except in obese people when only the medial end is visible. The medial end articulates with the first costal cartilage and the manubrium sterni. The gap between the two medial ends of the clavicles above the upper end of the manubrium is known as the suprasternal notch, and it usually lies around the level of the lower border of the body of the second thoracic vertebra. The lateral end of the clavicle projects to a variable degree above the level of the acromion with which it articulates. The acromioclavicular joint may be felt about 2–3 cm medial to the lateral border of the acromion. For descriptive and clinical convenience the clavicle is classically divided into thirds.

The medial third of the clavicle is convex forwards and gives attachment to the clavicular head of the sternocleidomastoid. Between the sternal and clavicular heads of the sternocleidomastoid is a hollow, behind which the internal juglar vein descends in front of the subclavian artery with the vagus nerve postero-medial and the phrenic nerve postero-lateral to the vein. Behind the clavicle, beneath the hollow, the subclavian vein joins the internal juglar vein to form the brachiocephalic vein. Here the upper part of the brachiocephalic vein is separated from the cervical pleura by the phrenic nerve and the internal thoracic artery. Just lateral to the point of junction, the subclavian vein runs behind the clavicle.

The junction between the medial and middle thirds is marked by the lateral limit of the origin of the sternocleidomastoid (*see also* Muscles of the Neck, p. 210). This represents the antero-inferior angle of the posterior triangle of the neck and the lateral limit of the cervical pleura.

The middle third overlies the brachial plexus and the subclavian artery. At the lower border of the clavicle, the subclavian artery and the brachial plexus cross the lateral border of the first rib to enter the axilla, at which point the subclavian artery becomes continuous with the axillary artery.

The junction of the middle and lateral thirds is marked by the anterior limit of the insertion of the trapezius which forms the postero-inferior angle of the posterior triangle of the neck.

The lateral third is concave forwards and overlies the shoulder joint laterally and the coracoid process medially.

The Scapula

The scapula lies on the back over the second to seventh ribs and helps to form part of the posterior wall of the axilla. The acromion forms the summit of the shoulder lateral to the acromioclavicular joint and the coracoid process lies below the lateral third of the clavicle.

The acromion is subcutaneous and is easily felt in the summit of the shoulder. The tip of the acromion is the most anterior point and is readily palpated about one finger's breadth lateral to the lateral end of the clavicle. The lateral and posterior borders of the acromion join at a sharp angle, known as the acromial angle, which is easily felt at the back of the shoulder and is useful for making measurements of the upper limb. In abduction the acromial angle may be obscured by the deltoid.

The crest of the spine of the scapula is easily felt as it slopes medially and downwards to meet the vertebral border of the scapula at the level of the spine of T3; a useful point to remember when trying to identify vertebral spines by number.

The vertebral border can be identified by firm palpation. It usually lies about three finger breadths from the median plane when the scapula is in the resting position, but may be only one finger's breadth from the median plane when the shoulders are braced backwards. Equally, when the shoulders are drawn forwards the scapula moves away from the median plane thus facilitating examination of the posterior chest wall and underlying lungs. When the hand is placed on the back of the head, the scapula is rotated so that the vertebral border corresponds roughly with the position and direction of the oblique fissure of the lung. This also serves to expose the back of the lower lobe of the lung to easier auscultation.

The coracoid process lies in the infraclavicular fossa and can be felt about 2–3 cm below the junction of the lateral and middle thirds of the clavicle. Since it is covered by the anterior border of the deltoid, deep palpation may be required. It is also easier to feel if the arm is abducted during palpation, when the coracoid process may be felt to move under the examining fingers. The brachial plexus and axillary artery lie close to the medial side of the coracoid process.

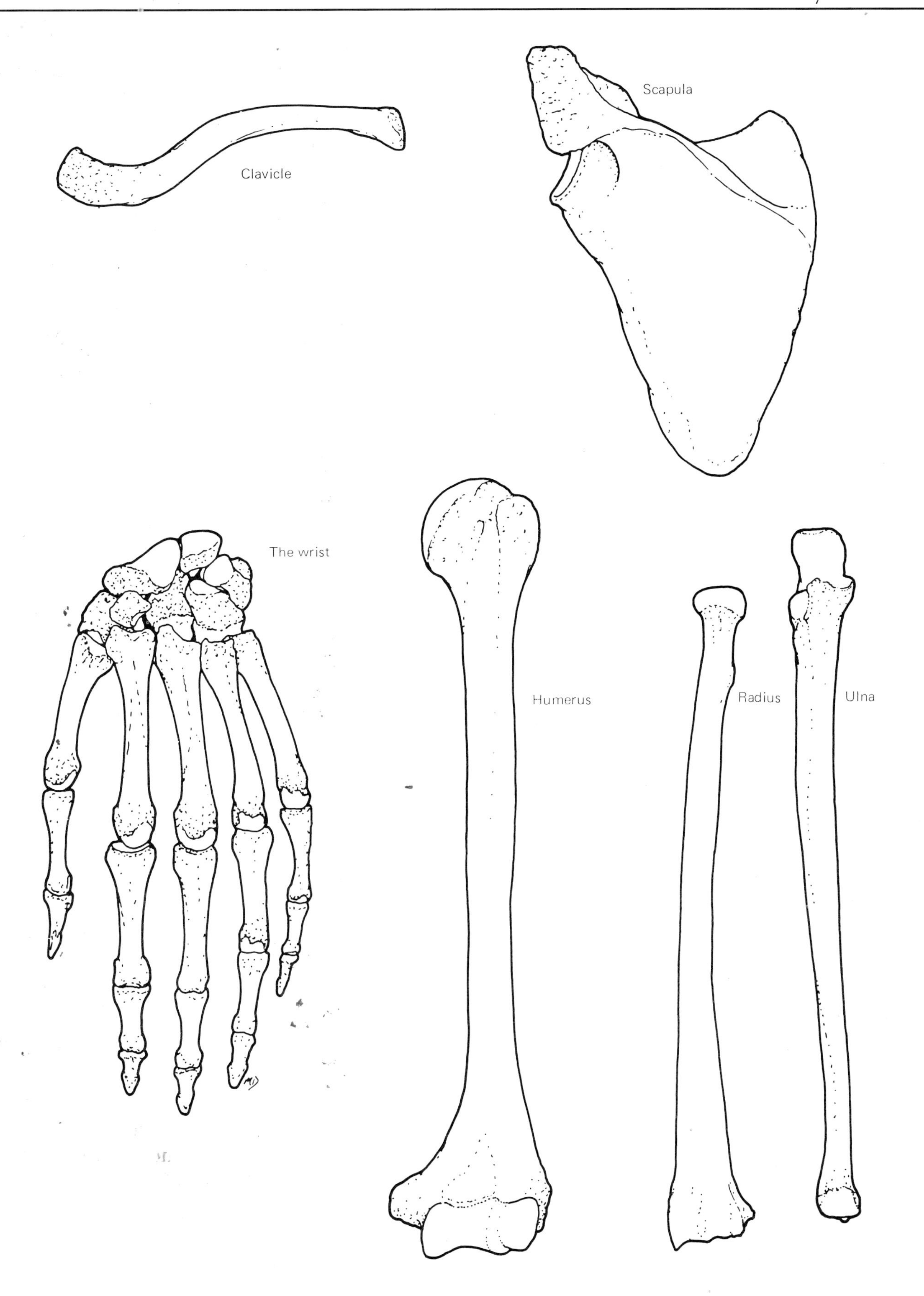
Clavicle
Scapula
The wrist
Humerus
Radius
Ulna

The Humerus

The greater tuberosity forms the most laterally palpable bony landmark of the shoulder region. It is covered by the deltoid muscle which gives the rounded appearance to the shoulder. If the head of the humerus becomes dislocated from the glenoid fossa, then the rounded appearance is lost and the acromion becomes the most laterally palpable point.

The head of the humerus may be felt indistinctly by upward palpation in the axilla, especially during movement.

The surgical neck of the humerus may be felt high up in the lateral wall of the axilla behind the coracobrachialis muscle. It lies about three finger breadths below the level of the acromion.

The shaft of the humerus may be palpated throughout its length. The deltoid tuberosity is on its lateral surface opposite the insertion of the coracobrachialis. Distally, the medial and lateral supracondylar ridges may be distinctly felt and the ulnar nerve may be compressed against the medial epicondyle.

The medial epicondyle is the easily palpable distal extremity of the medial supracondylar ridge. The ulnar nerve lies on its posterior surface and may be easily rolled between the finger and the bone. Here the unprotected nerve may be subjected to minor trauma and create the sensation of 'hitting the funny bone'.

The lateral epicondyle is not as prominent as the medial, but is still easily palpable.

The Ulna

The ulna is easily felt throughout its length from the olecranon to the head. Its posterior border is subcutaneous and forms a useful landmark in the forearm since it intervenes between the extensor and flexor musculature. Distally, the posterior surface of the ulna ends in a rounded prominence called the head.

The head of the ulna is the prominence seen on the ulna side of the wrist when the forearm is prone. As the forearm is supinated the head disappears.

The styloid process is not conspicuous. However, it may easily be identified by the following method. Lie the hand palm down on a table with the wrist in ulnar deviation but neither flexed nor extended. This highlights the tendon of the extensor carpi ulnaris on the ulnar side of the head of the ulna. Now, radially deviate the hand and notice that the tendon of the extensor carpi ulnaris runs in a groove between the head of the ulna and a smaller palpable bony prominence which is continuous with the subcutaneous border of the ulna. This is the styloid process.

The olecranon is subcutaneous and easily palpated. When the elbow is fully extended the top of the olecranon and the prominences of the two epicondyles of the humerus are in the same transverse plane. However, when the forearm is flexed they form the three corners of a nearly equilateral triangle. When the elbow is dislocated or when there is a fracture of the olecranon these relationships become distorted.

The Radius

The radius is only easily examined at its proximal and distal ends.

The head of the radius is readily felt about 2 cm distal to the lateral epicondyle of the humerus where it can be felt to rotate during movements of pronation and supination.

The styloid process is the distal prolongation of the lateral surface. It is the bone felt at the proximal part of the anatomical snuff-box.
The tips of the styloid processes of the radius and ulna should be examined and their levels compared. The tip of the radial styloid process is the more distal of the two by almost one finger's breadth. This relationship may become altered in certain fractures of the distal end of the radius.

The dorsal tubercle of Lister lies on the dorsal surface of the distal end of the radius in line with the cleft between the index and middle fingers. Its size is variable. It is used as a landmark in certain surgical incisions at the wrist.

The Carpus or Wrist

The wrist comprises eight small bones arranged in two rows. Parts of almost all the carpal bones can be felt between the tendons and the borders of the back of the wrist.

The scaphoid is very important clinically since it is commonly fractured in falls on the outstretched hand and these fractures may not always be readily visible on the radiograph.

With the hand bent towards the ulnar side, the scaphoid may be palpated in the floor of the anatomical snuff-box near the radial styloid process, but it moves with the hand whereas the process does not. Pressure on the floor of the anatomical snuff-box will evoke a painful response in cases of fracture of the scaphoid.

The tubercle of the scaphoid is the small prominence felt immediately proximal to the ball of the thumb in the angle between the ball of the thumb and the flexor carpi radialis tendon.

The ridge of the trapezium may be felt by applying firm pressure to the ball of the thumb immediately distal to the tubercle of the scaphoid. It is more easily felt when the wrist is extended.

The pisiform bone is identified as the knob at the ulnar border of the front of the wrist just proximal to the hypothenar eminence. The ulnar nerve and vessels lie close along its radial side.

The hook of the hamate is hidden in the proximal part of the ball of the little finger. If the examiner's thumb is drawn towards the centre of the palm for 2 cm from the pisiform bone and then pressed deeply into the hypothenar eminence, the hook of the hamate may be indistinctly felt.

The Metacarpal Bones

The metacarpal bones may be seen and felt for their full length on the back of the hand. They are not easily felt in the palm except in very thin hands. The heads of the metacarpal bones are the knuckles, and the extensor tendons overlie them.

The metacarpophalangeal joints are just proximal to the knuckles. On the palm they lie about 0.5 cm proximal to the distal transverse crease of the palm, or 'heart line,' which runs over the heads of the ulnar metacarpals.

The Phalanges

These are easily felt from end to end. The dorsal surfaces of the first phalanx of the thumb and the first two phalanges of each finger are covered with the extensor tendons, and a venous arch lies across the back of the first phalanx of each digit. The palmar surfaces are covered with fibrous flexor sheaths, and the distal phalanx of each digit is covered with a fibrous, fatty pad known as the finger pulp.

The digital nerves and vessels lie along their sides.

MUSCLES OF THE SHOULDER AND ARM

To demonstrate the voluntary action and contours of the muscles:

Muscles on the Back

Trapezius [**1**] forms the posterior border of the posterior triangle of the neck with its upper part which is brought into relief when the shoulders are shrugged against resistance. This may also be used as a test for the integrity of the accessory nerve. The middle and lower parts are demonstrated by bracing the shoulders back against resistance (*See* photograph and diagram in section on 'The Back', p. 9).

Teres major [**2**] together with the tendon of the latissimus dorsi, which wraps around its lower border, forms the lower border of the posterior wall of the axilla. It may be tensed by bringing the abducted arm downwards and backwards against resistance.

Latissimus dorsi [**3**] is most easily demonstrated by asking the subject to cough or breathe out against resistance, when it may be seen and palpated lateral to the inferior angle of the scapula. Alternatively, the abducted and laterally rotated arm may be adducted against resistance.

Subscapularis may be demonstrated when, with the elbow at right angles, the upper arm is medially rotated against resistance.

Supraspinatus lies under the middle part of the trapezius above the spine of the scapula; it is responsible for the first 15° of abduction when the arm is raised from the side.

Infraspinatus may be felt to contract in the angle between the lateral border of the trapezius and the posterior border of the deltoid when, with the elbow at right angles, the upper arm is rotated laterally against resistance.

Muscles on the Chest

Pectoralis major [**4**] lies on the front of the chest and in the anterior wall of the axilla. It may be made to contract by pushing the hands together in front of the chest or by pushing the hands hard onto the hips. When tense, the lower part of the muscle makes the anterior margin of the axilla more distinct and the upper part gives the infraclavicular fossa a distinct medial margin.

The infraclavicular fossa [**IF**] is bounded by the clavicle above, the deltoid laterally and the pectoralis major medially.

Muscles on the Arm

Deltoid [**5**] forms the fleshy, rounded elevation of the shoulder and may be demonstrated when the partially abducted arm is maintained in abduction against resistance.

Biceps [**6**] forms the fleshy elevation on the front of the arm. It is limited medially by the medial bicipital furrow [**MF**] and laterally by the lateral bicipital furrow [**LF**]. It may be demonstrated by flexing the supinated forearm against resistance when its tendon may be felt as a tense stout band extending well into the cubital fossa. The bicipital aponeurosis [**6a**] spreads in a medial direction from the tendon and its sharp, concave proximal border may be felt below the bend of the elbow. The rotatory action of the biceps may be demonstrated by noting the hardening of the muscle when the fully extended forearm is rapidly supinated.

Coracobrachialis [**7**] and the long head of biceps.

Brachialis [**8**] lies deep to the biceps and may be felt deeply on either side of the tapering distal third of the biceps. Its action is demonstrated by flexing the semi-pronated forearm against resistance.

Triceps forms the fleshy mass on the back of the arm and may be demonstrated by extending the elbow against resistance.

Long head of triceps [**9a**].

Lateral head of triceps [**9b**].

Part of triceps tendon [**9c**] forming the posterior limit of the lateral bicipital furrow.

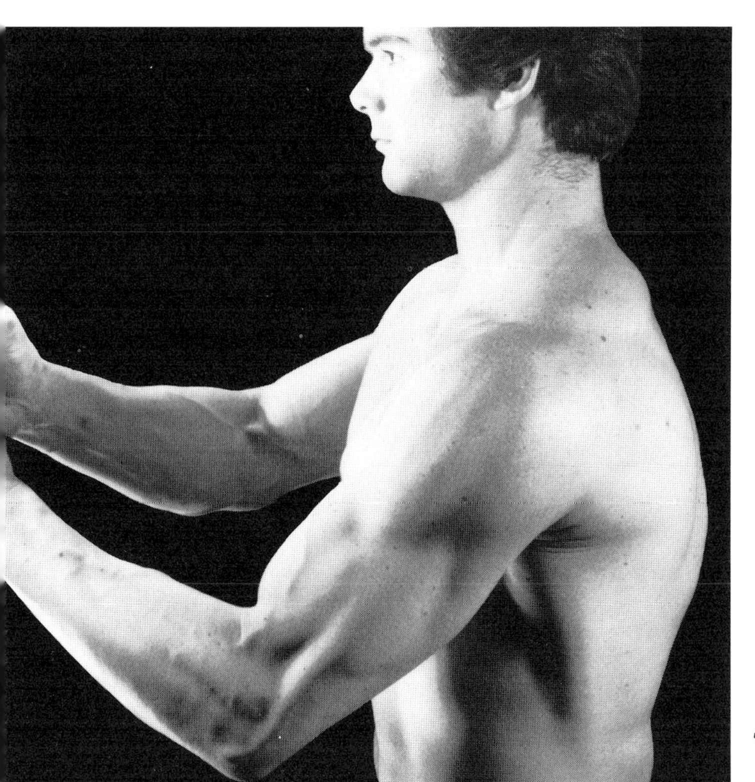

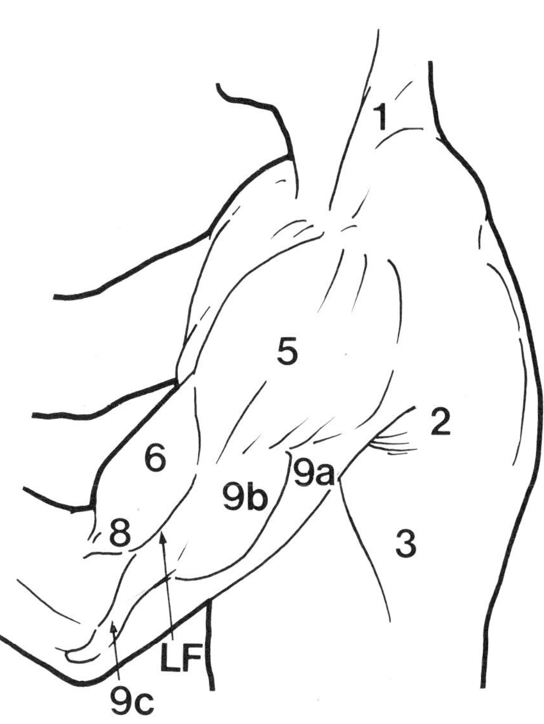
1
5
2
6
9a
9b
3
8
LF
9c

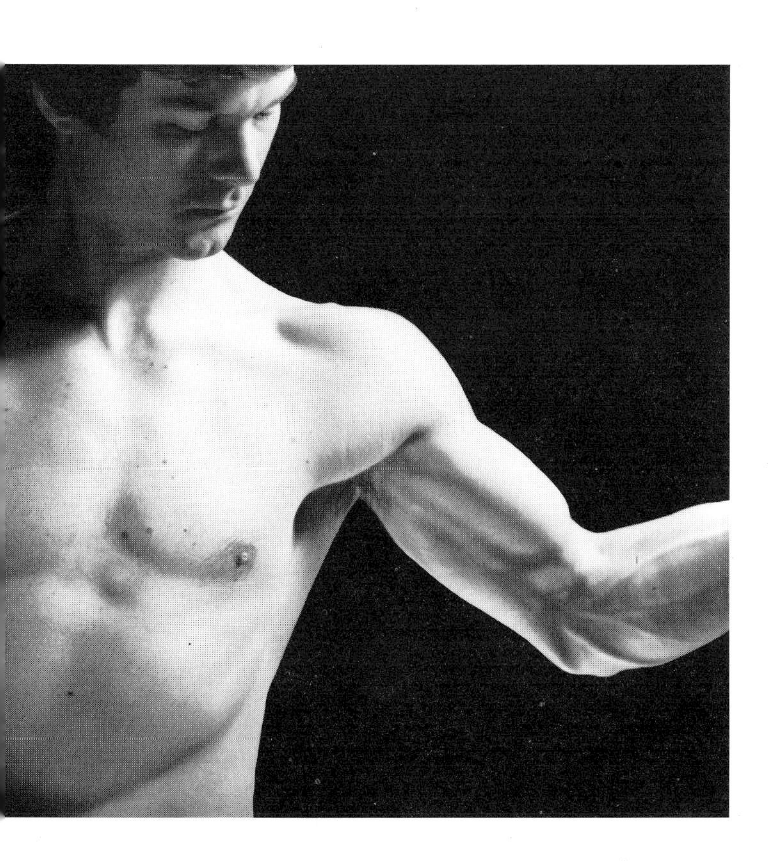

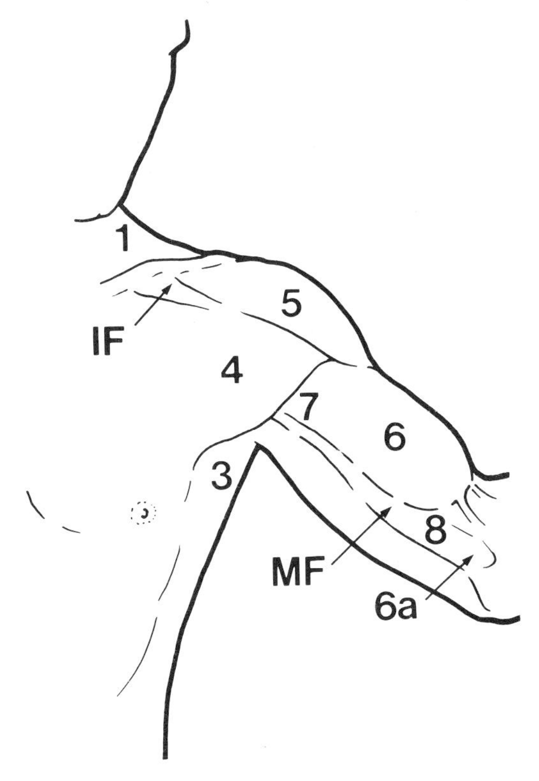
1
5
IF
4
7
6
3
8
MF
6a

THE AXILLA

If the arm is relaxed to loosen the skin and fascia of the armpit, the fingers can be pushed high into the axillary space.

On the medial wall the ribs can be felt through the serratus anterior. The long thoracic nerve runs down the medial wall just behind the mid-axillary line with a very slight inclination backwards. It is the only nerve in the body to lie on the external surface of the muscle it supplies, and here it may be subject to damage during a mastectomy. The damage manifests itself as an inability to brush the hair with the arm of the affected side because of inadequate scapular rotation. If the patient pushes on the outstretched arm, e.g. opening a swing door, the serratus anterior cannot hold the vertebral border of the scapula against the chest wall and it protrudes, as compared with the opposite scapula. This is called 'winging' of the scapula.

The anterior axillary fold [**1**] is formed by the pectoralis major.

The posterior axillary fold [**2**] is formed by the latissimus dorsi and the lower border of the teres major muscle.

The lateral wall is narrow because the muscles of the anterior and posterior folds insert close together on the humerus. Lying in the lateral angle are the short head of the biceps and coracobrachialis through which the surgical neck of the humerus can be felt. Their combined mass produces a rounded ridge which extends from the lateral wall of the axilla to the proximal part of the arm. In the groove behind this ridge the pulsations of the axillary artery and its continuation, the brachial artery, may be felt against the humeral shaft. Immediately behind the vessels, in the axilla, is the radial nerve which can be rolled against the bone. Indeed, so exposed is the nerve in this position that the inebriated individual supporting himself with his arm hanging over the back of a chair may wake up in the morning with a transient radial nerve palsy—'the Saturday night palsy'.

The Lymph Nodes of the Axilla

The lymph nodes of the axilla can be divided into separate groups, each draining a specific territory and palpable when enlarged due to inflammatory or malignant processes.

The anterior or pectoral nodes lie along the course of the lateral thoracic artery and, when enlarged, may be felt in the antero-medial angle of the axilla when the arm is relaxed and slightly abducted. They drain the lateral part of the breast and the upper quadrant of the abdominal wall.

The posterior or subscapular nodes lie along the course of the subscapular artery in the postero-medial angle of the axilla. They drain the back above the iliac crest.

The lateral group of nodes lies along the axillary vein in the lateral wall of the axilla, and drains the deep structures of the arm and the superficial lymphatics from the medial side of the upper limb which accompany the basilic vein.

The central group lies centrally and receives lymph from the preceding three groups, and directly from the tail of the breast.

The apical group lies high in the apex of the axilla and receives lymph from all the other groups, including the *infraclavicular group* of nodes in the deltopectoral groove which drains the lymphatics from the lateral side of the upper limb accompanying the cephalic vein.

REFLEXES OF THE ELBOW REGION

The triceps jerk (sixth and seventh cervical nerves). The examiner supports the subject's forearm allowing about 80° flexion at the elbow to keep the triceps tendon taut. A tap on the triceps tendon, just proximal to the olecranon, elicits an extensor movement at the elbow with an associated contraction and hardening of the triceps muscle.

The biceps jerk (fifth and sixth cervical nerves). The examiner supports the subject's forearm in a similar position as that for the triceps jerk. He then puts his thumb across the biceps tendon and strikes it with the patella hammer. This stretches the biceps tendon and results in a reflex contraction of the biceps.

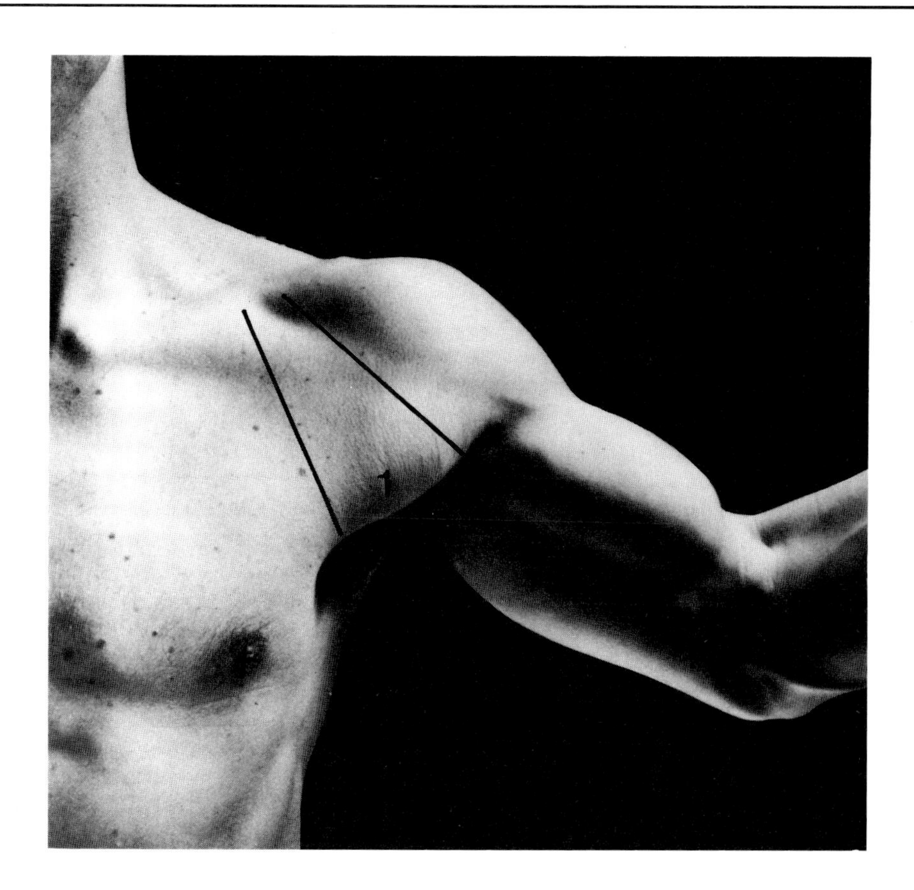

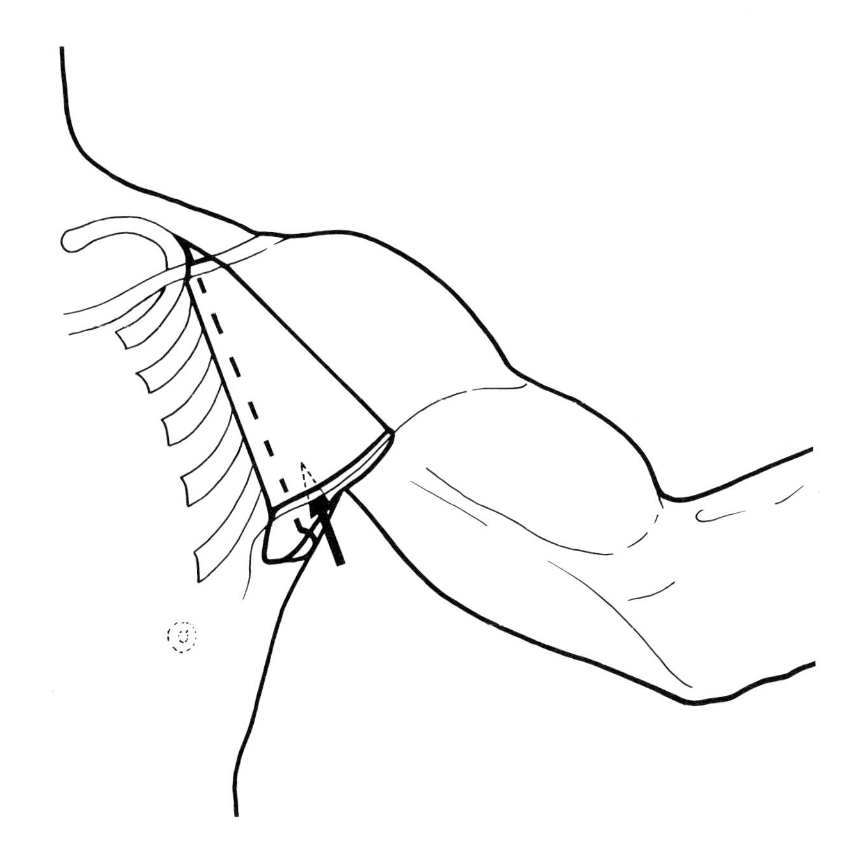

RADIOLOGY OF THE SHOULDER REGION

The shoulder joint is a complex radiographic area and the choice of views taken depends on the suspected pathology.

RADIOGRAPH A (top) This is an antero-posterior radiograph of the shoulder with the arm externally rotated and slightly abducted.

The Scapula

The cortex of the vertebral border [**1**] may sometimes be seen as an unbroken white line superimposed on the clavicle, ribs and lung. However, it is frequently invisible.

The bone in the infraspinous fossa [**2**] is very thin and radiolucent.

The axillary border of the scapula [**3**] through which many of the forces from the arm are transmitted is dense and marked trabecular patterns may be seen superiorly and inferiorly.

The acromion [**4**] is conspicuous. There is a secondary centre of ossification at the tip which unites with the main mass at about the age of 21 years.

The spine of the scapula [**5**] is continuous with the acromion and is seen as a long thin linear shadow extending to within 2 cm of the vertebral border.

The coracoid process [**6**] appears in scapular adduction as a ring shadow about 1 cm in diameter. It also exhibits a secondary centre at its tip which unites with the main mass about three years before the acromial tip.

The glenoid fossa [**7**] is seen here as an ellipse because the shoulders are braced backwards and the x-ray beam is pointing sideways at the fossa. The apparent gap between the head of the humerus and the glenoid fossa is normally 5–8 mm and is due to the fact that articular cartilage is radiolucent. Thus the apparent gap is a measure of the thickness of the articular cartilage and is called the *radiological joint space.*

If the shoulders are braced forwards, the glenoid fossa faces the x-ray beam end on and its cortical margins appear as an almost circular shadow.

The Humerus

The head of the humerus [**8**] around which the articular cortex is seen as a thin white line about 0.5 mm thick. It is continuous with the much thicker cortex of the shaft which may be 5–8 mm thick.

The trabecular pattern reveals a regular arrangement of horizontal and vertical lines which fade out until they are completely absent in the middle third of the humerus.

The greater tuberosity [**9**] shows a sparse trabecular pattern here which may disappear completely in old age, simulating an area of pathological bone resorption.

The lesser tuberosity [**10**].

The lateral end of the clavicle usually projects a little higher than the adjacent upper surface of the acromion. This relationship may be altered in dislocation of the joint or rupture of the acromioclavicular ligament.

RADIOGRAPH B (lower) This view of the shoulder has been included to illustrate the special relationship of the humeral head to the glenoid fossa and associated scapular structures. Several features can be clearly seen:

Coracoid process [**1**]
Acromion [**2**]
Acromioclavicular joint [**3**]
Head of humerus [**4**]
Bony rim of the shallow glenoid fossa [**5**]
The blade of the scapula seen end on [**6**]
The Lesser tuberosity [**7**]

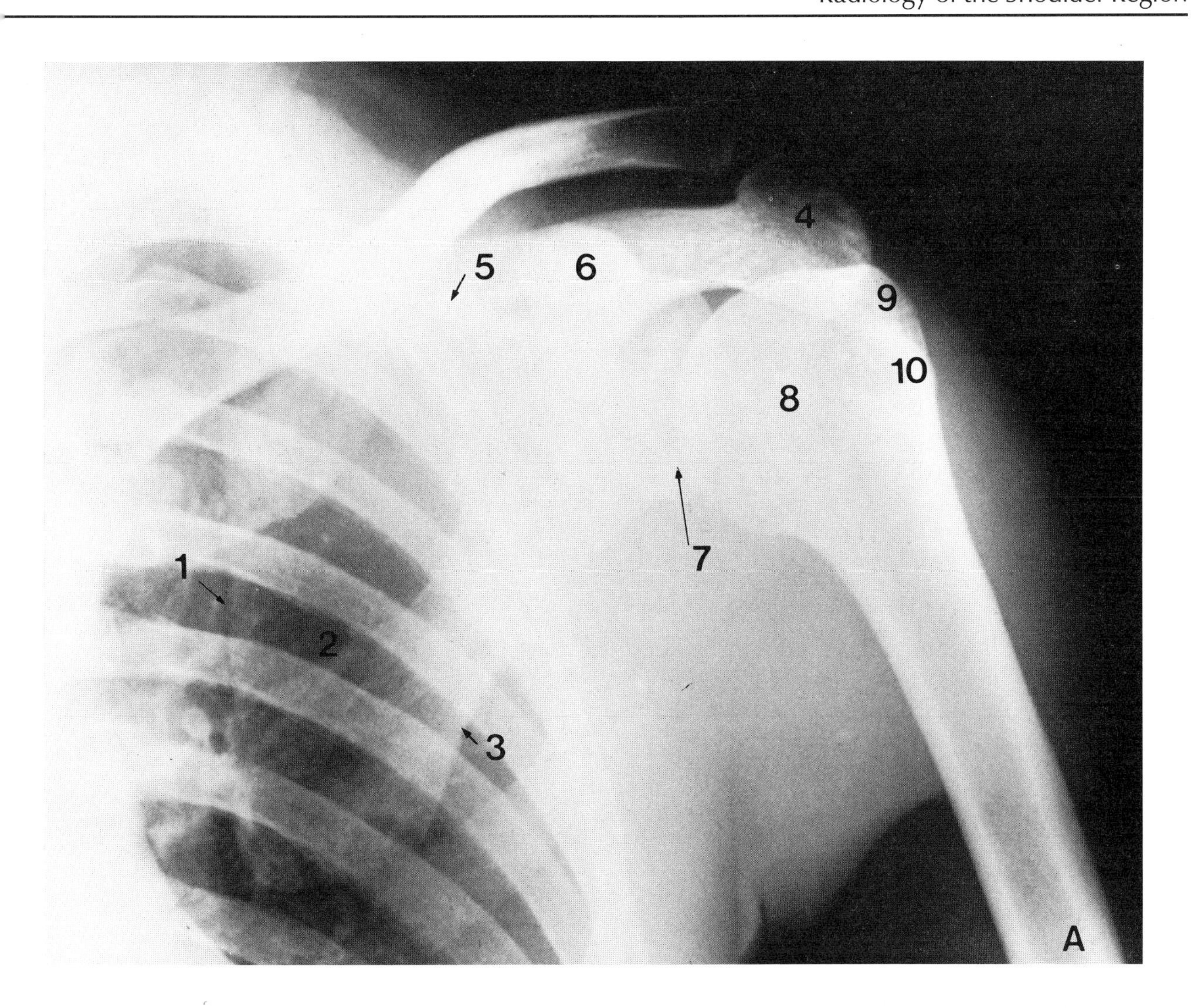

1
2
3
4
5
6
7
8
9
10
A

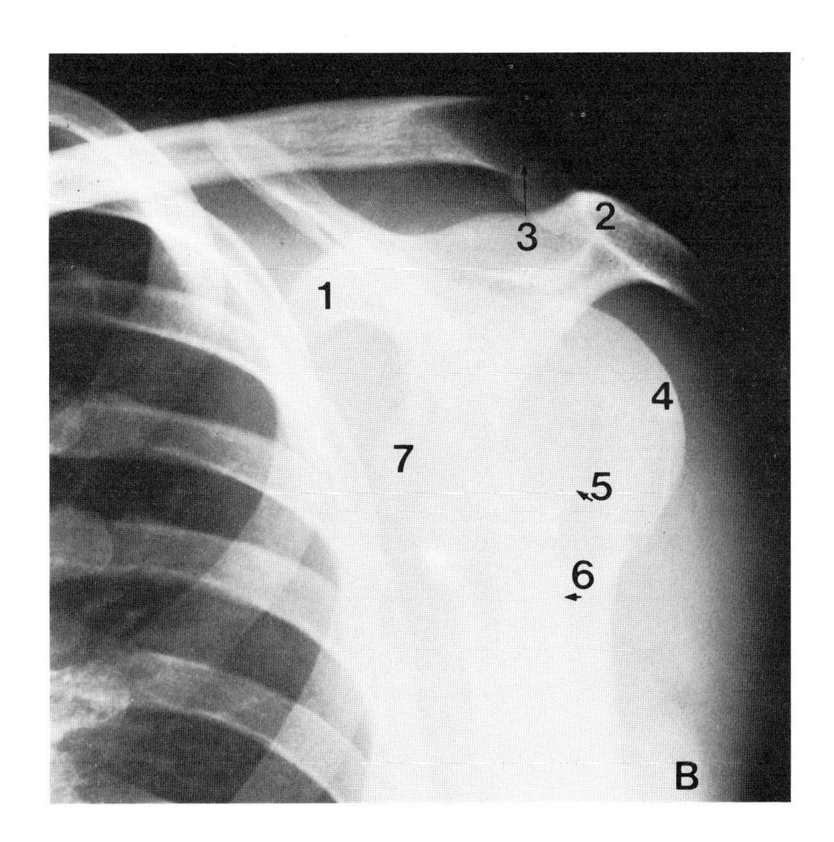

1
2
3
4
5
6
7
B

RADIOLOGY OF THE ELBOW REGION

Posterior View in Full Extension

Lateral epicondyle [1]
Medial epicondyle [2]
The superimposed olecranon and coronoid fossae produce a circular radiolucent zone with the olecranon superimposed over the lower half [3]
Coronoid process [4]
Joint space between coronoid and trochlea [5]
The compact bone around the head of the radius usually forms an unbroken white line [6]
Tuberosity of radius [7]

Lateral View

In this view the epicondylar shadows should be superimposed.

Supracondylar ridges [1]
The capitulum projecting anteriorly beyond the line of the anterior border of the humeral shaft for one-third of its thickness, to enable the forearm to be flexed beyond a right angle before the radial head comes into contact with the radial notch of the humerus [2]
Olecranon [3]
The coronoid process [4] partly overlaps the radial head [5]

Ossification in the Elbow and Forearm

Lower end of humerus. Dates of appearance

Capitulum— 2 years
Medial epicondyle— 5 years
Trochlea— 10 years
Lateral epicondyle— 12 years

The secondary centres for the lateral epicondyle, capitulum and trochlea fuse at puberty to form a single epiphysis which fuses with the main shaft at about the age of 16 in males, and 14 in females.

The centre for the medial epicondyle fuses with the main mass of bone at about the twentieth year.

Radius. Dates of appearance

Shaft— antenatal
Head— 5 years, fuses at 18 years
Distal end—2 years, fuses at 20 years

Ulna. Dates of appearance

Shaft— antenatal
Olecranon—10 years, fuses at 20 years
Distal end—5 years, fuses at 20 years

These dates are only approximate and vary by several years. Therefore, if a fracture is ever suspected around a joint of a youngster *always* x-ray both sides for comparison.

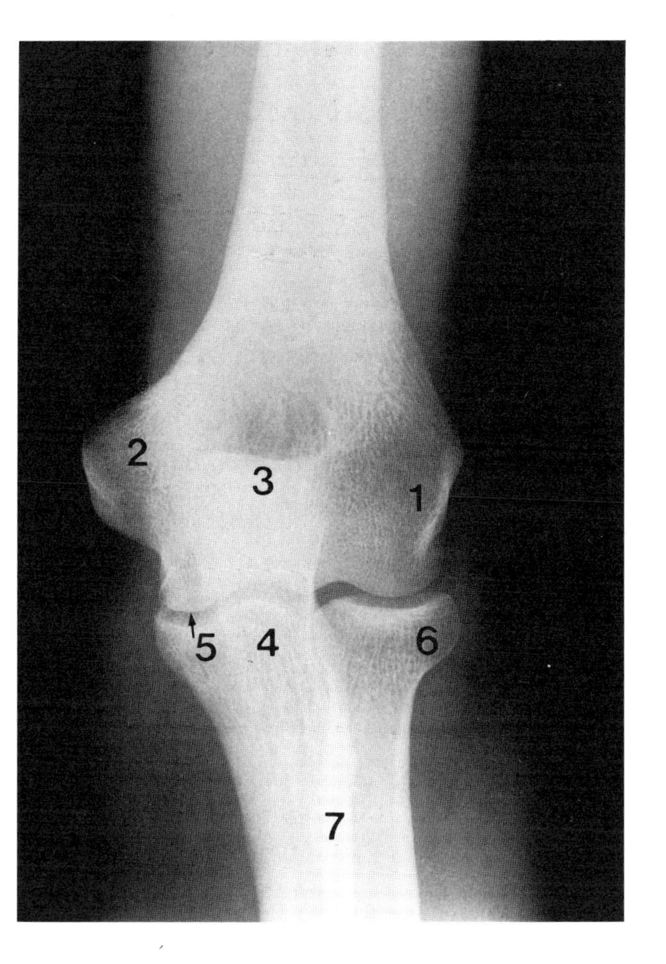

Posterior view

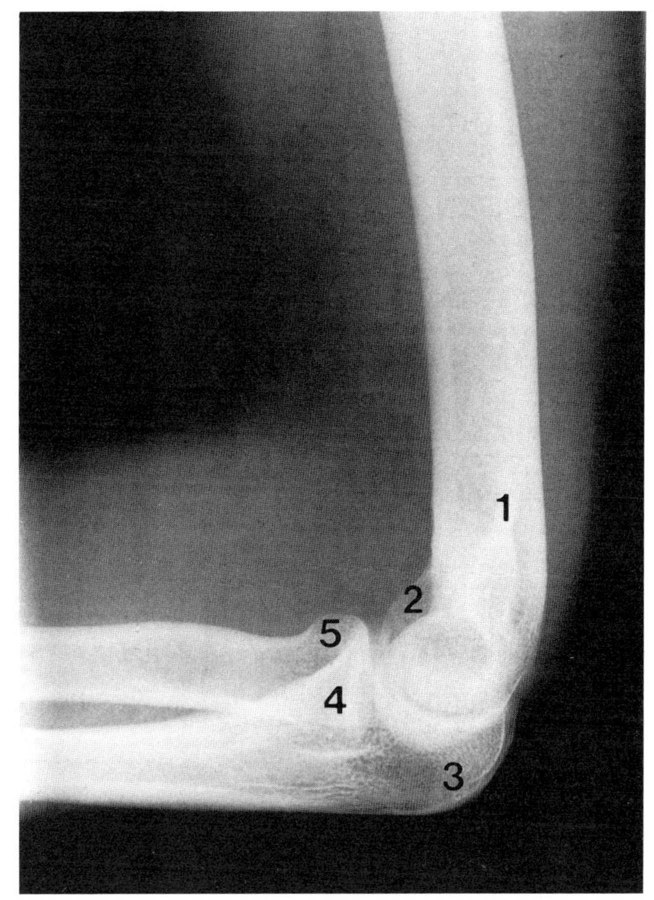

Lateral view

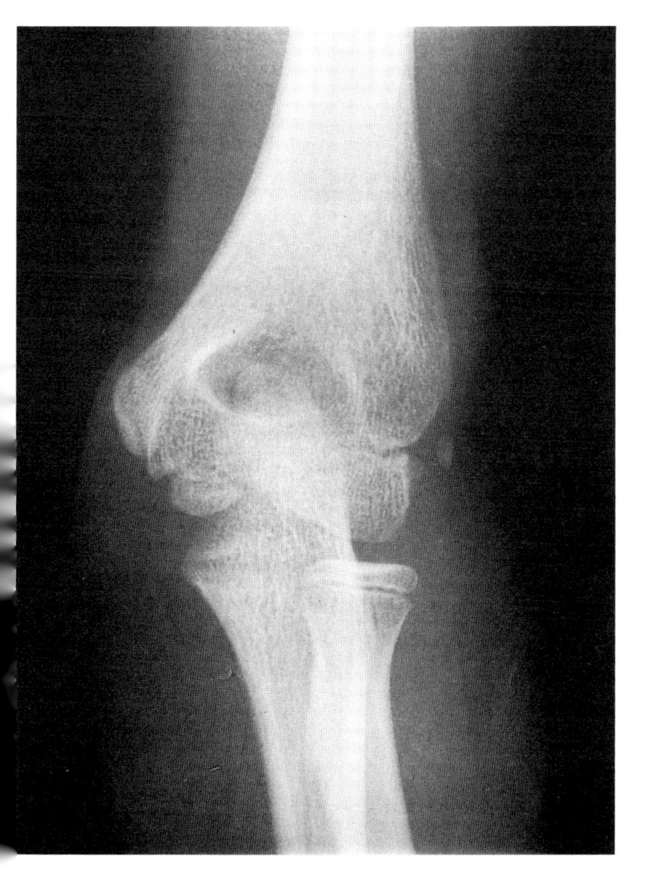

Posterior view

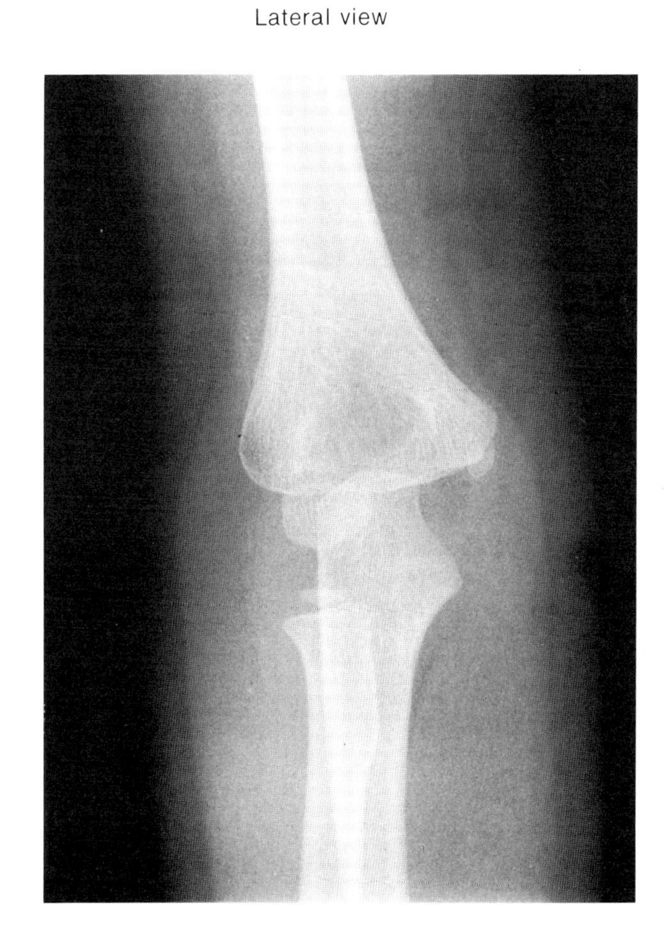

Posterior view

THE EXTENSOR SURFACE OF THE FOREARM AND HAND

Bony Landmarks

Lateral epicondyle [**A**]
Olecranon [**B**]
Subcutaneous border of ulna [**C**]
Head of ulna [**D**]
Styloid process of ulna [**E**]
Head of radius [**F**]
Base of second metacarpal [**G**]—the styloid process is particularly prominent

Soft Tissues

Brachioradialis [**1**] is demonstrated by flexing the elbow against resistance with the forearm midway between pronation and supination. In this position the muscle is seen as a conspicuous ridge running down the radial border of the forearm.

Extensor carpi radialis longus [**2a**].
Extensor carpi radialis brevis [**2b**]. The bellies of these two muscles may be displayed by extending the hand to the radial side against resistance with the elbow flexed. The belly of the longus is seen above the lateral epicondyle, the brevis just distal to it.

Extensor digitorum [**3**] is tested and displayed by keeping the fingers extended against a force which tends to flex them at the interphalangeal joints.

Extensor digiti minimi [**4**].

Extensor carpi ulnaris [**5**] is tested by dorsiflexing the hand to the ulnar side against resistance. Its tendon lies between the head of the ulna and the styloid process of the ulna at the wrist.

Anconeus [**6**].

Flexor carpi ulnaris [**7**] overlying the bulk of the flexor digitorum profundus.

Extensor pollicis brevis [**8**]. *Abductor pollicis longus* [**9**]. These two muscles are responsible for the oblique fleshy elevation on the back of the radial side of the wrist which is seen when the fist is clenched. Their tendons lie so close together in the radial side of the anatomical snuff-box, with the abductor on the radial side, that on the surface they appear as a single tendon.

The first dorsal interosseous muscle [**10**] may be displayed by firmly pressing the tips of the thumb and index finger together causing the fleshy elevation between the metacarpal bones of the thumb and index finger to harden.

Extensor indicis tendon [**11**].

Second dorsal interosseous [**12**].

Third dorsal interosseous [**13**].

Notice the position of the extensor expansion on the dorsum of the middle finger.

The Anatomical Snuff-box

This is brought into relief when the thumb is fully extended and the extensor tendons are drawn up on the radial side of the wrist. The 'snuff-box' is the depression between the extensor pollicis longus tendon, forming the ulnar or posterior border, and the tendons of the extensor pollicis brevis and abductor pollicis longus, forming the radial or anterior border.

The cutaneous branches of the radial nerve run over the tendon of the extensor pollicis longus and can be rolled against it.

Several bony points can be felt in the floor of the snuff-box. Proximally, the radial styloid process and distally, the base of the thumb metacarpal can easily be identified. In between these two, the trapezium and scaphoid can be felt (*See* Bony Landmarks of the Upper Limb, p. 114–15). The pulse of the radial artery may be felt as it lies on the scaphoid and trapezium in the floor of the snuff-box before entering the palm between the two heads of the first dorsal interosseous. In some individuals it is also possible to feel the movement of the tendon of extensor carpi radialis longus when the fist is clenched.

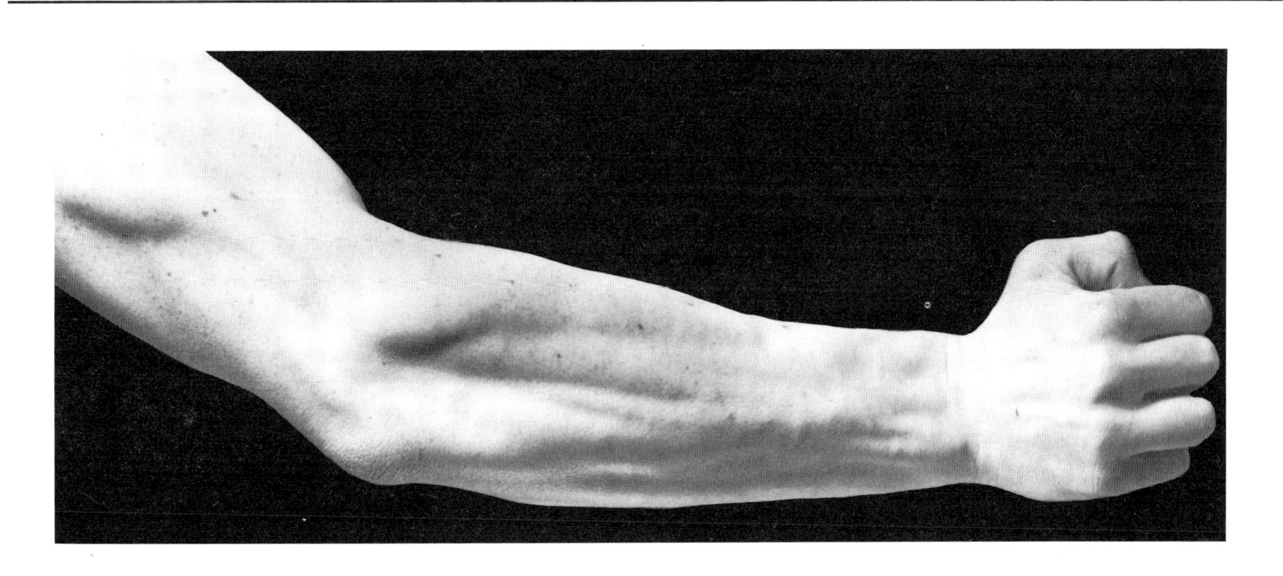

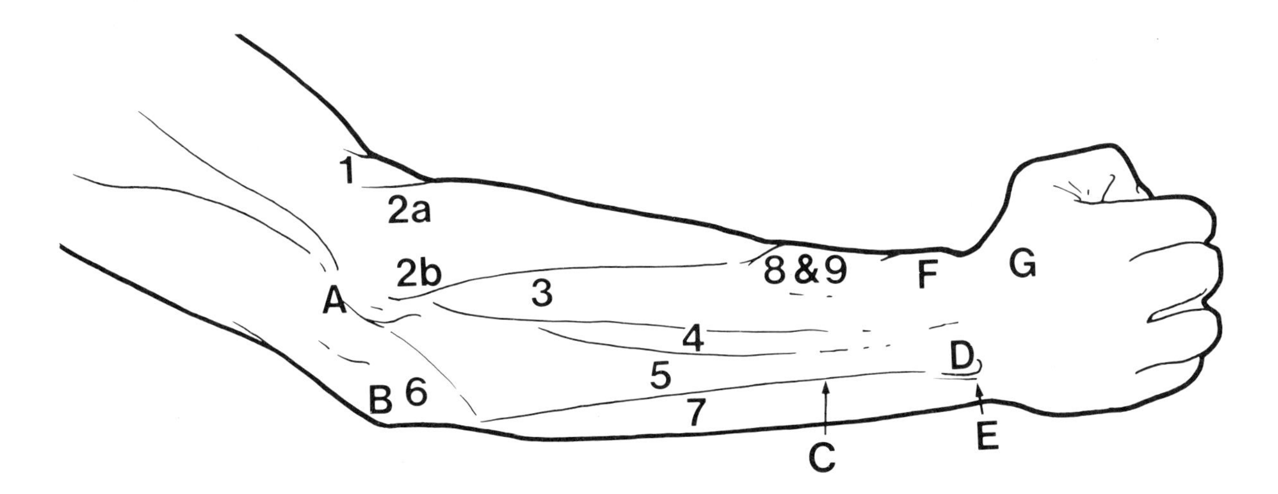
1
2a
2b
A
3
8 & 9
F
G
4
D
5
B 6
7
C
E

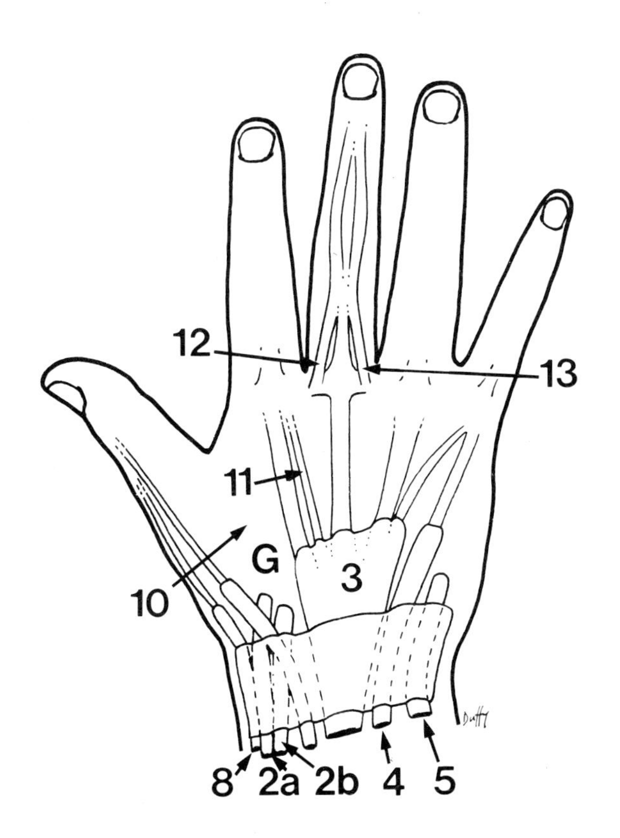
12
13
11
G
10
3
8
2a
2b
4
5

FLEXOR MUSCLES OF THE FOREARM

These muscles are responsible for grip and are therefore more bulky than the extensor muscles. They are arranged in two layers with five muscles in the superficial group and three in the deep group.

The Superficial Muscles

These all take origin, in part, from the common flexor origin at the medial epicondyle and their positions may be visualised as follows: with the heel of the hand over the opposite medial epicondyle and the palm on the forearm, the fingers point distally along the five muscles; the thumb for the pronator teres, the index finger for the flexor carpi radialis, the middle finger for the flexor digitorum superficialis, the ring finger for the palmaris longus and the little finger for the flexor carpi ulnaris. These muscles may be demonstrated by the following methods:

Pronator teres [**1**] runs from the medial epicondyle to the middle of the radial border of the forearm. It is tested by pronating the supinated forearm against resistance with the arm at the side.

Flexor carpi radialis [**2**] runs along a line from the medial epicondyle to the base of the second metacarpal. It is the most radial tendon brought into relief at the wrist by flexing the wrist against resistance.

Palmaris longus [**3**] is sometimes absent on one or both sides but, when present, it is brought into relief by flexing the wrist against resistance when it lies on the ulnar side of flexor carpi radialis. The median nerve then lies between the two on a slightly deeper plane. If the palmaris longus is absent then the middle finger tendon of the flexor digitorum superficialis which is usually deep to the palmaris longus comes into relief. If there is doubt, then alternate flexion of the wrist and middle finger should identify the tendon.

Flexor digitorum superficialis [**4**] is tested by flexing the fingers against resistance at the proximal interphalangeal joints. The tendon for the ring finger is felt at the wrist between the palmaris longus and the flexor carpi ulnaris when the closed fist is flexed against resistance.

Remember that the tendons of the middle and ring fingers lie superficial to the tendons of the index and little fingers at the wrist; thus even in the absence of the palmaris longus only the former two tendons can be brought into relief.

Flexor carpi ulnaris [**5**] runs along the ulnar border of the forearm to the pisiform bone and appears bulky because it overlies the flexor digitorum profundus. When the closed fist is flexed against resistance it is the tendon seen nearest the ulnar border at the wrist and is easily traced to the pisiform.

The ulnar artery and nerve lie in the depression between this tendon and the adjacent tendon of the flexor digitorum superficialis.

The Deep Muscles

Flexor digitorum profundus is the muscular mass felt hardening on the ulnar side of the arm when the fist is clenched. It is tested by the examiner holding the subject's fingers extended at the proximal interphalangeal joints while the subject flexes the distal interphalangeal joints.

Flexor pollicis longus is tested by flexing the thumb at the interphalangeal joint.

Pronator quadratus assists the pronator teres and lies deeply across the distal fifth of the forearm with its distal border level with the proximal wrist crease.

The Brachioradialis Reflex (sixth cervical nerve)

The arm to be tested is supported freely in front of the trunk by the subject's other arm. With the arm halfway between pronation and supination, a brisk tap on the brachioradalis tendon [**6**], just proximal to the radial head, should elicit a contraction of the muscle with resultant flexion at the elbow. The interphalangeal joints of the fingers sometimes also flex.

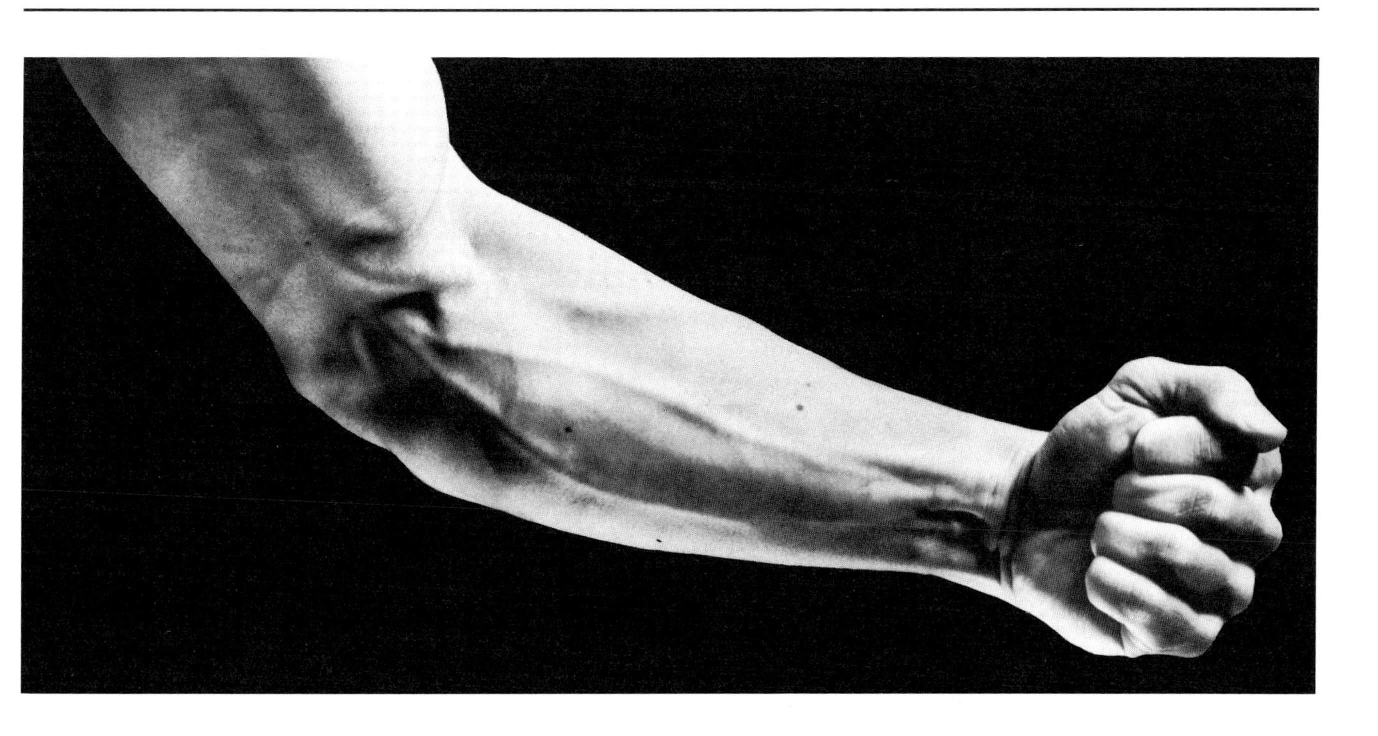

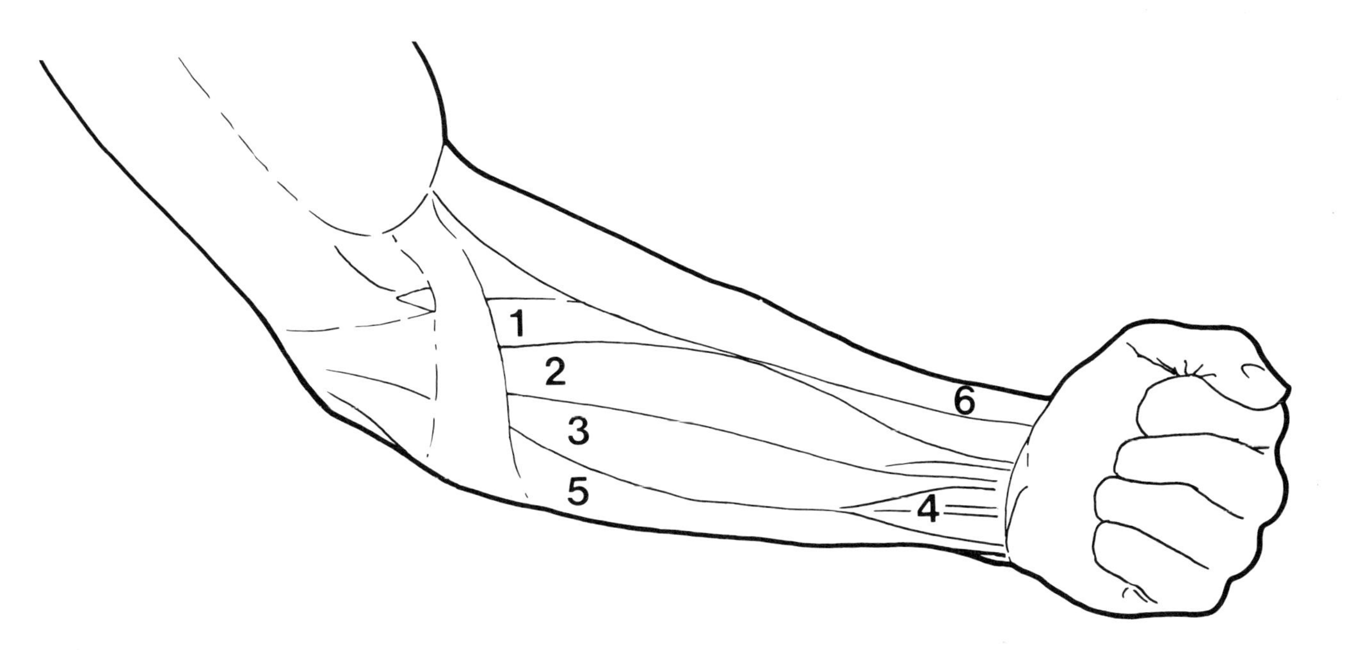
1
2
3
4
5
6

The Cubital Fossa

A good working knowledge of the cubital fossa and its overlying pattern of veins is essential if painful and unnecessary accidents are to be avoided during venepuncture.

The cubital fossa is bounded proximally by an imaginary line between the humeral epicondyles, laterally by the medial border of brachioradialis and medially by the lateral border of the pronator teres. Its apex lies about one-quarter of the way down the forearm.

The floor is formed by the supinator muscle laterally and the brachialis medially.

The roof is formed by skin, superficial fascia and deep fascia. The superficial fascia contains the important cephalic and basilic veins and their interconnecting channels—most commonly the median cubital vein (*See* The Veins of the Upper Limb, p. 136). The medial and lateral cutaneous nerves of the forearm lie on the medial and lateral sides of the basilic and cephalic veins respectively.

The contents of the cubital fossa are best related to the biceps tendon which forms the most easily identifiable landmark in the area. Medial to the biceps tendon all the deep structures are separated from the median cubital and basilic veins by the bicipital aponeurosis.

The brachial artery runs along the ulnar side of the tendon and, opposite the neck of the radius, divides into the radial artery which continues in front of the tendon, and the ulnar artery which passes downwards and medially away from the tendon under cover of the aponeurosis. However, in high division of the brachial artery, the ulnar artery frequently passes superficial to the aponeurosis and is then in danger during venepuncture, or intravenous injection of irritant drugs.

The median nerve lies medial to the brachial artery and then to the ulnar artery, but crosses in front of the latter as it leaves the fossa.

The radial nerve appears in the upper lateral angle of the fossa and its superficial terminal branch runs down the lateral border under the protection of the brachioradialis muscle.

The Flexor Retinaculum

The flexor retinaculum is attached to the ridge of the trapezium and the tubercle of the scaphoid laterally, and to the hamate and the pisiform bones medially. It measures about 2.5 cm in width and 2.0 cm proximo-distally. Its proximal limit corresponds to the distal wrist crease.

The Anterior Synovial Sheaths

Three synovial sheaths lie deep to the flexor retinaculum and extend a short distance proximally on the anterior aspect of the wrist about as far as the middle crease of the wrist. The sheaths are:

The common flexor sheath **[1]** for the flexor digitorum superficialis and the flexor digitorum profundus, sometimes called the ulnar bursa, extends almost as far distally as the distal transverse palmar crease, except for a single prolongation as far as the terminal phalanx of the little finger.

The sheath for the tendon of flexor pollicis longus **[2]** sometimes called the radial bursa, winds around the ulnar side of the flexor carpi radialis tendon and sheath, and is continued into the thumb as far as the terminal phalanx.

The long flexor tendons **[3]** to the index, middle and ring fingers have separate synovial sheaths, distinct from the common flexor sheath, which extend between the metacarpal head and the base of the terminal phalanx of each digit.

In about one-half of the population, the radial and ulnar bursae communicate at the site indicated by the arrow.

If these spaces become infected, e.g. following a splinter or penetrating injury, then infection of the index, middle and ring finger sheaths remains confined to the finger concerned. However, infection of the thumb or little finger sheaths may spread into the palm via the ulnar or radial bursa respectively, and from one bursa to the other through the frequent intercommunication between the two. Occasionally, sepsis may even spread into the forearm through the proximal extension of these bursae beneath the flexor retinaculum and the flexor digitorum profundus.

Flexor-carpi radialis sheath **[4]**

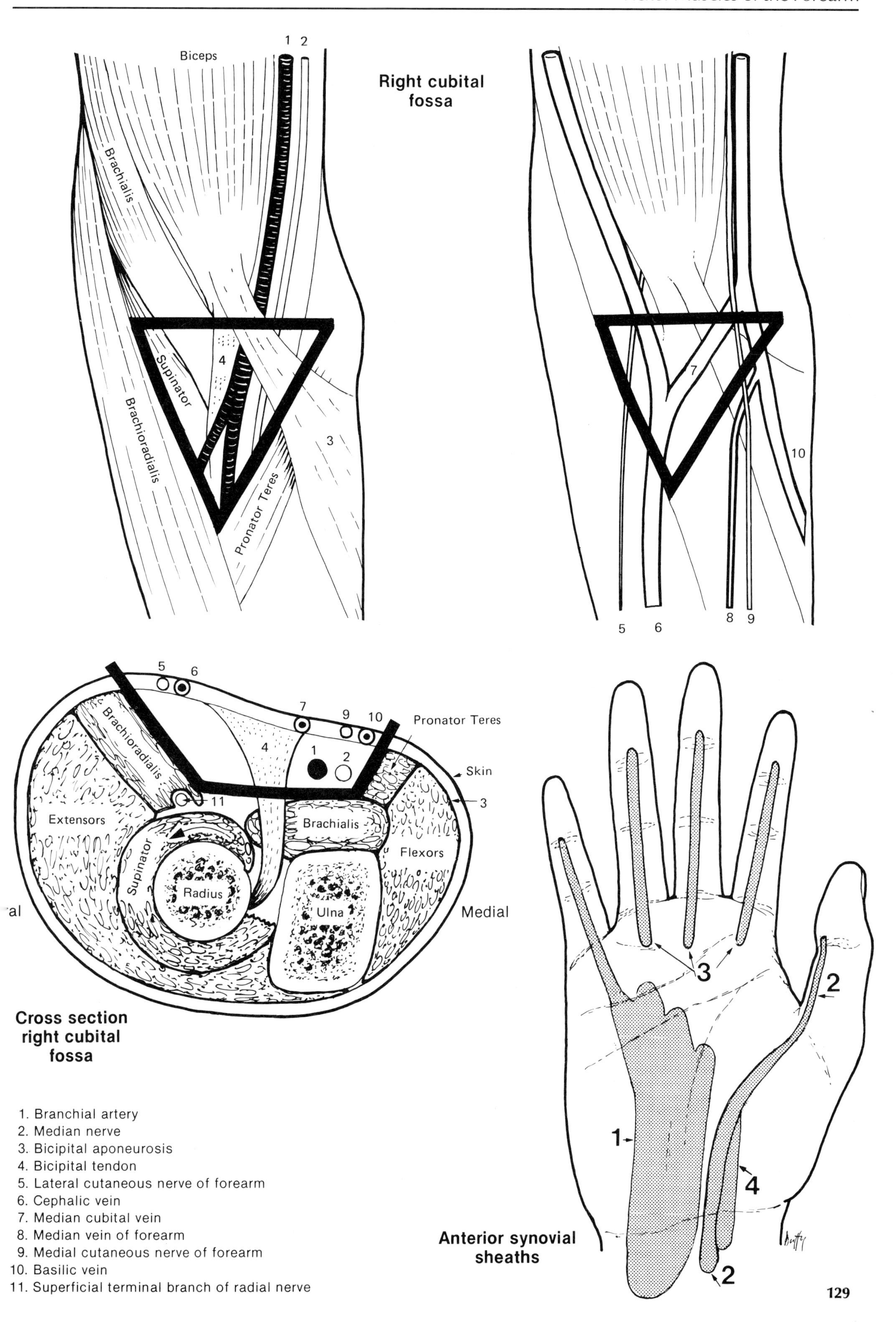

Right cubital fossa

Cross section right cubital fossa

Anterior synovial sheaths

1. Branchial artery
2. Median nerve
3. Bicipital aponeurosis
4. Bicipital tendon
5. Lateral cutaneous nerve of forearm
6. Cephalic vein
7. Median cubital vein
8. Median vein of forearm
9. Medial cutaneous nerve of forearm
10. Basilic vein
11. Superficial terminal branch of radial nerve

THE HAND

The Skin of the Hand

Three basic types of skin marking exist, all of which are well illustrated on the hands. These are tension lines, papillary ridges (also known as fingerprints or friction lines), and flexure creases.

Tension lines are the fine linear furrows seen on the back of the hand and between the wrist creases which form a network of small lozenge-shaped areas.

Papillary ridges or fingerprints are seen on the digits, palms and soles. The pattern of these epidermal ridges is determined by the arrangement and size of the underlying dermal papillae which interlock with the basal layer of the epidermis. In thin, hairy skin, as on the back of the hand, the papillae are scattered randomly, but in the thick, non-hairy skin of the hands and feet the dermal papillae are arranged in rows which throw the overlying epidermis into ridges. The individual patterns of these ridges constitute the fingerprints. Although slightly different in every individual, three basic patterns can be recognised—loops, whorls and arches. The openings of the sweat glands may be seen with a magnifying glass on the crests of the ridges.

Flexure creases are caused by the firm anchoring of the skin to the deep fascia by fibrous strands and are related to habitual joint movements. The fibrous strands quilt the fatty superficial fascia into pads which allow versatile and successful gripping.

The distal transverse crease or 'heart line' [**A**] runs across the heads of the four ulnar metacarpal bones and is accentuated by flexion of the metacarpophalangeal joints.

The radial longitudinal crease or 'life line' [**B**] is accentuated by opposition of the thumb when the thenar eminence rises into relief.

The proximal transverse crease or 'head line' [**C**] is a flexure crease anchoring the palmar skin between the heart and life lines.

The middle palmar crease [**D**].

Distal wrist crease [**E**].

Middle wrist crease [**F**].

Proximal wrist crease [**G**].

To Test the Muscles of the Hand

Adductor pollicis [**1**]. Try to hold a piece of paper against forcible removal between the thumb and the palmar aspect of the base of the index finger with the thumb nail at right angles to the palm.

Flexor pollicis brevis [**2**]. Flex the proximal phalanx against resistance while keeping the interphalangeal joint extended.

Abductor pollicis brevis [**3**]. Abduct the thumb at right angles to the palm against resistance.

Opponens pollicis [**4**]. Press the extended thumb towards the base of the little finger against resistance.

Abductor digiti minimi [**5**]. With the back of the hand and the fingers flat upon the table, abduct the little finger against resistance.

Flexor digiti minimi [**6**]. With the interphalangeal joints extended, flex the metacarpophalangeal joint of the little finger against resistance.

The interossei. The dorsal interossei spread the fingers apart and the palmar interossei bring them together.

The lumbricals. These can be assessed in conjunction with the interossei by their ability to flex the metacarpophalangeal joints with extension at the interphalangeal joints.

The muscles of the hand receive their segmental nerve supply from T1, the lowest root of the brachial plexus. Severe upward traction on the arm, such as may occur during a forcible breech delivery, may tear or damage this root. The result is then a claw-shaped hand due to the unmodified opposing actions of the long flexors and extensors—Klumpke's paralysis.

Loop

Whorl

Arch

Papillary ridges or fingerprints

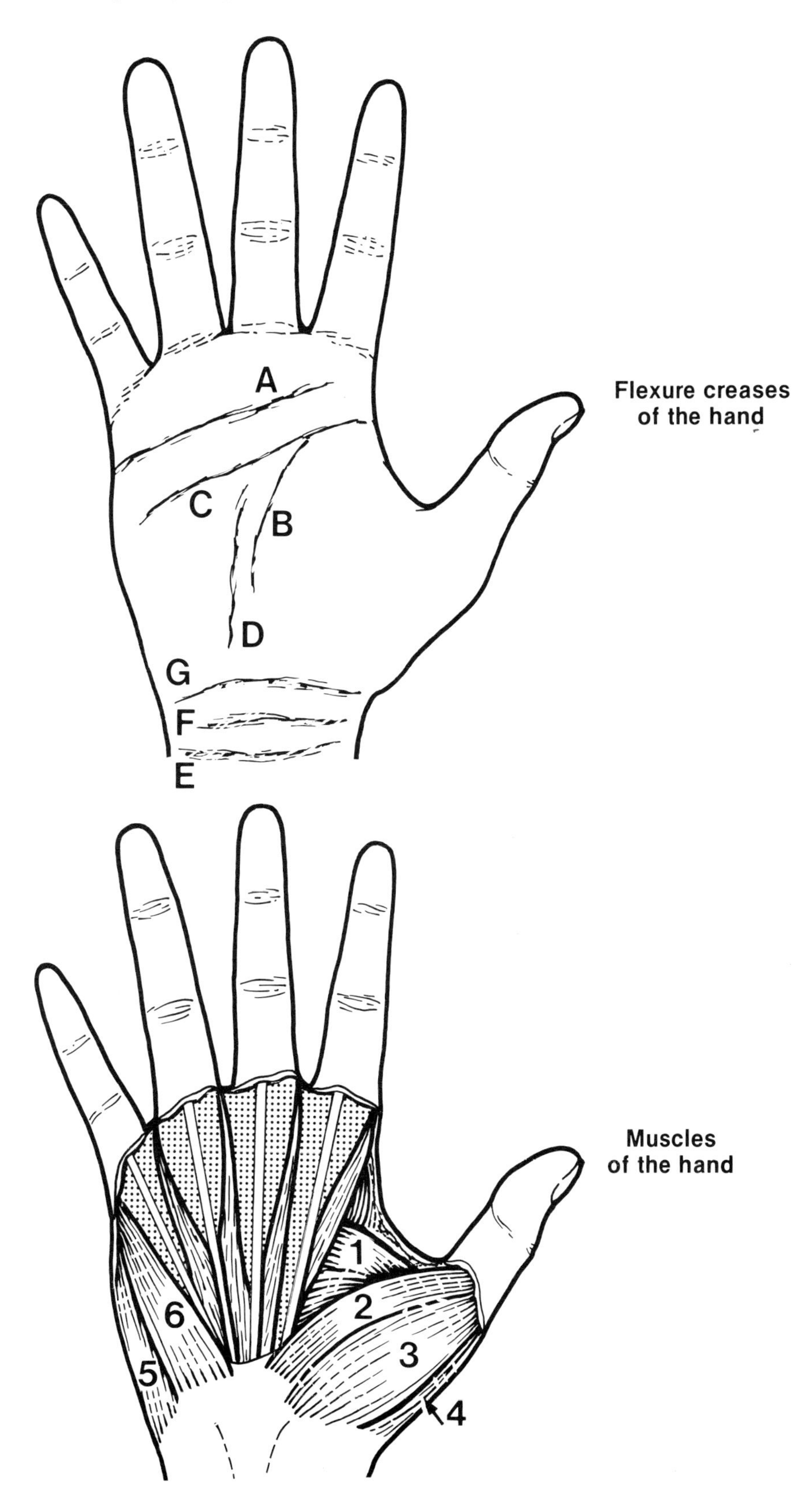

Flexure creases of the hand

Muscles of the hand

RADIOLOGY OF THE WRIST AND HAND

There are 29 bones visible on a radiograph of the wrist and hand. For most purposes only two views are required to evaluate any pathological condition; an anterior view and a lateral view. Occasionally oblique views may also be useful.

Anterior View

This view is taken with the patient resting his hand palm down on the x-ray plate. In the under 20-year-olds, in whom ossification may be incomplete, and in patients with suspected bone disease, pictures of both hands should always be taken for comparison, preferably on the same film.

Key to the Anterior View

Styloid process of the radius extending further than the styloid process of the ulna [**1**]
Styloid process of the ulna [**2**]
Scaphoid [**3**]
Lunate [**4**]
Triquetral [**5**]
Hamate—the hook of the hamate is seen end on and therefore appears as an oval white ring [**6**]
Pisiform [**7**]
Capitate [**8**]
Trapezoid [**9**]
Trapezium [**10**]
First metacarpal [**11**]
Sesamoid bones [**12**]
Proximal phalanx [**13**]
Middle phalanx [**14**]
Terminal phalanx [**15**]
Head of metacarpal [**16**]
Body of metacarpal [**17**]
Base of metacarpal [**18**]

Note that the radiological joint spaces of the radiocarpal and intercarpal joints are well defined, but the distal radioulnar joint space is obscured by overlapping shadows.

Lateral View

This view is particularly useful in patients with dislocations or fractures of the wrist. It is of limited value in evaluating the phalanges and metacarpals.

Key to the Lateral View

Base of first metacarpal [**1**]
Trapezium [**2**]
Trapezoid and hook of hamate superimposed [**3**]
Capitate [**4**]
Pisiform [**5**]
Tubercle of scaphoid [**6**]
Scaphoid [**7**]
Lunate [**8**]
Styloid process of radius [**9**]
Triquetral [**10**]
Hamate [**11**]

Anterior View in Adduction

Note the good view of the scaphoid.

Anterior view

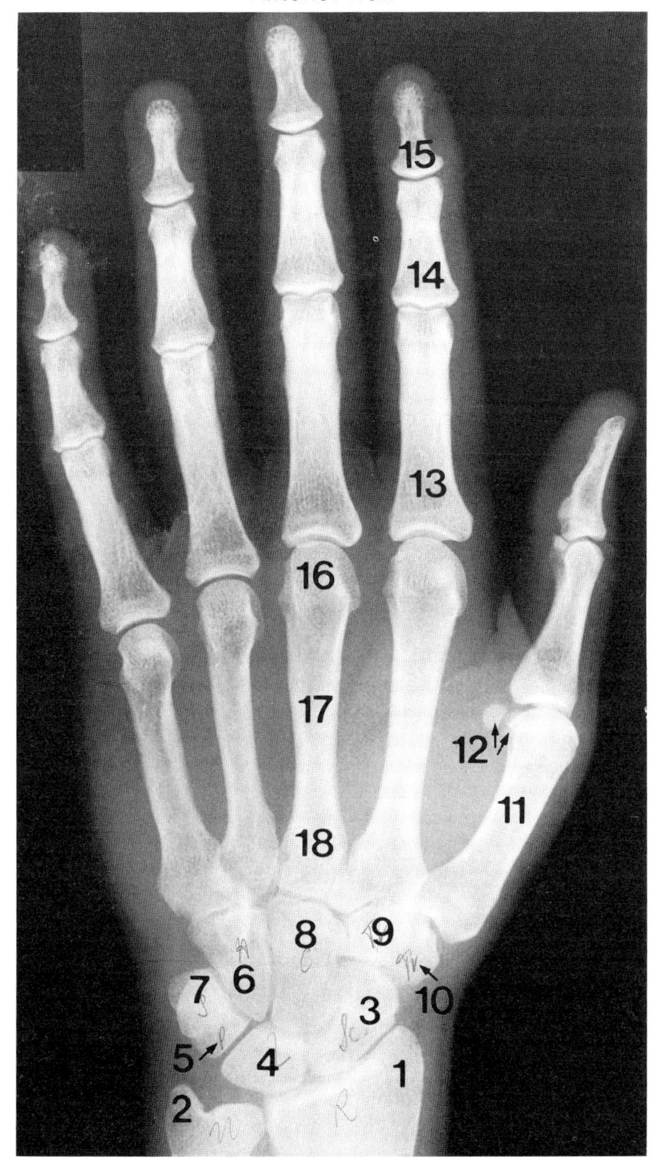

Lateral view

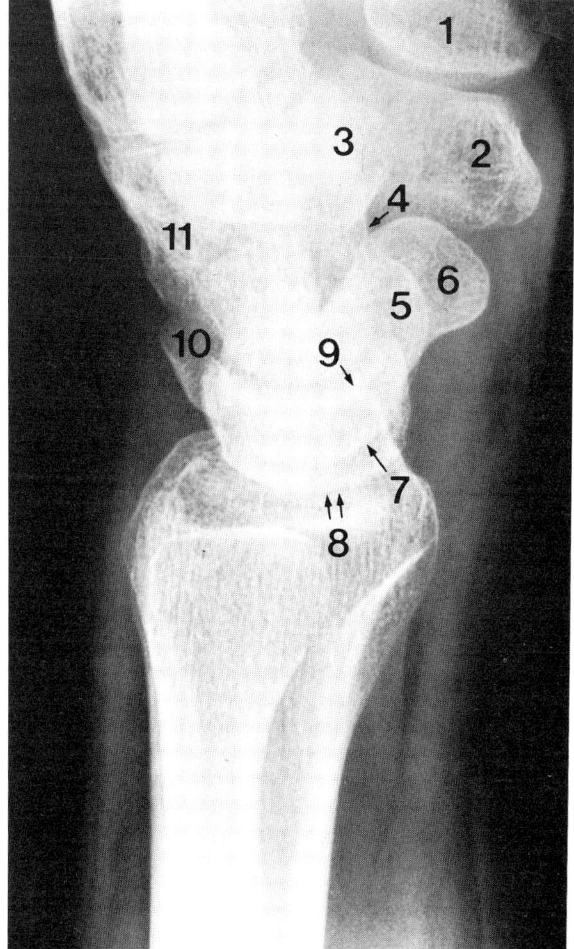

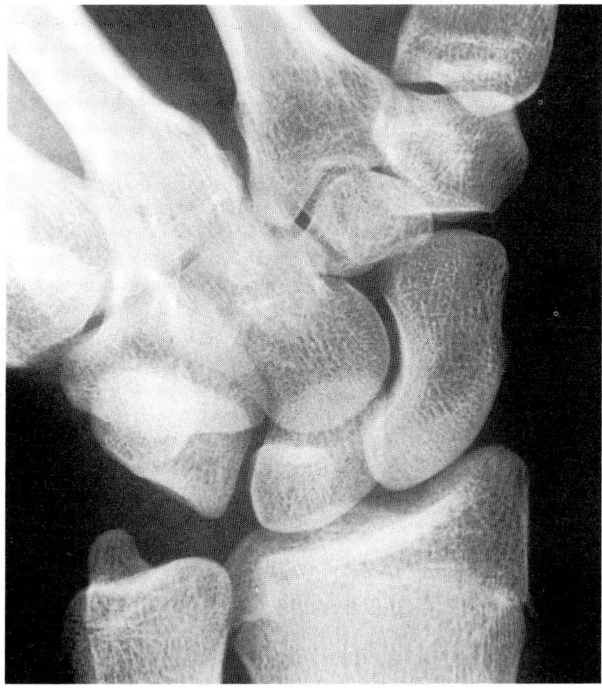

Anterior view in adduction

ARTERIES AND PULSES OF THE UPPER LIMB

The Axillary Artery [1]

The axillary artery begins as the continuation of the brachiocephalic artery at the lateral border of the first rib and ends by becoming the brachial artery at the lateral extremity of the posterior axillary fold which is formed by the teres major. It is divided into three parts by the pectoralis minor, but the length of each part depends on the position of the arm. For example, when the arm hangs naturally by the side, the first part is about 5 cm in length, but when the scapula is raised or rotated, as in abduction of the arm, the coracoid process, which receives the insertion of the pectoralis minor, moves medially and upwards, thus shortening the first part.

The course is represented by a line drawn from the midpoint of the clavicle and passing immediately below the coracoid process to the medial bicipital groove behind the coracobrachialis muscle. When the arm hangs by the side this line is concave downwards, but when the arm is raised it becomes concave upwards.

The pulse of the axillary artery is readily palpated in the lateral wall of the axilla in the groove behind the coracobrachialis muscle. This is a useful pressure point to control distal bleeding, although paraesthesiae may result from the inevitable pressure on the median, ulnar and radial nerves which are in close relation to the artery at this point.

The Brachial Artery [2]

This is the continuation of the axillary artery beyond the teres major. It may be marked by a line from the limit of the axillary artery behind the coracobrachialis muscle, along the medial bicipital groove to the tendon of the biceps muscle in the middle of the cubital fossa where it divides into the radial and ulnar arteries. In the cubital fossa, it lies beneath the bicipital aponeurosis which separates it from the median cubital vein, and here it may be subject to insult from inexpert venepuncture. Here, the median nerve lies about 2 cm medial to the artery.

High division of the brachial artery sometimes occurs and the artery divides into its terminal branches proximal to the cubital fossa. It may divide at any point between the axilla and the cubital fossa, in which case the two arteries descend side by side following the normal course of the brachial artery.

The brachial pulse may be felt along the whole course of the artery by compressing the artery against the humerus, i.e. lateral pressure proximally and dorso-lateral pressure distally. It is best felt just medial to the bicipital aponeurosis at the level of the medial epicondyle, and it is at this point that one listens for the Korotkoff sounds when measuring blood pressure.

The Radial Artery [3]

If the arm is placed in the semi-prone position and the brachioradialis tensed, then the course of the radial artery may be indicated by a slightly convex line beginning at the biceps tendon and running down the ulnar side of the tensed brachioradialis to a point just medial to the radial styloid process on the anterior aspect where the pulse may be felt, and thence obliquely across the anatomical snuff-box to the proximal end of the interval between the metacarpal bones of the thumb and index finger.

The radial pulses may be felt (1) against the distal border of the radius as mentioned above and (2) in the anatomical snuff-box.

The Ulnar Artery [4]

The ulnar artery may be represented by a line convex medially from the tendon of the biceps to the pisiform bone and thence to the hook of the hamate.

The superficial palmar arch [5] is formed mainly from the superficial branch of the ulnar artery and partly from the superficial palmar branch of the radial artery. It lies deep to the fascia. The distal convexity of the arch lies level with the cleft between the outstretched thumb and palm.

The deep palmar arch [6] is formed from the radial artery and the deep branch of the ulnar artery. It lies deep to the ulnar bursa on the bases of the metacarpals and on the interosseous muscles. Its distal convexity is 3 cm distal to the distal crease of the wrist.

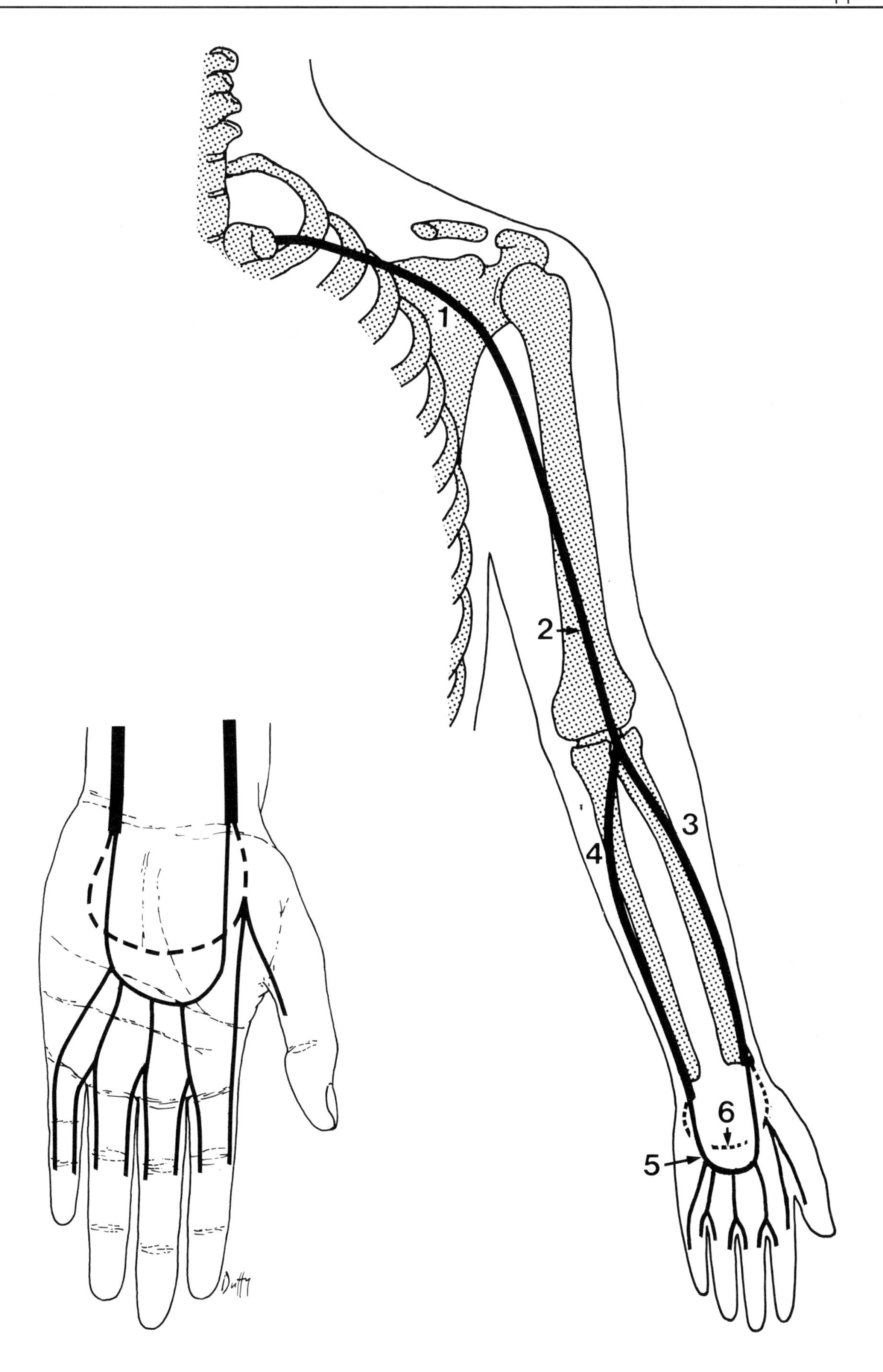
1
2
3
4
5
6
Duffy

VEINS OF THE UPPER LIMB

Apart from the axillary artery which is accompanied by a single vein, all of the arteries, both large and small, are accompanied by two venae comitantes.

The Axillary Vein

The axillary vein is the continuation of the basilic vein at the lower border of the teres major tendon in the posterior wall of the axilla. It ends by becoming the subclavian vein at the lateral border of the first rib. Its course is identical to the axillary artery to which it lies medially, and a little anteriorly in the proximal half, especially when the arm is abducted.

The Superficial Veins

These are arranged in irregular networks in the superficial fascia. They are connected with the deep veins by inconstant perforating veins which pierce the deep fascia. The blood is drained from the superficial system principally by the basilic and cephalic veins.
The main superficial channels are as follows:

[1] *The dorsal venous arch* would be better named the dorsal venous plexus, for its arch-like nature is seldom apparent. It lies on the back of the hand, its position and pattern being highly variable.

[2] *The basilic vein* arises from the ulnar side of the arch and ascends along the ulnar side of the distal half of the forearm before inclining forwards to pass in front of the medial epicondyle to enter the medial bicipital furrow. Opposite the insertion of the coracobrachialis it pierces the deep fascia to ascend along the medial side of the brachial vessels to become the axillary vein at the lower border of the teres major tendon. In the forearm it is usually easily visible, particularly in men.

[3] *The cephalic vein* arises from the radial end of the dorsal venous arch. Inclining forwards, it ascends to the radial side of the biceps tendon. It then inclines laterally and ascends along the lateral prominence of the biceps to reach the groove in front of the shoulder between the deltoid and pectoralis major muscles (the deltopectoral groove), where it gradually pierces the deep fascia. It ascends in the groove to the level of the coracoid process where it turns medially between the pectoralis major and the pectoralis minor, pierces the clavipectoral fascia and ends in the axillary vein at a point just below the middle of the clavicle.

This vein is commonly used for intravenous cannulation at the wrist where it can usually be felt, if not seen, just dorsal to the radial styloid process when the superficial veins are engorged with the aid of a tourniquet. Occasionally, it may also be necessary to cannulate this vein where it lies in the deltopectoral groove.

[4] *The median cubital vein* is a short, wide vein, useful for venepuncture, which runs upwards and medially across the bicipital tendon and aponeurosis, the latter separating it from the brachial artery. It joins the basilic vein just above the medial epicondyle. Its relationship to the brachial artery and median nerve should always be borne in mind during venepuncture.

[5] *The anterior median vein of the forearm* when present, runs up the middle of the front of the forearm and may join the basilic or cephalic vein, or may divide at the cubital fossa into the *median cephalic and median basilic veins* [**6**].

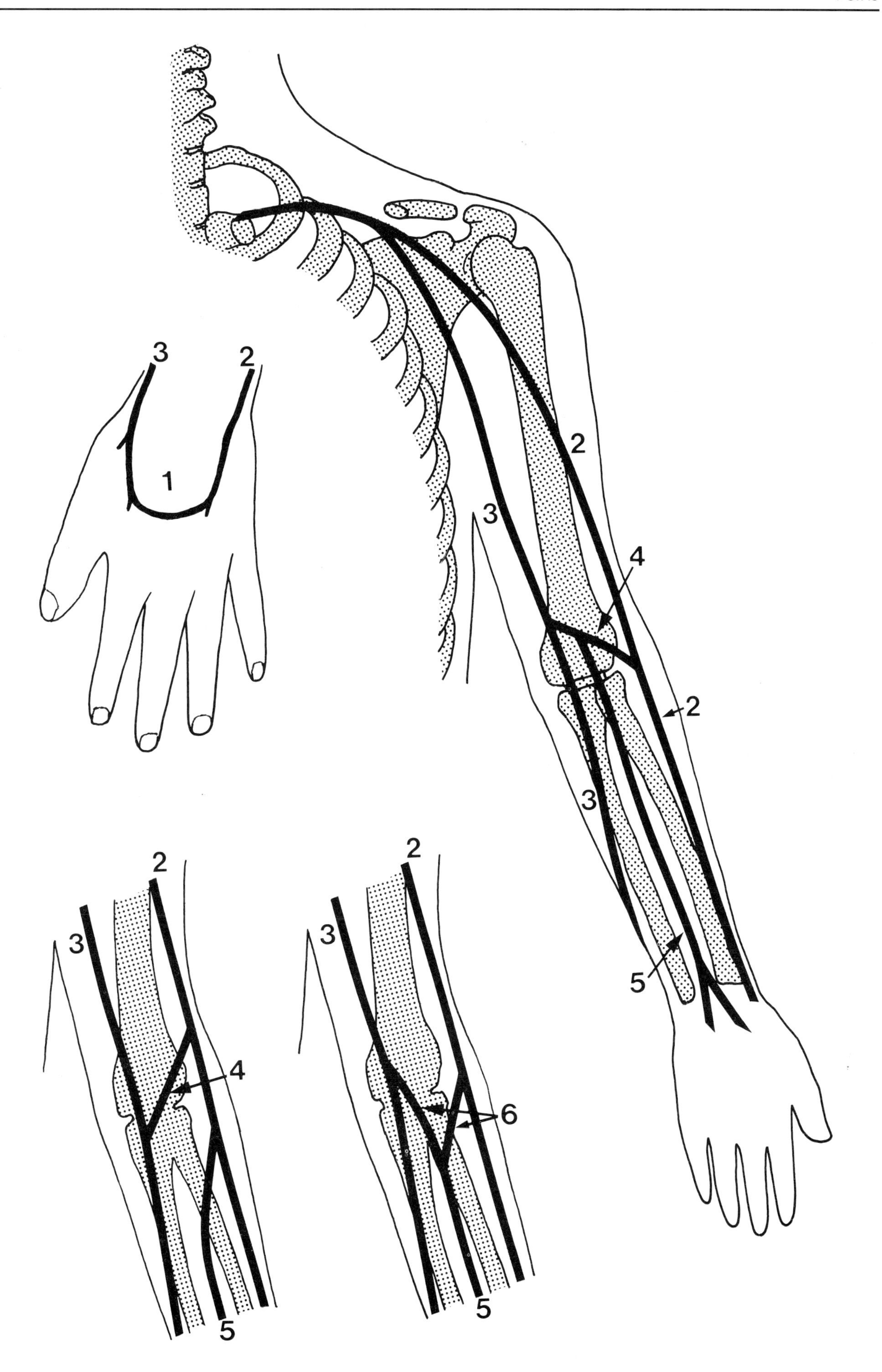
3
2
1
2
3
4
2
3
5
2
3
4
5
2
3
6
5

CUTANEOUS NERVES AND DERMATOMES OF THE UPPER LIMB

Cutaneous Nerves

Acromial branch of supraclavicular nerve from the cervical plexus [**1**]
Upper lateral cutaneous nerve of the arm [**2**]
Lower lateral cutaneous nerve of the arm [**3**]
Lateral cutaneous nerve of the forearm [**4**]
Radial nerve [**5**]
Median nerve [**6**]
Ulnar nerve [**7**]
Medial cutaneous nerve of the forearm [**8**]
Intercostobrachial nerve [**9**]
Third intercostal [**10**]
Posterior cutaneous nerve of the forearm [**11**]
Posterior cutaneous nerve of the arm [**12**]

Dermatomes—Some General Considerations

In the developing embryo, the body wall is supplied segmentally or metamerically by pairs of spinal nerves. Towards the end of the fourth week of intrauterine life the limb buds develop as lateral outgrowths from the body wall. As the limb elongates, the skin that envelopes it is drawn out from the trunk over the developing limb.

Study of the trunk dermatomes near the root of the limb reveals that many of the dermatomes are now missing; they lie peripherally along the limb. The elongation process results, inevitably, in dermatomes from discontinuous spinal levels lying adjacent to one another. The line separating them is known as an *axial line*. Each limb, therefore, has two axial lines—one on the ventral surface, sometimes known as the anterior axial line and one on the dorsal surface, sometimes called the posterior axial line.

Normally there is considerable overlap between sequential dermatomes, and thus, damage to a single sensory nerve root will produce considerably less sensory loss than might be expected. However, due to the discontinuous nature of adjacent dermatomes on either side of the axial lines there is *no* sensory overlap across axial lines which is useful to remember when investigating cutaneous paraesthesia or analgesia of segmental origin. Therefore, the skin should always be tested on either side of the axial lines when sensory changes are suspected.

Pre-Axial and Post-Axial Borders

The pre- and post-axial borders of a limb represent the most cranial and caudal borders, respectively, of the developing limb bud. Hence they represent the lines of junction between the ventral and dorsal surfaces of the growing limb. Thus, they also serve to mark the junction between the flexor and extensor compartments of the limb.

In adult life they are conveniently marked by superficial veins:

Pre-axial border
Cephalic vein in the arm
Long saphenous vein in the leg

Post-axial border
Basilic vein in the arm
Short saphenous vein in the leg

It is interesting to note that the hairs on the limbs diverge to either side of the axial border. This is best noticed when the limbs are wet.

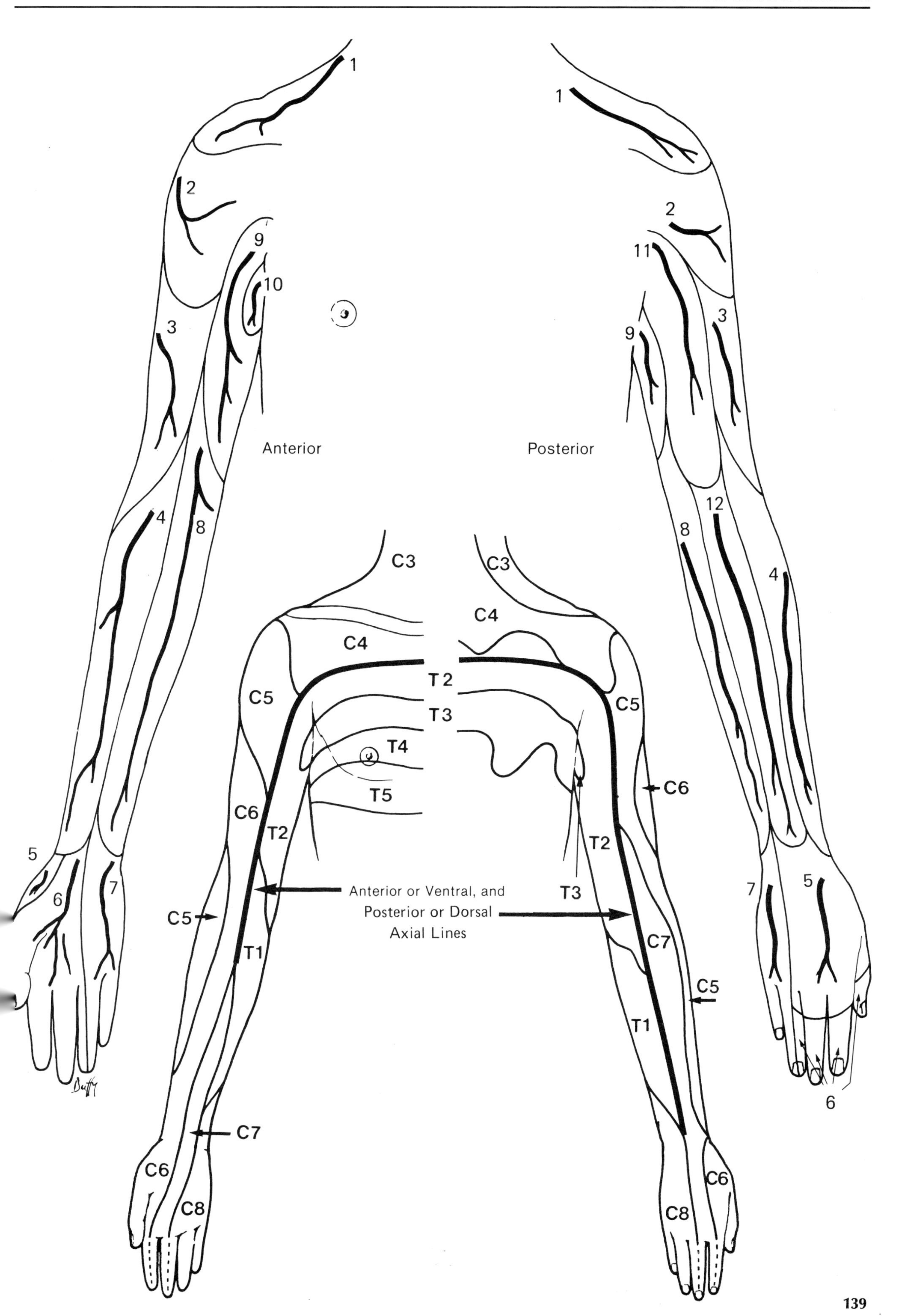

Anterior
Posterior
1
2
3
4
5
6
7
8
9
10
11
12
C3
C4
C5
C6
C7
C8
T1
T2
T3
T4
T5
Anterior or Ventral, and
Posterior or Dorsal
Axial Lines

THE MEDIAN NERVE

Root Value
C(5), 6, 7, 8, T1

Course and Surface Marking

The median nerve is formed from two limbs or heads, one from the medial cord and one from the lateral cord, of the brachial plexus, opposite a point a little below and medial to the coracoid process. The two limbs unite in the lateral wall of the axilla, antero-lateral to the axillary artery which they clasp. From here the nerve is marked by a line running down the medial bicipital furrow to a point in the cubital fossa at the angle between the proximal border of the bicipital aponeurosis and the pronator teres.

The nerve can be palpated in the medial bicipital furrow when it is put on the stretch by abducting the arm. Initially, the nerve lies antero-lateral to the artery but, opposite the insertion of the coracobrachialis, it crosses in front of the artery to lie on its medial side. In the cubital fossa it lies about 0.5–1 cm medial to the brachial artery.

From this point, it passes deep to the bicipital aponeurosis and is marked by a line to a point at the wrist immediately to the ulnar side of the tendon of the flexor carpi radialis.

Note that, although the nerve is superficial in the axilla and arm, it lies deep in the forearm passing in turn deep to the flexor carpi radialis, the palmaris longus and the radial head of the flexor digitorum superficialis. At the wrist it lies deep to the palmaris longus between the flexor digitorum superficialis and the ulnar bursa medially, and the flexor carpi radialis laterally. It enters the palm deep to the flexor retinaculum.

In the hand, the digital branches lie nearer the palmar surface than the corresponding digital arteries.

Muscles Supplied

All the flexor muscles of the forearm except the flexor carpi ulnaris and the ulnar side of the flexor digitorum profundus.

The thenar muscles and the two radial lumbricals.

Area of Cutaneous Supply

The palmar cutaneous branch commences in the forearm a short distance above the flexor retinaculum. It passes over the retinaculum and supplies the skin over the thenar eminence.

The palmar digital branches arise after the nerve has passed beneath the flexor retinaculum, and these supply the palmar aspect of the $3\frac{1}{2}$ radial digits, and as far back as the terminal interphalangeal joints on the dorsal surfaces of the same digits.

Effects of Section

The nerve is commonly damaged at the wrist or elbow.

Division at the wrist results in loss of function of the thenar muscles (except the adductor pollicis) and the two radial lumbricals. But, possibly more importantly, it results in the loss of sensation over the thumb and adjacent $2\frac{1}{2}$ fingers, in addition to variable loss over the radial side of the palm.

If the nerve is divided at the elbow, as it may be in a supracondylar fracture of the humerus, then there is additional serious muscle impairment. Pronation is lost and flexion is considerably weakened and accompanied by ulnar deviation because flexion is now dependent solely on the flexor carpi ulnaris and the ulnar half of the flexor digitorum profundus.

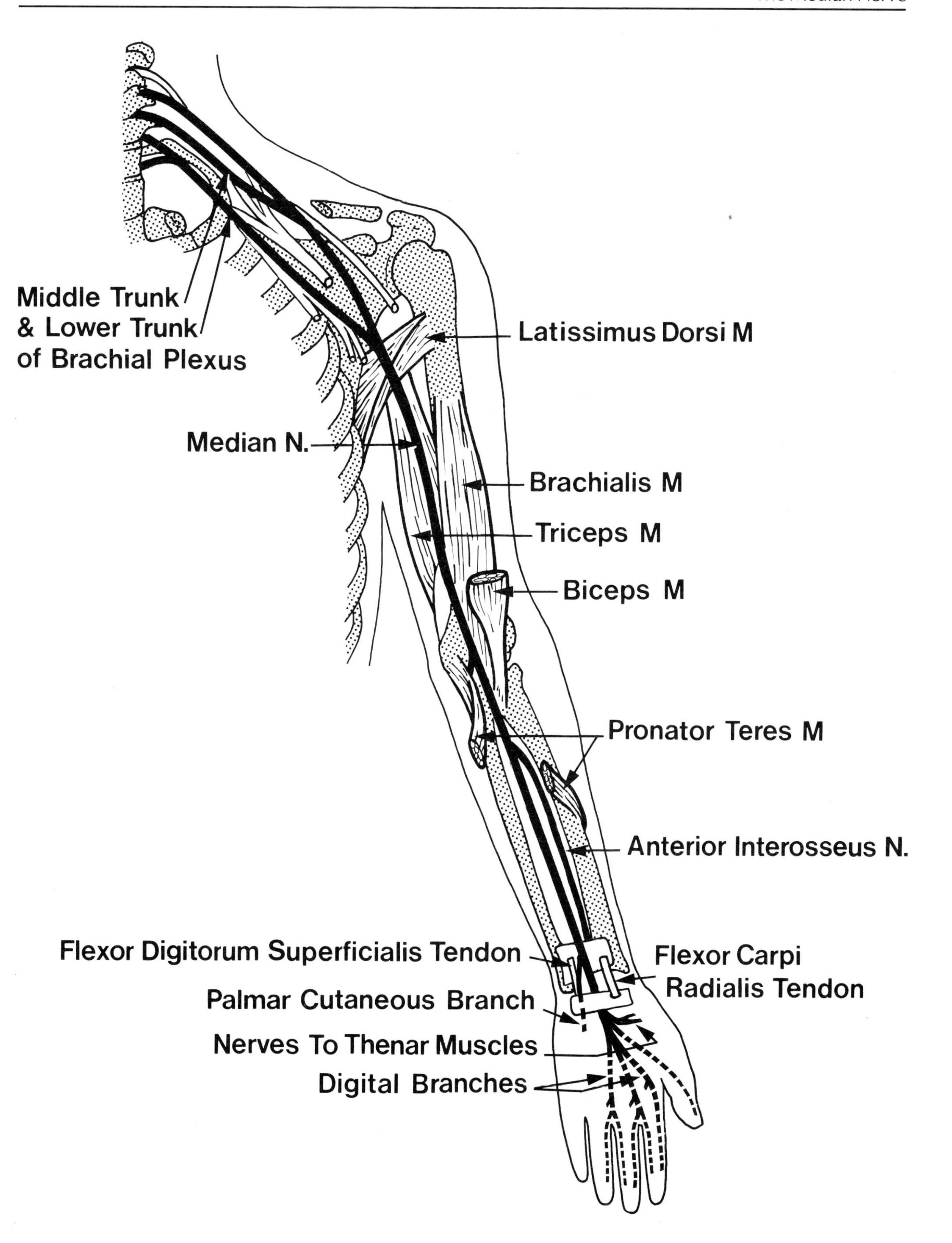
Middle Trunk
& Lower Trunk
of Brachial Plexus
Latissimus Dorsi M
Median N.
Brachialis M
Triceps M
Biceps M
Pronator Teres M
Anterior Interosseus N.
Flexor Digitorum Superficialis Tendon
Flexor Carpi
Radialis Tendon
Palmar Cutaneous Branch
Nerves To Thenar Muscles
Digital Branches

THE ULNAR NERVE

Root Value
C8, T1

Course and Surface Marking

With the arm in the anatomical position, the course of the ulnar nerve may be represented by a line drawn from a point on the lateral wall of the axilla, behind the swelling for the coracobrachialis muscle, down to a point immediately behind the medial epicondyle.

In the forearm, its further course may be represented by a line drawn from the point behind the medial epicondyle to the pisiform bone.

Muscles Supplied

Flexor carpi ulnaris (C7 C8 T1).
The ulnar side of flexor digitorum profundus (C7 C8).
All the muscles of the hand, except those of the thenar eminence and the first two lumbricals (T1).

Area of Cutaneous Supply

There are three cutaneous branches:

The dorsal branch arises about 5 cm above the wrist; it passes distally and backwards deep to the flexor carpi ulnaris to supply cutaneous sensation to the medial side of the back of the wrist and hand as far as the distal interphalangeal joints of the little finger and the ulnar side of the ring finger.

The palmar cutaneous branch arises around the middle of the forearm and runs down to supply the skin over the medial side of the palm.

The superficial terminal branch supplies the skin of the medial side of the hand. It divides into digital branches which can be compressed against the hook of the hamate bone, and which supply the palmar surfaces of the little finger and the ulnar half of the ring finger, extending on to the dorsal surfaces to include the nail bed and skin as far back as the distal interphalangeal joint.

Effects of section

The ulnar nerve is most commonly damaged at the elbow or wrist. If damaged at the wrist all the small muscles of the hand, except the muscles of the thenar eminence and the first two lumbricals, are paralysed. This results in the unmodified opposing actions of the long flexors and extensors. Normally, the interossei and lumbricals are responsible for flexion at the metacarpophalangeal joints and they assist the long extensor at the interphalangeal joints. When the ulnar nerve is damaged all the interossei are paralysed resulting in extension of the metacarpophalangeal joints and some flexion at the interphalangeal joints. There is also loss of abduction and adduction of the fingers. Paralysis of the two ulnar lumbricals to the ring and little fingers, and the loss of their extensor action at the interphalangeal joints lead to a more noticeable flexion deformity of these fingers. The flexion deformity is less in the index and middle fingers because of their intact lumbricals. The sensory loss does *not* include the area of supply of the dorsal cutaneous branch.

If, however, the nerve is damaged at the elbow, the flexion of the ring and little fingers at the interphalangeal joints is not as bad as when the injury is at the wrist because the ulnar part of the flexor digitorum profundus is also paralysed. In addition, the paralysis of the flexor carpi ulnaris results in radial deviation on flexion of the wrist.

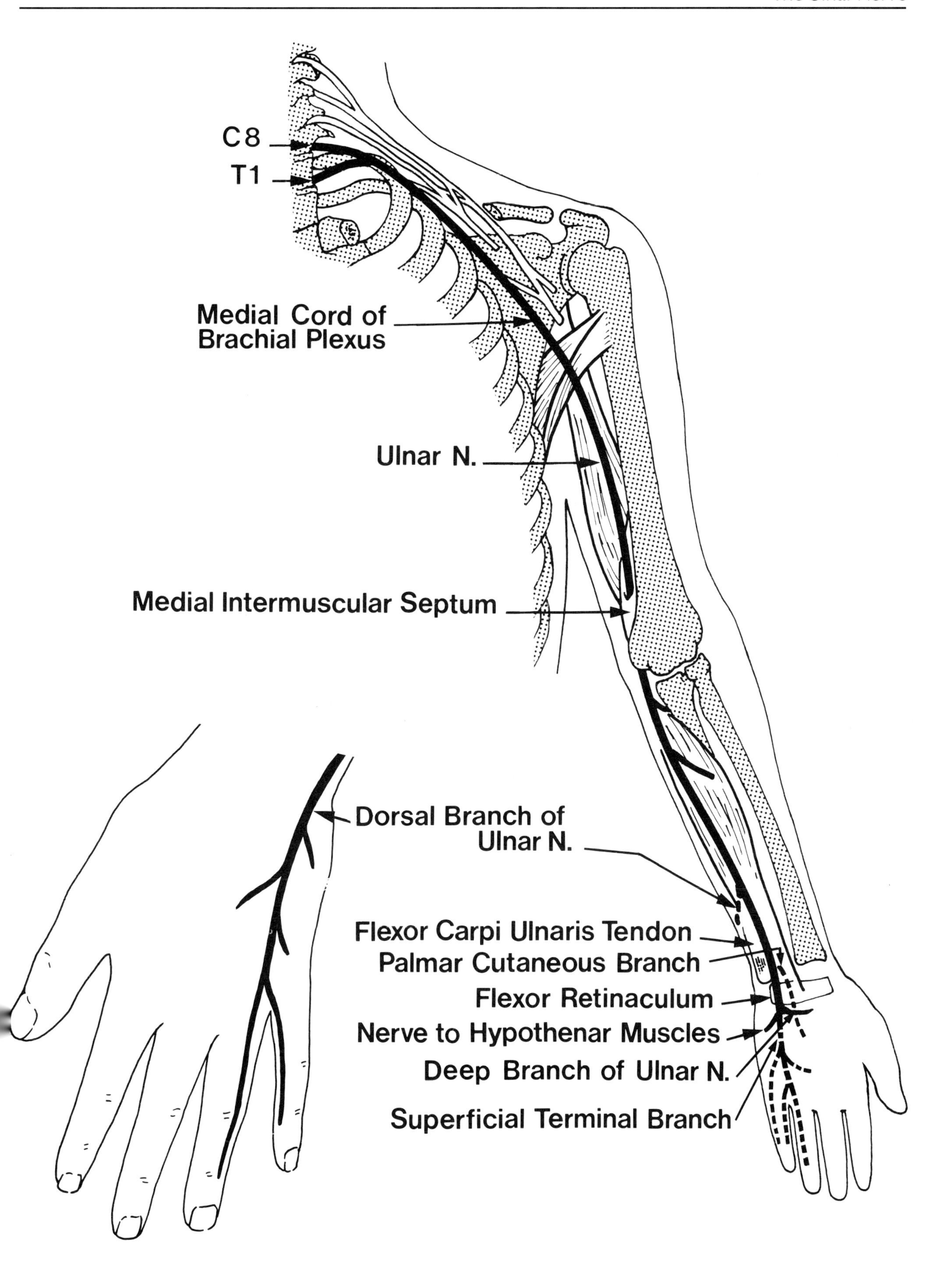
C8
T1
Medial Cord of Brachial Plexus
Ulnar N.
Medial Intermuscular Septum
Dorsal Branch of Ulnar N.
Flexor Carpi Ulnaris Tendon
Palmar Cutaneous Branch
Flexor Retinaculum
Nerve to Hypothenar Muscles
Deep Branch of Ulnar N.
Superficial Terminal Branch

THE RADIAL NERVE

Root Value
C(5), 6, 7, 8

Course and Surface Marking

The course of the radial nerve may be mapped on the arm by a line drawn across the posterior aspect of the arm, from the junction of the posterior wall of the axilla with the arm, to a point at the junction of the middle and distal thirds of a line between the acromion and the lateral epicondyle. From there the course is indicated by a line drawn obliquely downwards and forwards to the front of the lateral epicondyle where the nerve divides into its two terminal branches, about 1 cm lateral to the biceps tendon.

The superficial terminal branch is indicated by a line running from the front of the lateral epicondyle to a point on the brachioradialis tendon, 7 cm above the wrist.

The deep terminal branch is indicated by a line from the same point in front of the lateral epicondyle to a point on the dorsum of the wrist, 1 cm distal to the midpoint of a line between the head of the ulna and the dorsal tubercle of Lister on the radius, where the nerve ends as a pseudoganglion from which filaments are distributed to the ligaments and articulations of the carpus.

Muscles Supplied

Radial nerve
Three heads of the triceps C6 C7 C8
Brachioradialis C5 C6
Extensor carpi radialis longus C5 C6

Deep terminal branch
Extensor carpi radialis brevis C6 C7
Supinator C6 C7
Extensor carpi ulnaris C7 C8
Extensor digitorum C7 C8
Abductor pollicis longus C7 C8
Extensor pollicis longus C7 C8
Extensor pollicis brevis C7 C8
Extensor indicis C7 C8

Area of cutaneous supply

The posterior cutaneous nerve of the arm arises in the axilla and supplies the skin on the back of the arm as far down as the olecranon.

The lower lateral cutaneous nerve of the arm supplies the skin on the side of the arm between the elbow and the insertion of the deltoid.

The posterior nerve of the forearm arises in the arm at about the level of the insertion of the deltoid and supplies the skin over the posterior aspect of the forearm between the elbow and the wrist.

The superficial terminal branch divides into digital branches which supply the dorsal surfaces of the $3\frac{1}{2}$ radial digits as far as the distal interphalangeal joints.

Effects of section

Surprisingly, because of nerve overlap, division of the radial nerve produces only a very small area of anaesthesia over the dorsum of the hand between the first and second metacarpals.

The motor effects of section depend on the level. If the section is high, i.e. in the axilla, then the triceps will be paralysed. If the insult is more distal after the branches to the triceps have emerged, e.g. the spiral groove, then only wrist drop will ensue due to the loss of the wrist extensors.

Damage to the interosseous nerve (deep terminal branch of the radial nerve) leaves the extensor carpi radialis longus intact, as it is supplied from above the terminal division of the radial nerve, and this muscle alone is powerful enough to maintain extension at the wrist.

THE AXILLARY NERVE

This nerve supplies the deltoid and a small area of skin over its insertion. The nerve may be damaged in, either dislocation of the shoulder, or in reduction of a shoulder dislocation and it therefore carries medico-legal significance. Before reducing a dislocated shoulder check the area of sensation to establish the integrity of the nerve.

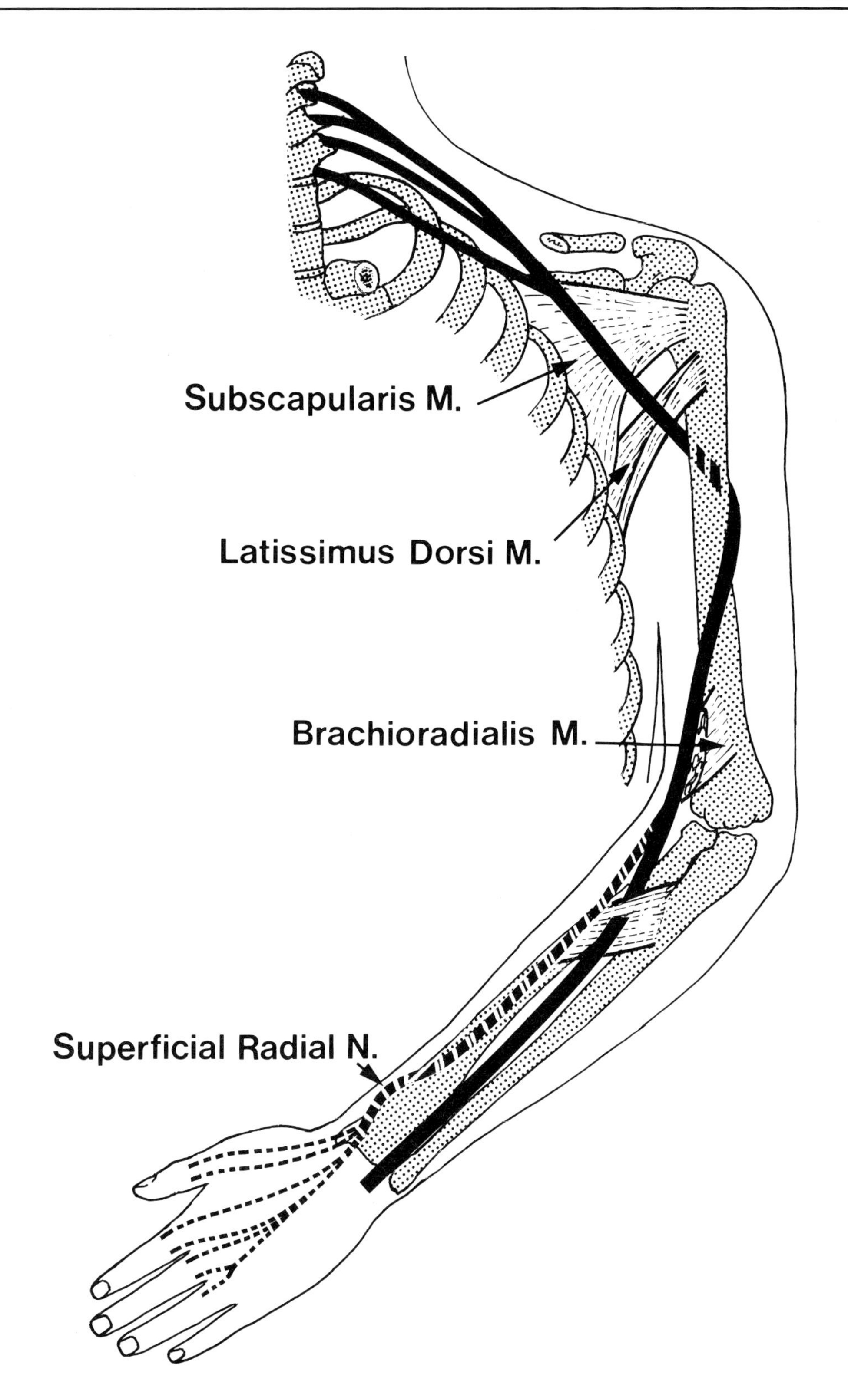
Subscapularis M.
Latissimus Dorsi M.
Brachioradialis M.
Superficial Radial N.

9. The Lower Limb

BONY LANDMARKS

The Hip Bone

The hip bone consists of the ilium, the ischium and pubis, which, together with the sacrum, form the pelvic girdle.

The ilium is largely hidden under the cover of the gluteal muscles and the tensor fasciae latae. However, the iliac crest is easily palpable along its entire length. Each crest ends in front at the anterior superior iliac spine, and behind at the posterior superior iliac spine. The highest part of the iliac crest is felt towards the back and is level with the gap between the spines of L3 and L4.

The posterior superior iliac spine is the bone which can be felt in the dimple just above the buttock, and it lies at the level of the second sacral vertebra and the middle of the sacroiliac joint. A line drawn from this point to the tip of the coccyx represents the lateral margin of the lower part of the sacrum and the coccyx.

The posterior inferior iliac spine lies about 2.5 cm below the dimple from the superior spine. It lies at the posterior or inferior end of the sacroiliac joint and is therefore the point where that joint is most superficial.

The anterior superior iliac spine is easily felt, especially in the sitting position when the sartorius muscle is relaxed.

The iliac tubercle is a small prominence which may be felt on the outer surface of the iliac crest about 5 cm behind the anterior superior iliac spine. It lies at the level of the spine of the fifth lumbar vertebra.

The ischium lies deeply in the buttock under the cover of the gluteus maximus and the origin of the hamstring muscles.

The ischial tuberosity may be palpated in the lower part of the buttock. This is most easily done by pushing the fingers upwards to the buttock along a line between the medial and posterior aspects of the thigh. When the hip is flexed, as in the sitting position, the ischial tuberosity emerges from under the cover of the lower border of gluteus maximus, and is then only separated from the skin by a bursa and a pad of fat. The ischial tuberosities take the weight of the body in the sitting position.

The pubis—the body of the pubis may be felt alongside the pubic symphysis. The pubic crest is the upper border of the body and may be palpated through the rectus abdominus muscle and sheath.

The pubic tubercle lies at the lateral end of the pubic crest about 2.5 cm from the upper border of the symphysis. It is not always prominent and may sometimes only be recognised as the point where the bony resistance of the pubic crest changes to the yielding resistance of the inguinal ligament. It may be fairly easily palpated in the male by invaginating the scrotum with the examining finger, and feeling the tubercle of the upper border on the body of the pubis.

The conjoined ramus of the pubis and the ischium may be felt on either side between the medial side of the thigh and the perineum. It extends from the pubic region to the lower end of the ischium and is covered only with skin and fascia.

The Femur

The head of the femur lies within the acetabulum of the hip (innominate) bone. The centre of the head is deep to a point one thumb's breadth below the midpoint of the inguinal ligament. Although it lies deeply, its movements may be felt on deep palpation below the middle of the inguinal ligament when the thigh is rotated about its axis in the extended position.

The greater trochanter is the prominence felt and seen in front of the hollow on the side of the hip. It is the prominence of the greater trochanter which is responsible for the shapely appearance of the mature female pelvis.

Because the hip joint is surrounded by bulky muscles, fractures of the femur may be difficult to diagnose. The greater trochanter is the only part of the proximal end of the femur which can be felt; therefore, its relationship to bony points on the hip bone is of great importance.

Nelaton's line in a normal subject, with the lower limbs extended and the feet together, is a line drawn around the side of the buttocks between the anterior superior iliac spine and the ischial tuberosity which passes across the upper border of the greater trochanter.

Bryant's triangle is often used to compare the level of the greater trochanter on one side with that of the greater trochanter on the other. It is constructed by drawing three lines on the recumbent subject:

From the anterior superior iliac spine to the greater trochanter

A vertical line downwards from the anterior superior iliac spine

A horizontal line from the greater trochanter to meet the vertical line. The length of this horizontal line should be the same on both sides.

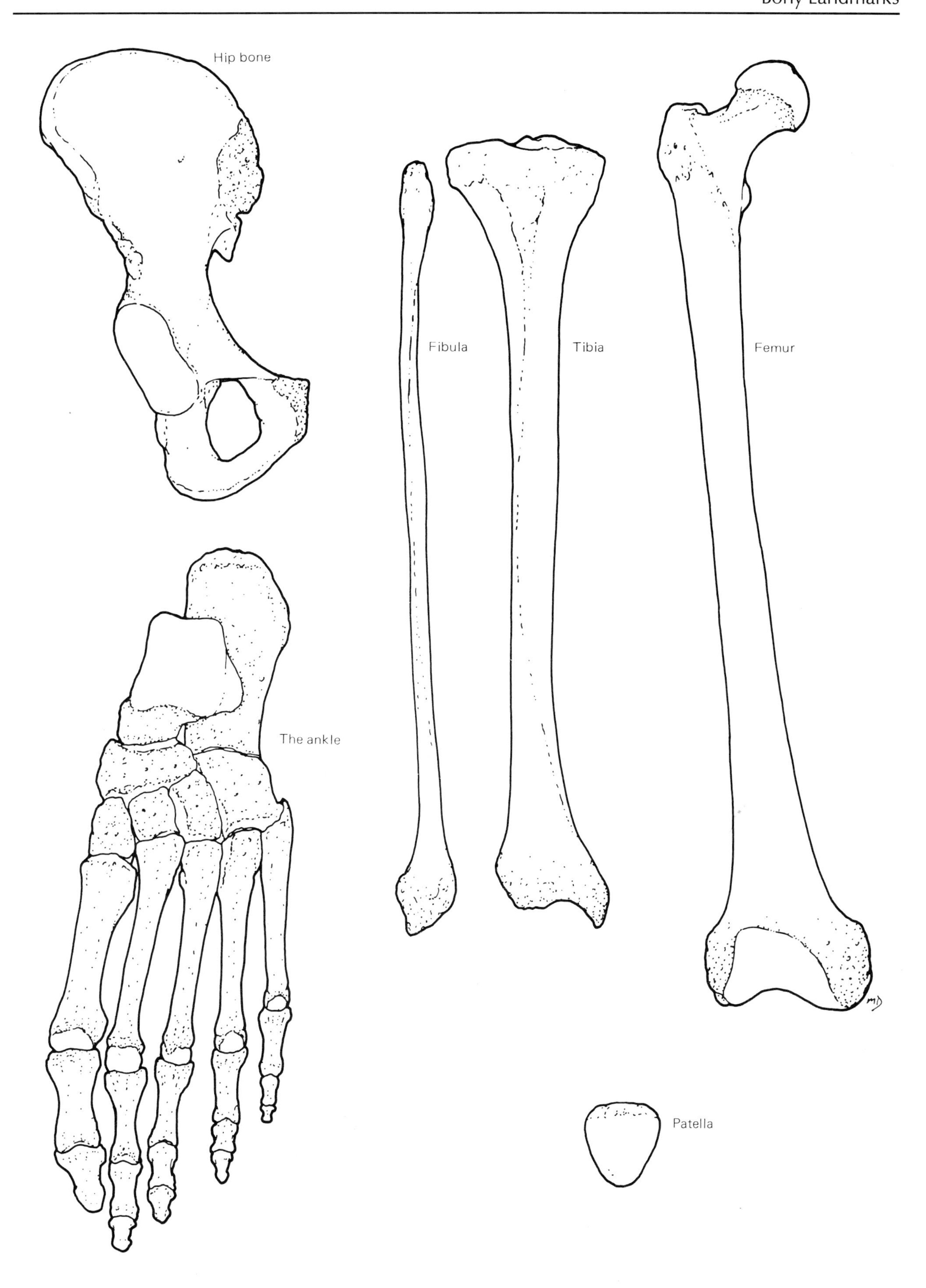
Hip bone
Fibula
Tibia
Femur
The ankle
Patella

Note that, in the normal subject the centre of the hip joint, the upper margin of the symphysis pubis, and the tip of the coccyx all lie in approximately the same plane in the erect posture.

The lesser trochanter of the femur can sometimes be felt by deep palpation just above the medial part of the gluteal fold when the thigh is extended and rotated medially.

The lateral condyle of the femur is superficial. It may be felt on the lateral aspect of the knee through the iliotibial tract and the aponeurotic expansion from the vastus lateralis muscle. Immediately posterior to the iliotibial tract is the fibular collateral ligament. Behind the fibular collateral ligament the biceps may be seen to descend across the most posterior part of the lateral condyle. When the knee is flexed the biceps slips backwards off the condyle.

The medial condyle of the femur can be felt superficially through the aponeurotic expansion from the vastus medialis. The sartorius muscle can be felt as a soft fleshy band crossing the posterior part of the side of the condyle. Just behind the sartorius is the tendon of gracilis, the saphenous nerve and accompanying vessels between the two. As the knee is flexed, the sartorius and the gracilis slip backwards off the condyle.

The adductor tubercle may be felt by pushing the thumb downwards along the medial side of the thigh until it is caught on the tubercle, the upper most part of the medial condyle. When the knee is bent at right angles, the tendon of adductor magnus, which lies in the groove between the sartorius and the vastus medialis, may be traced to the adductor tubercle. This tubercle marks the position of the epiphyseal line on the medial side. When the knee is flexed the anterior margins of the condyles may be felt as prominent ridges at the sides of the patella. The patella surface of the femur may, at the same time, be felt above the patella through the quadriceps tendon. When the knee is extended the latter surface is concealed by the patella.

The Patella

The patella is the large sesamoid bone lying in the quadriceps tendon in the anterior wall of the knee joint replacing part of the capsule. When the knee is flexed the position of the patella is fixed but, when the knee is relaxed in extension, the patella may be moved from side to side and slightly upwards and downwards. The distal end of the patella is one finger's breadth above the plane of the knee joint.

The plane of the knee joint lies between the femoral and tibial condyles. It is almost horizontal and is easily felt as a groove on either side of the knee beneath the femoral condyles. This groove also marks the position of the menisci.

The Tibia

The anterior border and medial surface of the tibia may be felt from end to end. The anterior margin, or border of the shin, is sharp and easily felt. Proximally, it commences at the tuberosity of the tibia and distally it inclines in a medial direction to become continuous with the anterior margin of the medial malleolus.

The tuberosity of the tibia lies about 5 cm below the patella in the flexed knee. It consists of a smooth proximal part and a rough distal part. The ligamentum patellae can be felt stretching between the distal part of the patella and the smooth proximal part of the tuberosity. When kneeling, the weight of the body rests on the rough part of the tuberosity, the distal part of the patella and the intervening ligamentum patellae.

Condyles of the tibia can be felt anteriorly on either side of the ligamentum patellae, especially when the joint is flexed. They are easily traced laterally but, posteriorly, they are hidden by muscles.

The medial malleolus is subcutaneous. The tip of the medial malleolus lies about 1.25 cm proximal to the level of the tip of the lateral malleolus.

The Fibula

The head of the fibula is felt just below the posterior part of the side of the lateral epicondyle of the tibia. It lies at the same level as the tibial tuberosity. The common peroneal nerve may be rolled between the finger and the back of the fibular head, although the nerve is separated from the bone by the proximal end of the soleus muscle.

The neck of the fibula lies immediately below the lateral side of the head. It may be palpated through the proximal part of the peroneus longus. The common peroneal nerve may be rolled against the lateral aspect of the neck of the fibula by the examining finger. The nerve divides into its terminal divisions between the bone and the peroneus longus muscle and it is especially subject to injury in fractures of the neck of the fibula.

The body of the fibula comes to lie subcutaneously in its distal quarter, where its triangular subcutaneous surface is felt above the lateral malleolus between the peroneus tertius muscle in front, and the peroneus brevis muscle behind. The lateral malleolus is easily palpable and lies at a lower level than the medial malleolus.

The Ankle Joint

This is formed between the talus, the distal end of the tibia and the two malleoli. The cavity of the joint may be traced as follows:

Anteriorly it lies opposite the angle between the front of the leg and the dorsum of the foot about 1.25 cm above the level of the distal end of the medial malleolus.

Posteriorly the cavity is at a slightly lower level than in front.

On either side the cavity reaches down to the lower part of each malleolus.

The Calcaneus

The calcaneus is the posterior pillar of the arch of the foot and forms the prominence of the heel. To the middle of its posterior surface, the tendocalcaneus or the Achilles tendon is inserted.

The lateral and medial tubercles may be felt on either side of the heel as blunt protuberances, on deep palpation, about 2.5 cm in front of the posterior surface of the heel and about 2.5 cm deep to the level of the sole of the foot.

The peroneal tubercle may be felt on the subcutaneous lateral surface of the calcaneus between the tendons of the peroneus longus and peroneus brevis, about one finger's breadth below the lateral malleolus.

The sustentaculum tali may be felt only as an indistinct horizontal bony ridge about 2.5 cm vertically beneath the medial malleolus.

The Talus

This bone is, to a large extent, hidden between the two malleoli. In the normal anatomical position, the medial side of the head of the talus may be palpated in the interval between the medial malleolus and the tubercle of the navicular bone. If the foot is now plantar-flexed and inverted, the interval between the medial malleolus and the tuberosity of the navicular bone is diminished, and the medial side of the head of the talus recedes from the surface. At the same time, the upper lateral part of the head can be felt as a marked prominence about 2.5 cm in front of the lateral malleolus; the articular surface of the body of the talus comes forward beneath the tibia and, in many cases, the anterolateral part of this surface can be felt as a sharp prominence between the prominence for the head of the talus and the lateral malleolus.

The Navicular Bone

The tuberosity of the navicular bone is probably the best landmark on the medial side of the foot. It may be felt 3.75 cm below and in front of the medial malleolus, about midway between the heel and the ball of the foot. If shoes are worn it is about level with the lip of the shoe.

The surface markings of the bone itself correspond to a strip, about the breadth of one finger, running laterally from the tuberosity to an imaginary line connecting the second interdigital cleft with the midpoint of the ankle joint.

The Medial Cuneiform

This bone provides the bony resistance felt on the medial side of the foot just in front of the tuberosity of the navicular bone.

The Ball of the Foot

This is formed by the head of the first metatarsal and its two associated sesamoid bones, which lie one on either side of the head. The common tendon is formed by the split insertion of the flexor hallucis brevis.

The base of the fifth metatarsal is the prominence which may be felt on the lateral border of the foot, about midway between the posterior surface of the heel and the tip of the little toe.

The Transverse Tarsal Joint

This is a collective term for the *calcaneocuboid and talonavicular joints.* Despite the fact that they are two separate joints they lie very nearly in the same transverse plane, and are often spoken of as one joint. Its transverse plane may be represented by a line drawn across the foot from a point just behind the tuberosity of the navicular bone to a point one finger's breadth behind the tuberosity of the fifth metatarsal.

The Tarsometatarsal Joints

These lie along an oblique line drawn from the tuberosity of the fifth metatarsal to a point 3.75 cm in front of the tuberosity of the navicular bone (or midway between the ball of the foot and the medial malleolus).

THE BUTTOCK, POSTERIOR AND LATERAL THIGH

Bony Landmarks

Iliac crest [**A**]
Greater trochanter [**B**]
Ischial tuberosity [**C**]
Patella [**D**]
Head of fibula [**E**]

Soft Tissues

The buttocks are separated from each other by the gluteal cleft [**F**] which extends as high as the fourth or third piece of the sacrum. Here the cleft forms the inferior angle of an inverted flattened triangle, between the two posterior inferior iliac spines, which overlies the upper part of the body of the sacrum.

Inferiorly, the buttock is limited by the gluteal fold [**G**] which runs horizontally *across* the inferior border of the gluteus maximus. Thus, the fold is *not* formed by the inferior border of the gluteus maximus.

Gluteus maximus [**1**] overlaps the ischial tuberosity with its lower margin in the standing position but, on sitting, the muscle slides up behind the tuberosity leaving it free to bear the body's weight.

Gluteus medius and *Gluteus minimus* [**2**] together constitute the major hip abductors and may be demonstrated by asking the subject to stand on one limb. The ipsilateral gluteus medius and minimus should contract and tilt the whole pelvis in an attempt to stabilise the centre of gravity. This tilting is most easily seen by watching the contralateral gluteal fold which should rise.

In some conditions, such as paralysis of the hip abductors, dislocation of the hip, and fracture of the neck of the femur, abduction of the hip is compromised, thus, when the patient is asked to stand on the affected limb, the normal tilting of the pelvis does not occur. This is known as Trendelenberg's sign.

The hamstrings are formed by:
Biceps femoris [**3**]
Semimembranosus [**4**]
Semitendinosus [**5**] lying on the semimembranosus to form the medial wall of the popliteal fossa.

The semitendinosus and the biceps form the bulk of the back of the thigh, but they separate above the knee to form the upper borders of the popliteal fossa.

The semimembranosus may be felt beneath the semitendinosus tendon in the supero-medial wall of the popliteal fossa. When sitting with the knee flexed, it may be felt as the bulky muscle mass between the fingers in the popliteal fossa with the thumb on the medial aspect of the lower thigh.

To test the hamstrings ask the subject to lie face downwards and flex the knee against resistance.

Tensor fasciae latae [**6**] forms the rounded elevation below and lateral to the anterior superior iliac spine. It may be represented by a band three fingers wide extending downwards and backwards from the anterior superior iliac spine to a level 5 cm below the greater trochanter where it becomes continuous with the iliotibial tract.

The iliotibial tract [**7**] runs from the tubercle of the iliac crest to the lateral condyle of the tibia. When the knee is flexed, its posterior border may be seen in the distal part of the thigh as the anterior boundary of a longitudinal groove which separates it from the biceps.

Vastus lateralis [**8**]

Rectus femoris [**9**]

Lateral head of gastrocnemius [**10**]

The popliteal fossa. Notice how, in the living, the popliteal fossa appears almost trapezoid or triangular because the strong popliteal fascia binds the two heads of gastrocnemius together almost obliterating the lower angle.

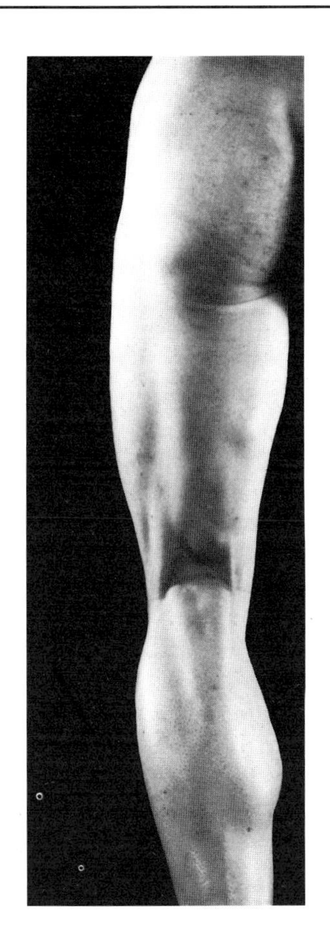

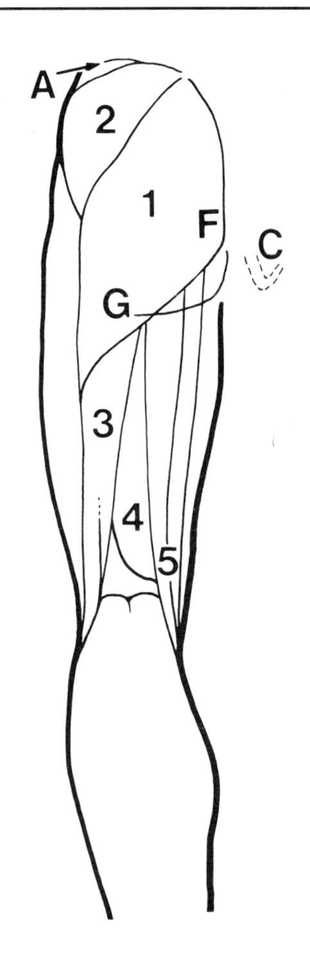

A
2
1
F
C
G
3
4
5

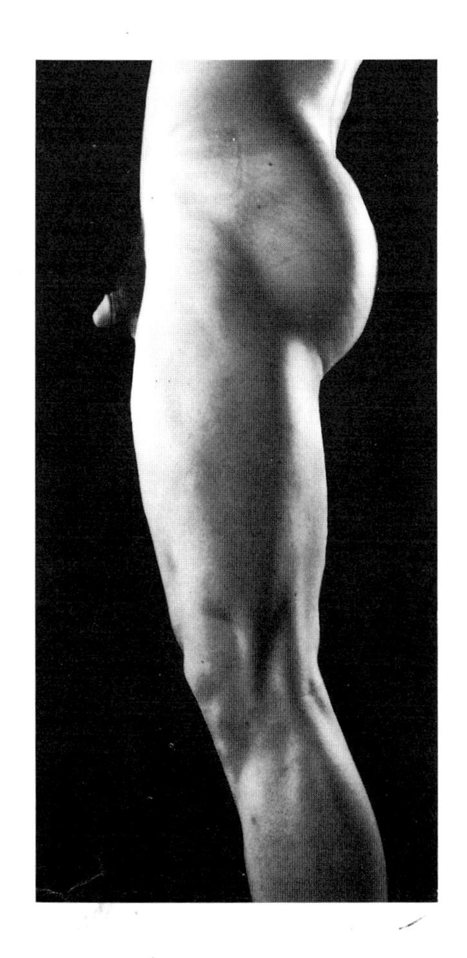

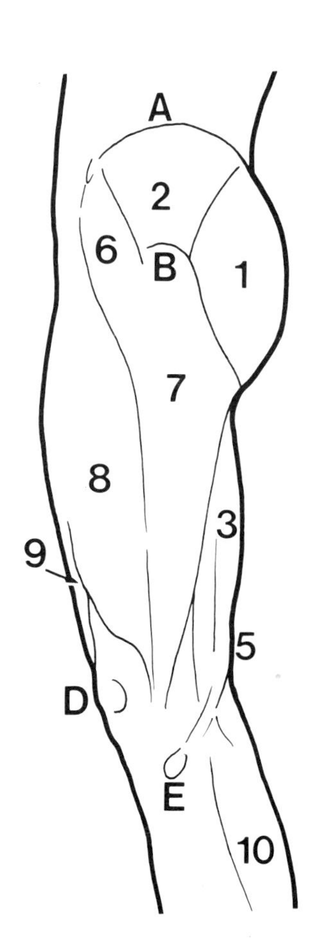

A
2
6
B
1
7
8
3
9
5
D
E
10

THE ANTERIOR AND MEDIAL ASPECTS OF THE THIGH

Bony Landmarks

Anterior superior iliac spine [**A**]
Patella [**B**]
Tibial tuberosity [**C**]

Soft Tissues

The quadriceps femoris consists of four individual muscles which form the major part of the muscle bulk on the front of the thigh.
The four muscles are:
Rectus femoris
Vastus medialis
Vastus lateralis
Vastus intermedius

The rectus femoris [**1**] may be easily seen as a ridge running down the front of the thigh when the extended leg is raised from the ground from a sitting position. In some individuals it is possible to see both the long head [**1L**] and the reflected head [**1R**] which meet at an acute angle at the junction of the upper and middle thirds of the thigh.

The vastus medialis [**2**] is a large muscle, the lower part of which forms the bulge above and medial to the patella.

The vastus lateralis [**3**] lies on the lateral side of the thigh. Its anterior border is slightly overlapped by the rectus femoris, and its lateral aspect is covered by the iliotibial tract. In its lower part, it forms an elevation above and lateral to the patella, which is less noticeable and more proximal than the elevation formed by the vastus medialis on the medial side.

The vastus intermedius is almost entirely hidden by the other three muscles.

Testing the quadriceps femoris is done by extending the knee against resistance.

The ligamentum patellae [**4**] is seen in thin subjects as a ridge running from the patella to the tibial tuberosity. Notice the infrapatellar fat pads on both sides [**4a**].

The knee jerk is demonstrated through the ligamentum patellae, the tendon of insertion of the quadriceps femoris. Ask the subject to cross his legs and relax the upper leg. If the ligamentum patellae is now struck briefly with a patella hammer, the quadriceps muscles will reflexly contract in an attempt to extend the knee. This reflex is known as the 'knee jerk' or 'quadriceps reflex' and is used to test the continuity of the reflex arc which passes through the third and fourth lumbar segments.

Sartorius (the tailor's muscle) [**5**] is represented by a line running from the anterior superior iliac spine to the posterior part of the medial femoral condyle and then forwards to the proximal part of the shaft of the tibia. It is a strap muscle with parallel fibres and is therefore easily identified at operation through incisions in the front of the thigh.

Demonstration of the sartorius is by flexing the hip when sitting, keeping the knee straight. In many subjects the sartorius will stand out as a long ridge running from the anterior superior iliac spine to the medial femoral condyle. This ridge is more obvious in its proximal third where it may be gripped between finger and thumb. Distally, it may be identified as a soft longitudinal elevation towards the posterior part of the medial femoral condyle. If the knee is now flexed the sartorius may be felt to slip off the condyle into the medial boundary of the popliteal fossa, its aponeurosis extending straight down to the flexed tibia. Functionally, its action may be demonstrated by asking the sitting subject to cross his legs so that, for example, the right ankle rests on the left knee with the right leg horizontal (the tailor's position). This demonstrates the action of the right sartorius.

The medial border of adductor longus [**6**] forms the medial boundary of the femoral triangle. It may be felt as a distinct ridge when an attempt is made to adduct the knees against resistance in the sitting position.

Gracilis is not seen here but, on adduction, it forms a palpable elongated band from the pubic symphysis to the posterior part of the medial condyle of the femur behind the sartorius. It then passes downwards and forwards to the proximal part of the tibial shaft.

The femoral triangle [**F**] is bounded by the inguinal ligament above, the medial border of the sartorius laterally, and the medial border of the adductor longus medially.

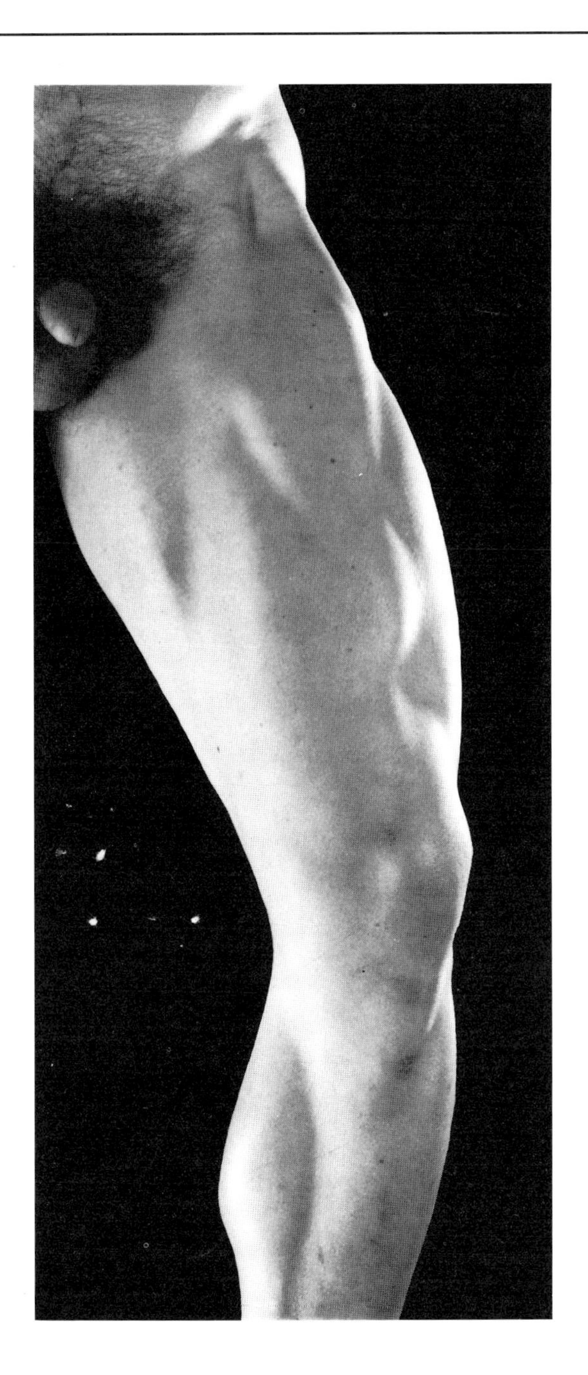

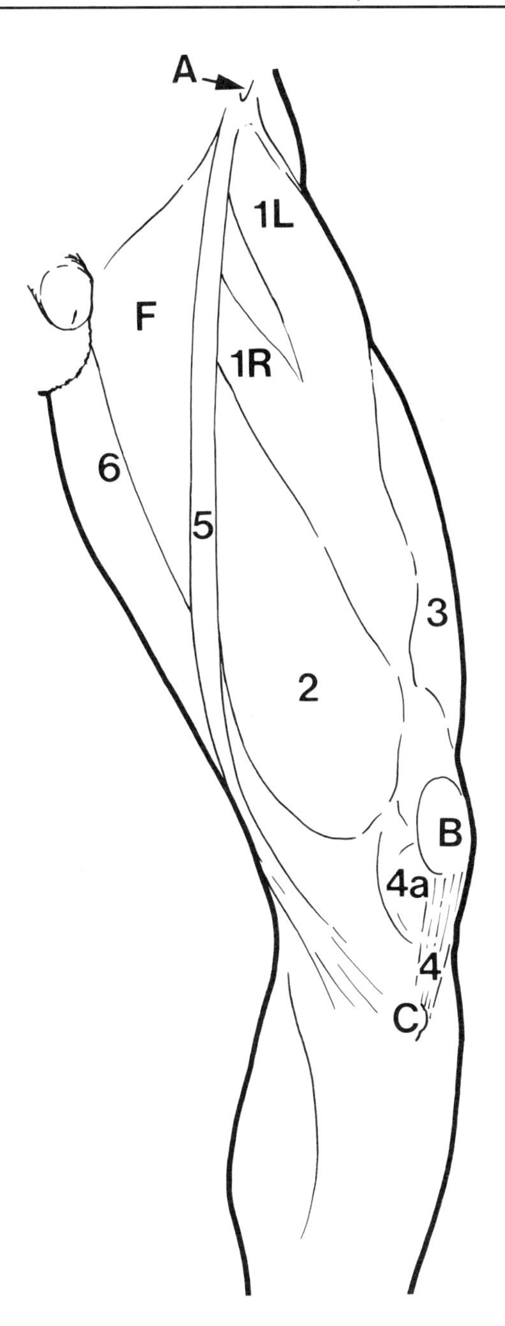
A
1L
F
1R
6
5
3
2
B
4a
4
C

RADIOLOGY OF THE HIP REGION

The hip joints are usually radiographed in the antero-posterior position with the heels slightly separated and the toes directed medially. In this position the femoral necks are parallel to the film. The field should include the lower lumbar spine, the sacrum and coccyx, the sacroiliac joints, both hip joints, the symphysis pubis and the femoral necks, as well as all the bones of the pelvis.

Sacrum [**1**]
Sacroiliac joint [**2**]
Ilium [**3**]
Ischium [**4**]
Pubis [**5**]
Pubic symphysis [**6**]
Anterior superior iliac spine [**7**]
Anterior inferior iliac spine [**8**]
Obturator foramen [**9**]
Coccyx [**10**]
Head of femur [**11**]
Cortical bone of acetabulum [**12**]
Radiological joint space [**13**]. The apparent gap between the two white lines representing the cortical bone of the femoral head and acetabulum, respectively, is known as the *radiological joint space*. It represents the thickness of the radiolucent hyaline cartilage lining the acetabulum and covering the femoral head. In this joint it is usually 4–7 mm
Fovea capitis [**14**]. Site of attachment of the ligamentum teres
Neck of femur [**15**]
Shaft of femur [**16**]. The angle between the neck and the shaft is usually between 120–140°, being more open in males because the pelvis tends to be narrower
Greater trochanter [**17**]
Lesser trochanter [**18**]
Shenton's Line [**19**]. In a normal individual, a line drawn as shown, along the upper margin of the obturator foramen, along the inferior margin of the neck of the femur and extending down the medial side of the shaft of the femur describes a smooth curve. This curve is unaffected by small changes in position, but in fractures and dislocations of the femur its continuity may become greatly distorted

Hip in Child Aged Six Years

Centre for the head of the femur which is present at birth [**20**]
Iliopubic limb of triradiate cartilage which disappears at around 16 years [**21**]
Site of fusion of ischiopubic rami [**22**]
Centre of ossification for the greater trochanter which appears between two and four years [**23**]

Ectopic Calcification

Phleboliths [**P**]. Small adherent thrombi in the pelvic veins commonly become calcified and are therefore seen on the radiograph. They tend to cast a circular shadow which sometimes appears laminated.

Lymph nodes [**N**]. Lymph nodes are also a common site of ectopic calcification. They tend to be larger, less regular in shape, and more mottled than phleboliths.

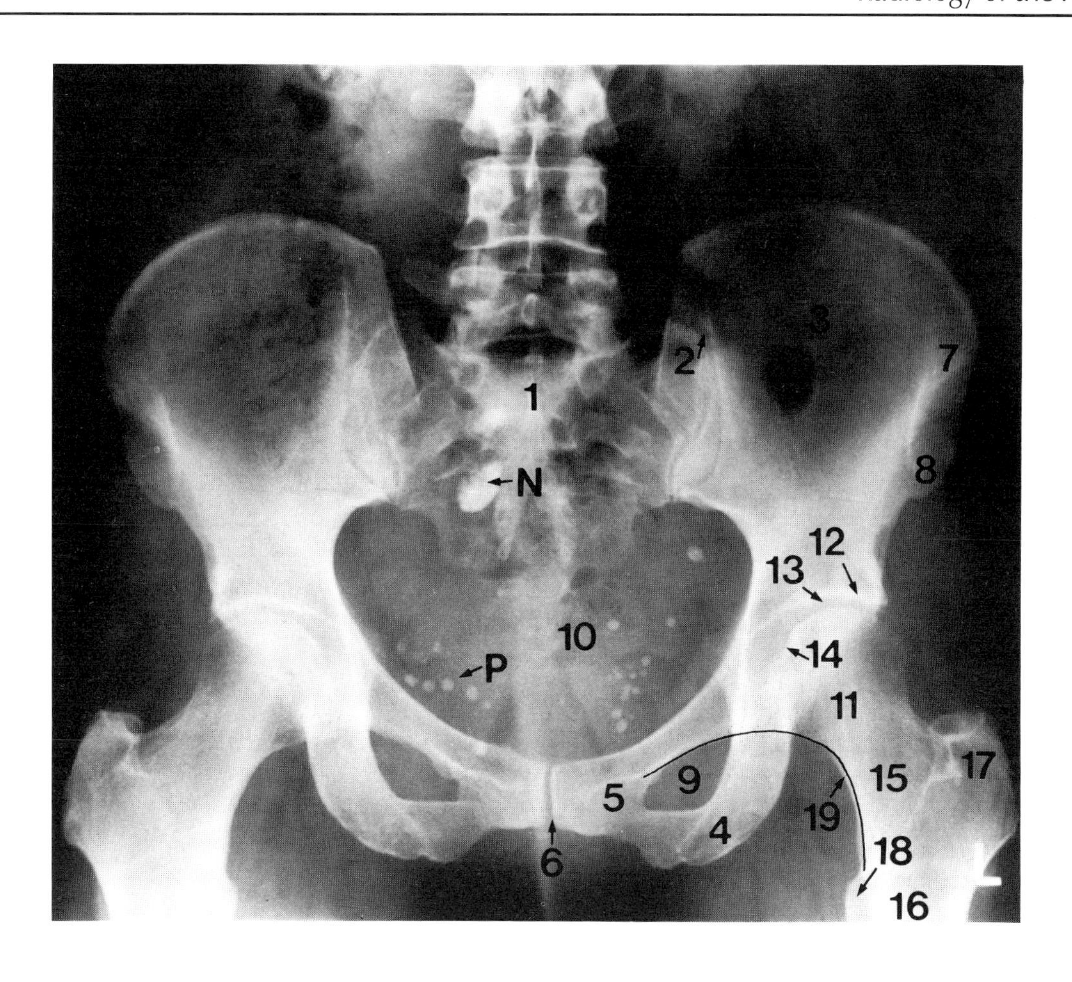

1
2
3
7
8
N
12
13
14
10
P
11
5
9
15
17
4
19
6
18
16
L

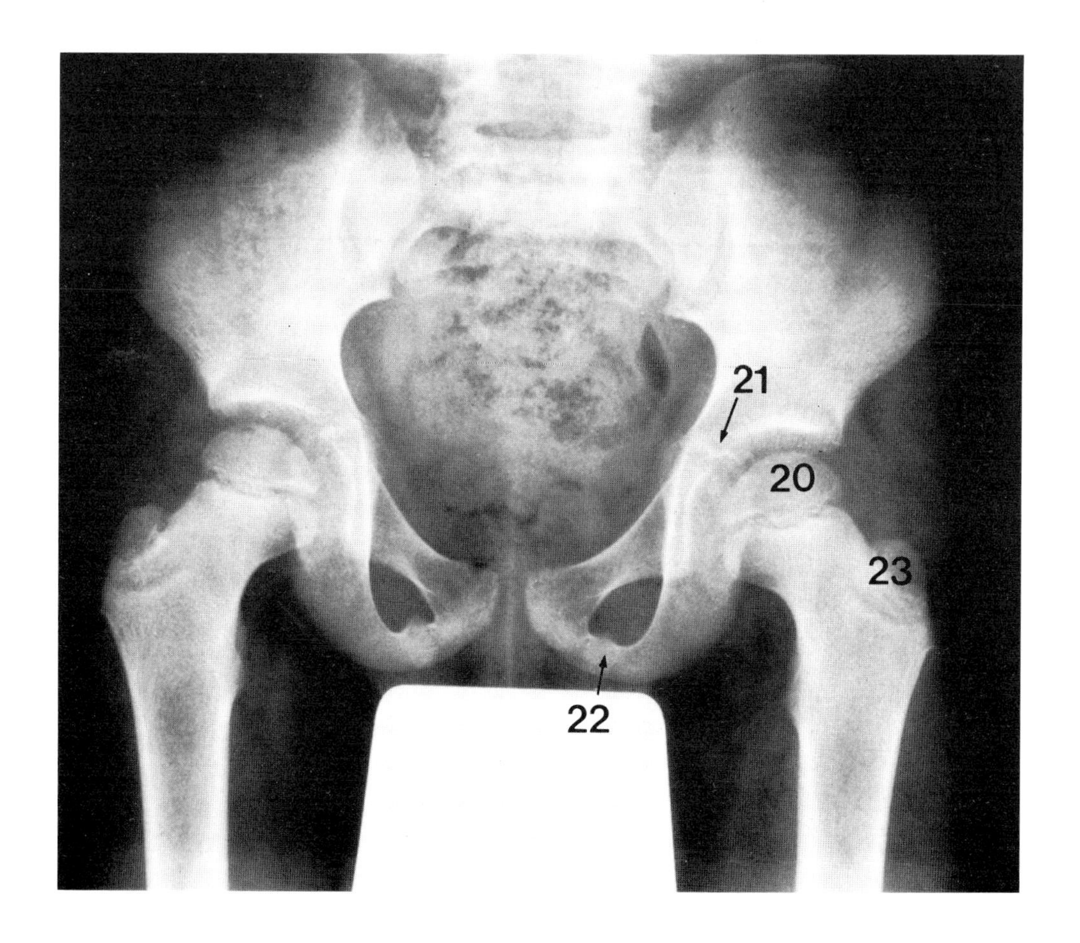

21
20
23
22

THE FLEXED KNEE

Medial Aspect

Bony Landmarks

Patella [**A**]
Medial femoral condyle [**B**]
Medial tibial condyle [**C**]
The joint line [**D**] felt as a groove between the femoral and tibial condyles. The groove is less distinct on the medial side than on the lateral side.
Tibial tuberosity [**E**]

Soft Tissues

The long head of rectus femoris [**1**]

The reflected head of rectus femoris [**2**]

Vastus medialis [**3**] forming a conspicuous fleshy bulge just above the medial femoral condyle

Sartorius [**4**] is not always easily seen. It may help to ask the subject to adduct and medially rotate his thigh against resistance. (*See* also The Anterior and Medial Thigh, p. 152.) Notice how, in this position, it has slipped off the medial femoral condyle

The adductor mass [**5**] comprises:
  Adductor longus
  Gracilis
  Adductor magnus
  The adductor magnus tendon may be felt if the palm is placed over the front of the thigh with the fingers deep to the posterior aspect of the vastus medialis while the thigh is adducted against resistance. It may be traced distally to the adductor tubercle on the proximal side of the medial condyle

The bulky tendon of the semimembranosus [**6**]

The thin tendon of the semitendinosus [**7**] which lies superficially on the bulk of the semimembranosus and on the popliteal side of the thicker tendon of the semimembranosus. (*See* also The Posterior and Lateral Thigh, p. 150)

The medial head of the gastrocnemius [**8**] which is larger than the lateral head

Soleus [**9**] peeping out from under the gastrocnemius. It is usually much easier to see this on the lateral side

The long saphenous vein [**10**]

Lateral Aspect

Bony Landmarks

Patella [**A**]
Lateral femoral condyle [**B**]
Lateral tibial condyle [**C**]
The joint line [**D**] felt as a groove between the tibial and femoral condyles. The medial and lateral grooves mark the positions of the medial and lateral menisci
Tibial tuberosity [**E**]
Head of fibula [**F**]

Soft Tissues

Vastus lateralis [**1**]

The iliotibial tract [**2**] forming the anterior boundary of a groove between the bulk of the vastus lateralis and the tendon of biceps femoris

Tendon of biceps femoris [**3**]

Semitendinosus [**4**]

Semimembranosus [**5**] in the medial wall of the popliteal fossa

Gastrocnemius [**6**]

Peroneus longus [**7**]

Tibialis anterior [**8**]

The common peroneal nerve may sometimes be seen superficially in thin people, in the angle between the biceps tendon and the peroneus longus and overlying the lateral head of the gastrocnemius. It can be rolled against the neck of the fibula. Notice that, as it crosses the neck of the fibula, it is about the same height as the bumper of a car which makes it particularly susceptible to damage in pedestrian road traffic accidents.

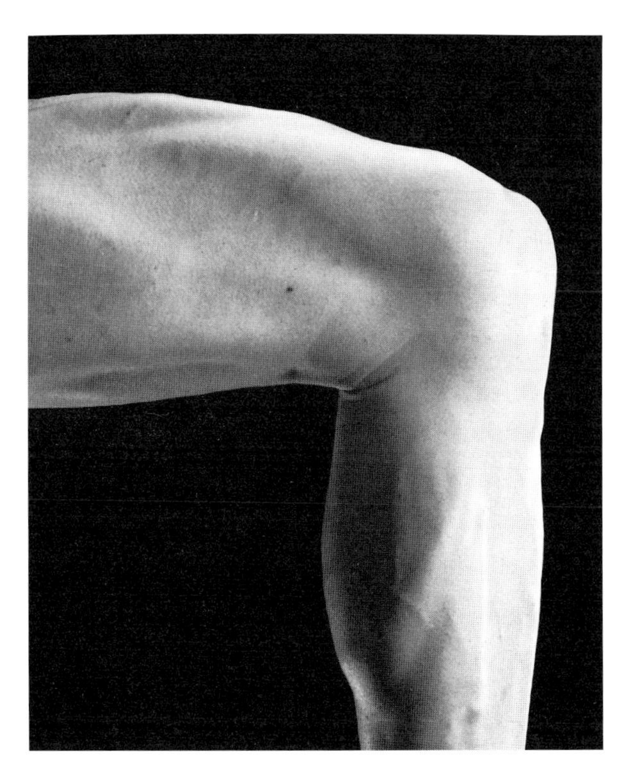

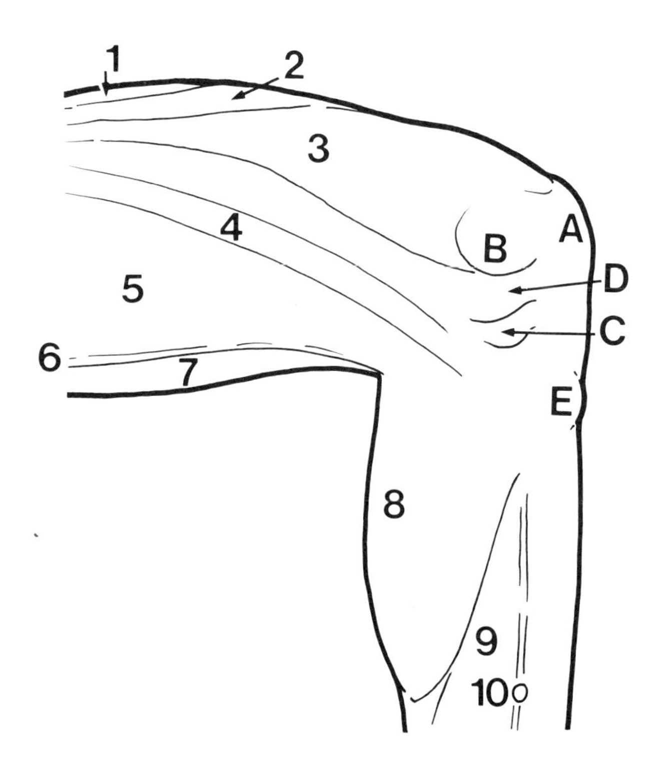

Medial aspect

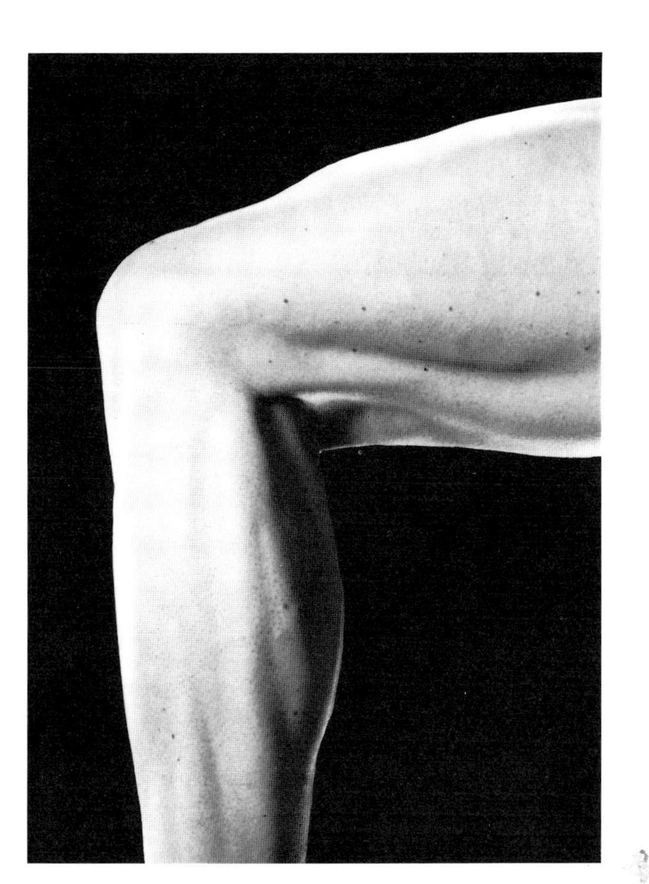

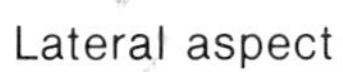

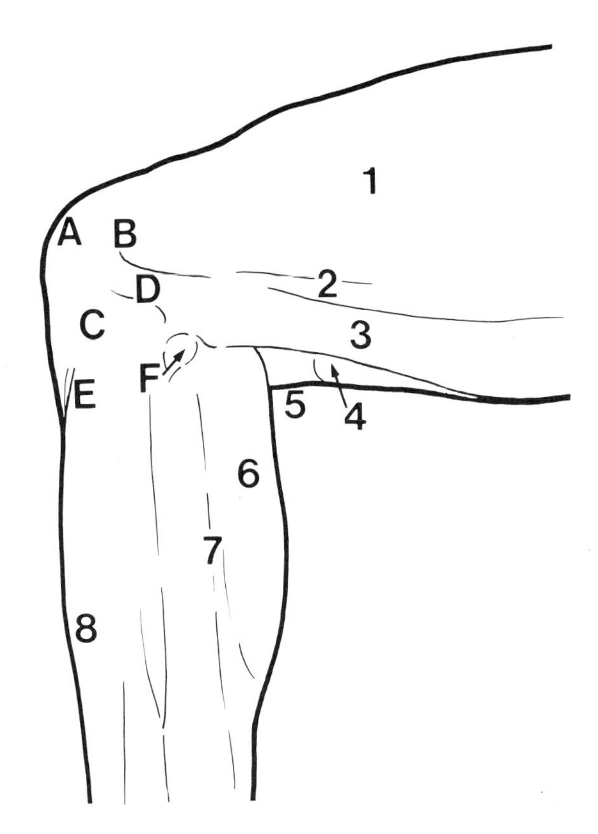

Lateral aspect

RADIOLOGY OF THE KNEE

The views usually taken of the knee include an antero-posterior view in full extension and a lateral view with the knee flexed to about 30°.

Antero-Posterior View

Patella [**1**]
Medial femoral condyle [**2**]
Lateral femoral condyle [**3**]
The radiological joint space [**4**] which represents the thickness of the radiolucent articular cartilage, is normally 3–5 mm wide and should be the same in both knees
Lateral tibial condyle [**5**]
Head of fibula [**6**]
Neck of fibula [**7**]
Intercondylar eminence [**8**]
Intercondylar notch [**9**]

The trabecular pattern can be seen to run mainly in a vertical direction. Notice also how the thick cortex of the femoral shaft gives way to the thin articular cortex which can be seen as a thin white line over the articular surface of the femoral condyles.

Lateral View

Patella [**1**]
The femoral condyles [**2**]. Because of the differing curves of the medial and lateral condyles, and the larger antero-posterior diameter of the lateral condyle, the two condyles cannot be fully superimposed. If the posterior margins are superimposed, it can be seen that about one-quarter of each condyle extends posterior to the line of the posterior aspect of the femoral shaft above
Sometimes a small bony shadow due to a sesamoid bone, the fabella, [**F**] in the lateral head of gastrocnemius may be seen just posterior to the femur. It is then also visible on the antero-posterior radiograph superimposed on the lateral femoral condyle
Intercondylar eminences [**3**]
Tibial tuberosity [**4**]
Head of fibula [**5**]

Knee of Child Aged 12 Years

Antero-posterior and lateral views. The outstanding feature in these radiographs is the presence of radiolucent cartilaginous growth plates between the metaphyses and epiphyses of the long bones.

Flared femoral metaphysis [**1**]
Distal femoral epiphysis [**2**] which has been present since birth
Radiolucent cartilaginous growth plates [**3**]
Patella [**4**] which ossifies from several centres in about the third year
Proximal tibial epiphysis [**5**] which appears at or immediately after birth
Tibial shaft [**6**]
Tibial tuberosity [**7**] which ossifies in early adolescence
The epiphyses unite with the main mass of bone in the late teens or early twenties and the site of fusion is often marked by a white line on the adult radiograph.

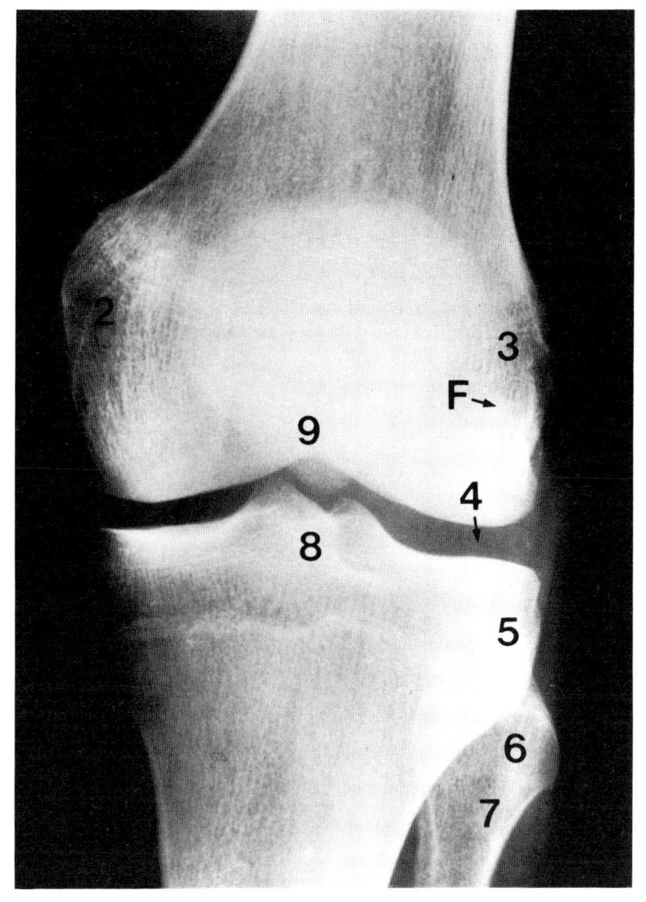

Anteroposterior view

Lateral view

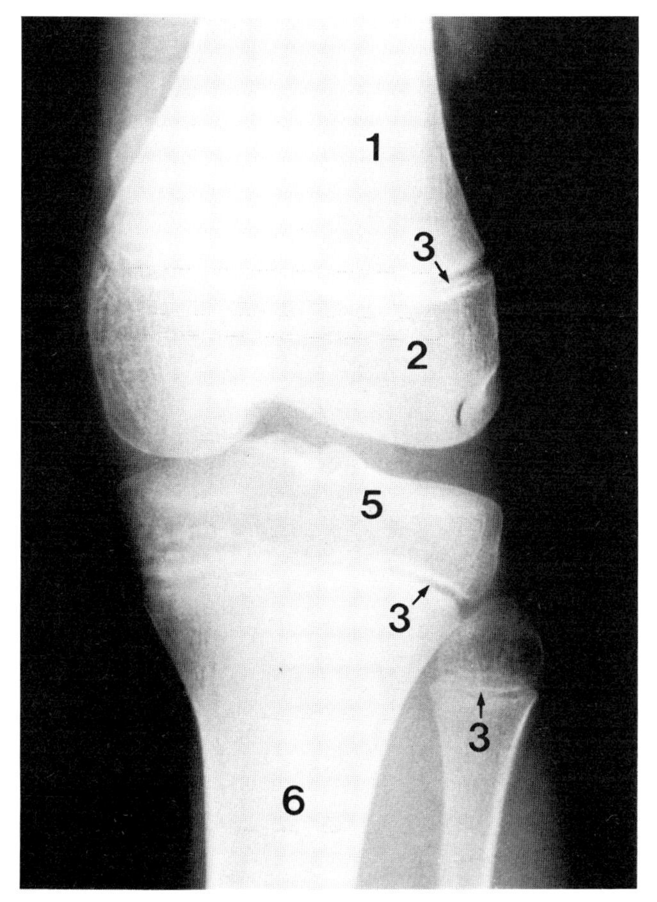

Anteroposterior view

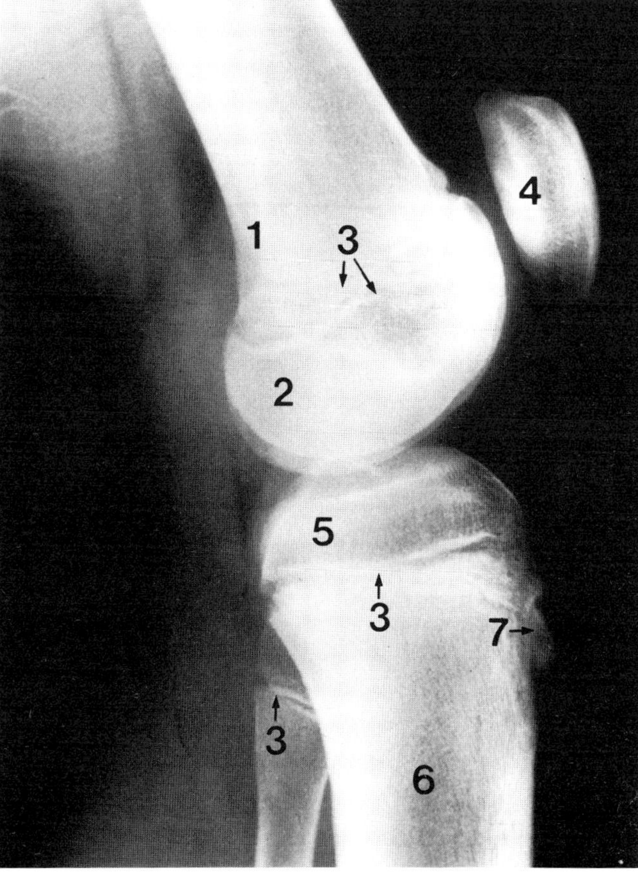

Lateral view

THE LATERAL ASPECT OF THE ANKLE, LEG AND FOOT

Bony Landmarks

Patella [**A**]
Head of fibula [**B**]
Lateral malleolus [**C**]
Peroneal tubercle [**D**]
Calcaneus [**E**]
Tuberosity of fifth metatarsal [**F**]

Soft Tissues

Iliotibial tract [**1**]

Biceps femoris [**2**]

Gastrocnemius [**3**] arises from two heads which unite to form the inferior angle of the popliteal fossa. The medial head is slightly larger and reaches a lower level than the lateral head. This muscle may be tested by plantar flexing the foot with the knee extended.

Soleus [**4**] lies deep to gastrocnemius but, when in action, it may be seen to peep out from under the gastrocnemius particularly on the lateral side where it forms a noticeable swelling behind the upper two thirds of a line between the head of the fibula and the lateral malleolus.

Plantaris runs opposite a line from the back of the lateral femoral condyle to the medial aspect of the Achilles tendon at its insertion.

The posterior intermuscular septum [**5**] is marked by a groove between the soleus and the peroneal muscles.

Tibialis anterior [**6**] may be felt as the fleshy mass protruding beyond the anterior margin of the middle third of the tibia when the medial side of the foot is dorsiflexed against resistance.

The peroneus longus [**7**] lies opposite a line commencing at the head of the fibula and running down behind the lateral malleolus and then forwards to the depression behind the tuberosity of the fifth metatarsal and finally running obliquely across the sole of the foot to the base of the first metatarsal.

The tendon of the peroneus longus [**7a**] overlying the tendon of the peroneus brevis.

Peroneus brevis [**8**] begins at the junction of the middle and lower thirds of the fibula and runs behind the lateral malleolus to the tuberosity of the fifth metatarsal.

The bellies of the peroneus longus and the peroneus brevis may be felt to contract on the lateral side of the leg when the foot is dorsiflexed or everted against resistance. The peroneus longus is felt in the upper part of the leg and the peroneus brevis in the lower third.

The synovial sheaths of the peroneal tendons begin about three finger breadths above the lateral malleolus. Initially, it is single before dividing on the calcaneus into two separate sheaths which enclose the tendons as far as their insertions.

Extensor digitorum longus [**9**] runs from the front of the head of the fibula to the front of the ankle joint, midway between the malleoli. Here the tendon of the peroneus tertius branches off, just prior to the division of the common tendon, into slips which pass to the lateral four toes. The common tendon and its digital slips are brought into relief when the foot is dorsiflexed. The tendon of the peroneus tertius is highlighted when the foot is everted.

Extensor digitorum brevis [**10**] forms the fleshy mass on the lateral side of the dorsum of the foot immediately lateral to the peroneus tertius tendon. It should be studied since it is often mistaken for a traumatic swelling or an abscess in a person with a painful ankle or foot. When the toes are spread apart and extended its tendons are thrown into relief and its fleshy belly hardens.

Achilles tendon [**11**] is formed from the tendons of the gastrocnemius and the soleus.

The ankle jerk is performed by striking the Achilles tendon with a patella hammer. This elicits a reflex contraction of the gastrocnemius and soleus via the first and second sacral segments (S1, 2).

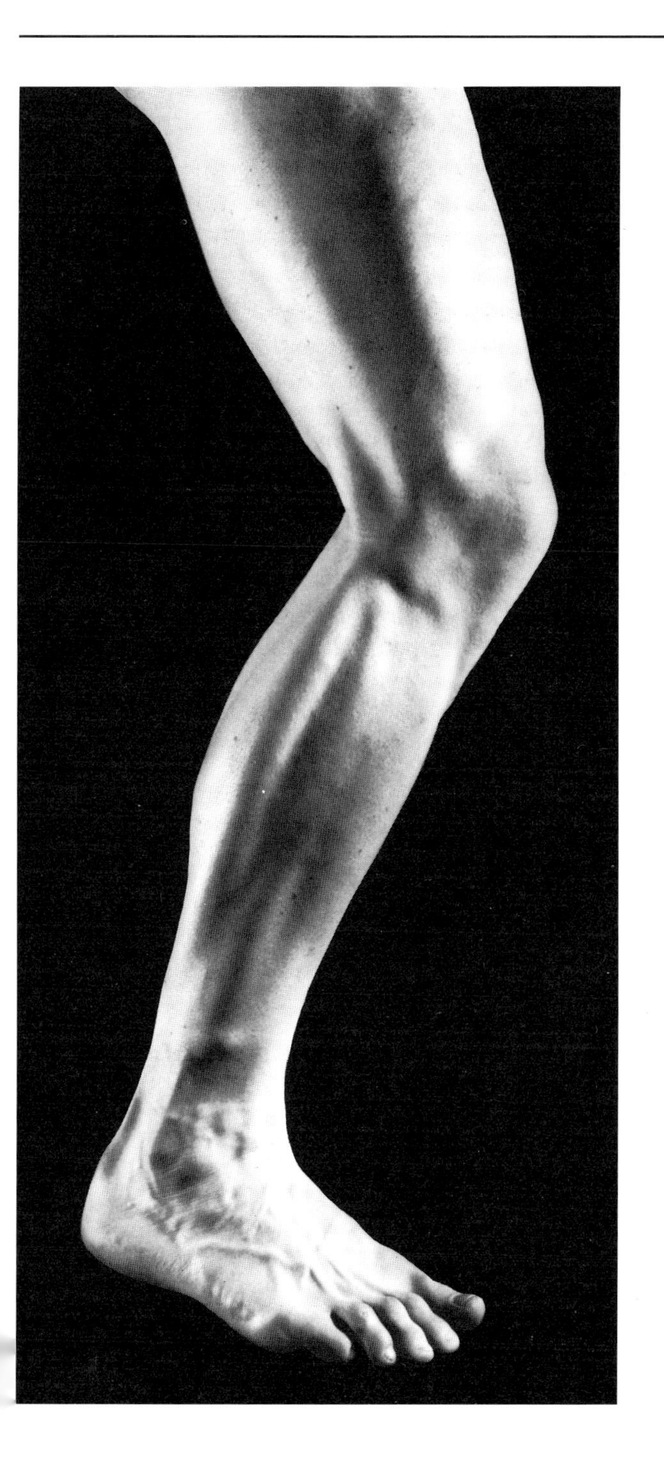

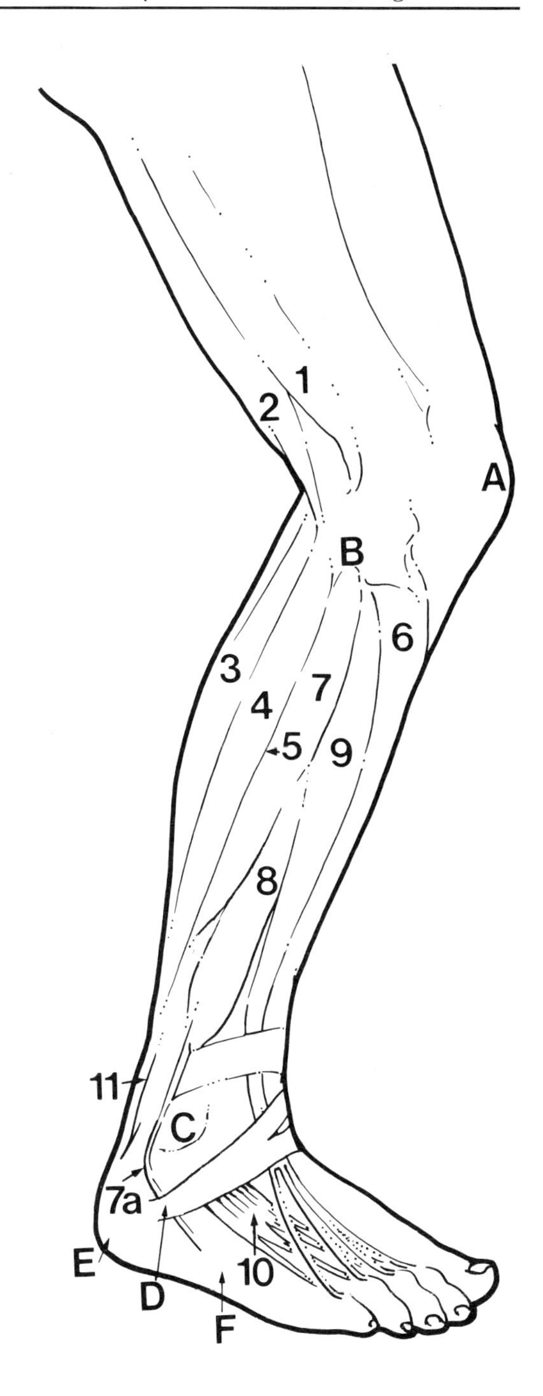
1
2
A
B
6
3
7
4
5
9
8
11
C
7a
E
D
10
F

THE ANTERIOR AND MEDIAL ASPECT OF THE ANKLE

Bony Landmarks

Medial malleolus [**A**]
Lateral malleolus [**B**] extending about 1.25 cm lower than the medial malleolus and also lying on a more posterior plane
Medial calcaneal tubercle [**C**]
Head of talus [**D**]

Soft Tissues

Superior extensor retinaculum [**1**] is a band about 2.5 cm broad stretching from the tibia to the fibula, its inferior border being about 2.5 cm above the plane of the ankle joint.

The inferior extensor retinaculum [**2**] is Y-shaped. The stem arises from the superio-lateral surface of the head of the calcaneus just behind the fleshy swelling of the extensor digitorum brevis. The upper limb passes to the front of the medial malleolus. The lower limb sweeps over to the navicular tuberosity and the medial border of the medial cuneiform.

The flexor retinaculum [**3**] straps down the tibialis posterior and the flexors as they pass from the back of the leg to the foot. It is a fan-shaped thickening of deep fascia passing between the medial malleolus and the lower medial margin of the calcaneus. Its upper border is ill defined and runs between the medial malleolus and the tendocalcaneus. The lower border is opposite a line drawn from the distal end of the medial malleolus to the medial calcaneal tubercle.

Extensor digitorum longus [**4**] sharing a common synovial sheath with the *peroneus tertius* [**4a**].

Extensor hallucis longus. [**5**] When this muscle contracts to dorsiflex the foot the big toe and its tendon can be felt above the ankle medial to the extensor digitorum as it raises a ridge of skin on the dorsum of the foot. Its synovial sheath begins about 2.5 cm above the ankle joint and ends just beyond the middle of the first metatarsal bone.

Tibialis anterior [**6**]

Tibialis posterior [**7**] the tendon of which escapes from under the cover of the flexor digitorum 6–7 cm above the heel, close to the medial margin of the tibia. The tendon winds around the back of the medial malleolus to insert mainly into the tuberosity of the navicular bone. The tendon rises into relief against resistance.

Tibialis anterior and tibialis posterior together form the chief inverters of the foot.

Flexor digitorum longus [**8**] has its tendon midway between the medial edge of the Achilles tendon and the sharp margin of the medial malleolus. The synovial sheath begins at the medial margin of the tendocalcaneus, about two finger breadths above the heel, and extends almost as far as the tuberosity of the navicular bone. Separate sheaths for the digital slips commence near the heads of the metatarsal bones and extend as far as the joints between the second and third phalanges.

Flexor hallucis longus. [**9**] At the level of the medial malleolus the tendon is just anterior to the medial margin of the Achilles tendon. From there the tendon courses obliquely across the talus to run beneath the sustentaculum tali and then forwards to the great toe. The synovial sheath begins about one thumb's breadth above the heel and encloses the tendon as far as the medial cuneiform. A second sheath commences near the head of the first metatarsal and encloses the tendon as far as the interphalangeal joint.

Abductor hallucis [**10**] is seen in some feet as a fleshy mass extending across the instep from the medial calcaneal tubercle as far as the ball of the great toe.

Notice the dorsal venous arch on the anterior view, with the long saphenous vein passing 2.5 cm in front of the medial malleolus.

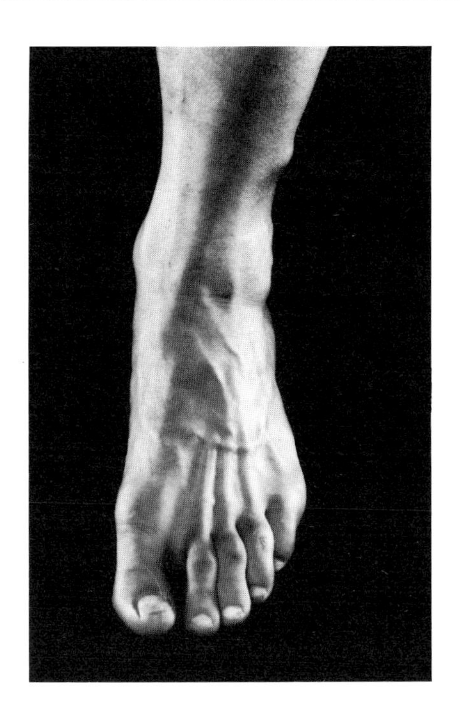

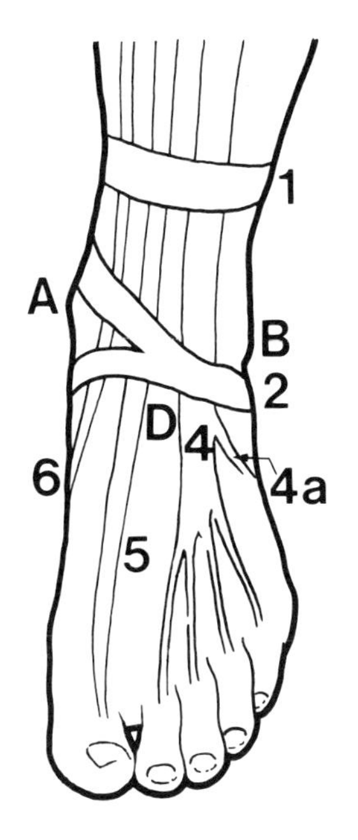
1
A
B
2
D
4
4a
6
5

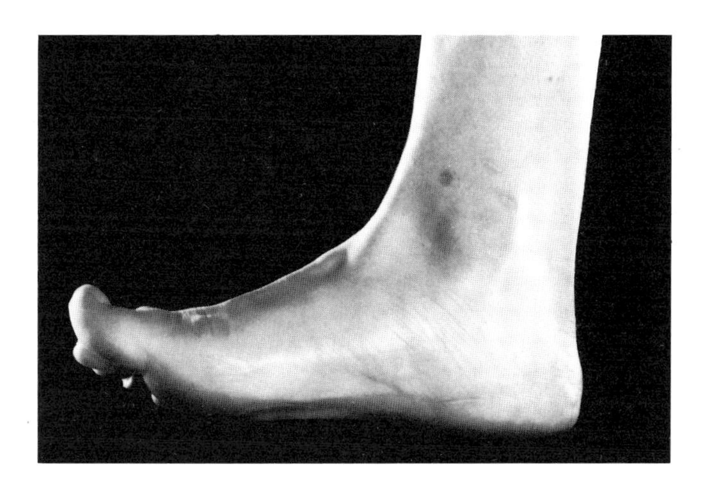

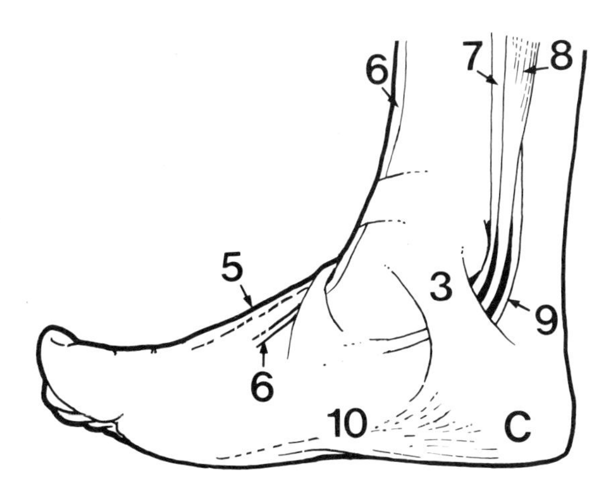
6
7
8
5
3
9
6
10
C

RADIOLOGY OF THE ANKLE AND FOOT

The Foot

Dorsi-plantar and oblique views are usually taken

First metatarsal [**1**]
Sesamoid bones [**2**]
Cuneiform bones [**3**]
Superimposed medial and intermedial cuneiform bones [**4**]
Navicular bone [**5**]
Head of talus [**6**]
Body of talus [**7**]
Cuboid [**8**]
Calcaneus [**9**]
Medial malleolus [**10**]
Lateral malleolus [**11**]
Fifth metatarsal [**12**]

The Ankle

Antero-posterior and lateral views are usually taken. The antero-posterior view is taken with the foot pointing slightly medially in order to rotate the shadow of the lateral malleolus away from the talofibular joint to demonstrate the ankle mortise clearly.

Oblique antero-posterior view
Fibula [**1**]
Tibia [**2**]
Lateral malleolus [**3**]
Inferior tibiofibular joint [**4**]
Talofibular joint [**5**]
Medial malleolus [**6**]
Body of talus [**7**]

Lateral view
Tibia [**1**]
Fibula [**2**]
Lateral malleolus [**3**]
Body of talus [**4**]
Calcaneus [**5**]
Navicular bone [**6**]
Cuboid [**7**]

Lines of Force

Cancellous bone consists of multiple trabeculae arranged along the direction of stress passing through a bone. These trabeculae appear as fine white lines on the radiograph, and are absent in areas of minimal stress. They can be seen to be linked together by other cross-bracing trabeculae for added stability. In the lateral radiograph of the ankle they appear to form an arch between the heel and ball of the foot with the highest point at the ankle.

The transmission of force is also well illustrated by the trabecular pattern of the hip region, particularly along the medial side of the neck of the femur.

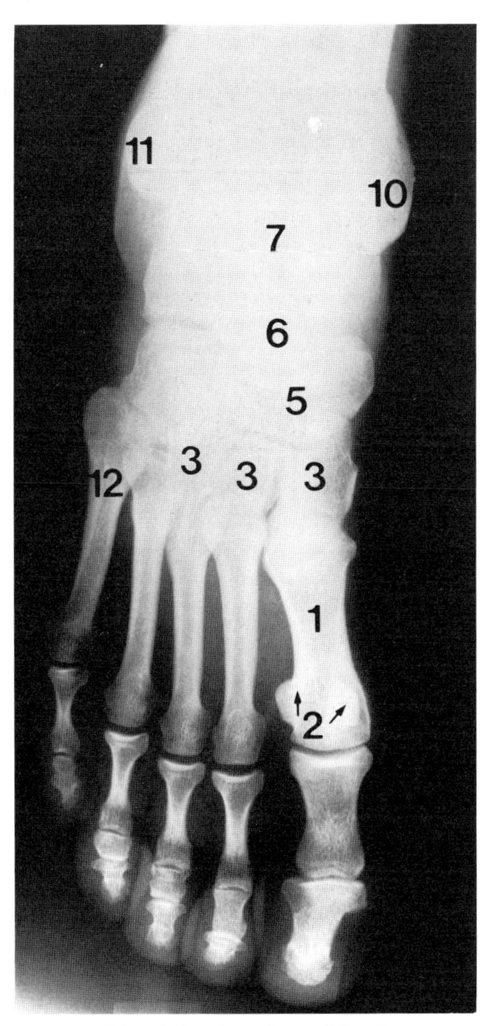

Dorsiplantar view of foot

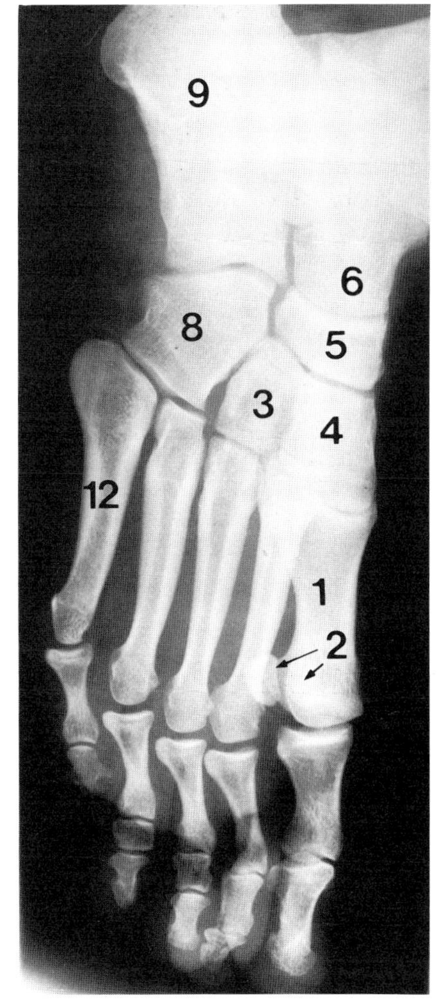

Oblique view of foot

Anteroposterior view of ankle

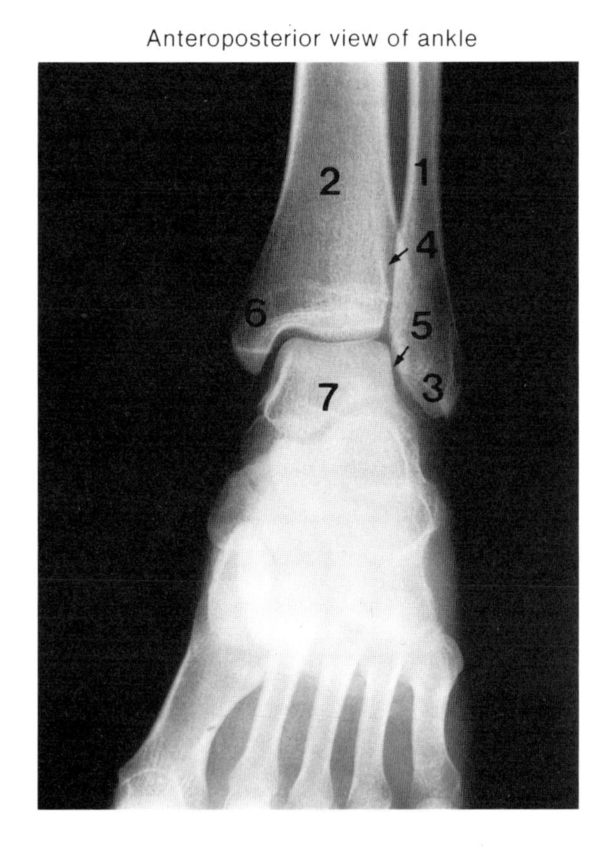

Lateral view of ankle

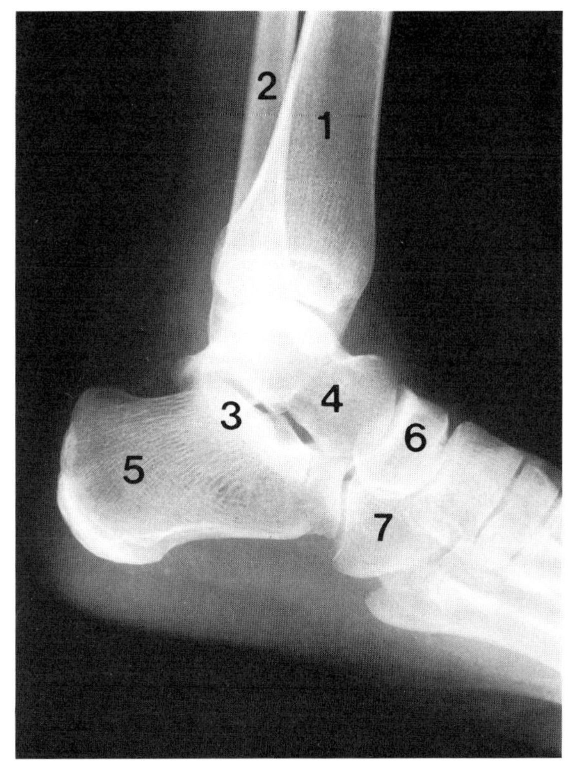

MAJOR ARTERIES OF THE LOWER LIMB

The femoral artery [**1**] may be represented by the upper two-thirds of a line running from the femoral point to the adductor tubercle when the thigh is slightly flexed and laterally rotated. The upper third of this line represents the course of the artery in the femoral triangle and the middle third the artery in the subsartorial canal. The upper 3.75 cm are enclosed in the femoral sheath.

The pulse of the femoral artery is easily felt by compressing the artery at its origin against the superior pubic ramus through the psoas major which lies in the floor of the femoral triangle. This is also a useful pressure point for controlling distal bleeding.

Profunda femoris artery [**2**]

The popliteal artery [**3**] is the continuation of the femoral artery once the latter has entered the posterior compartment of the thigh by piercing the adductor magnus. It enters the upper end of the popliteal fossa from the medial side, but quickly gains the midline of the fossa and then passes vertically downwards to the lower border of the popliteus muscle where it bifurcates into the anterior and posterior tibial arteries. This point of bifurcation is at the level of the tibial tuberosity.

The popliteal pulse may be felt by asking the subject to lie on his back and bend his knee to a right angle to relax the tense popliteal fascia which roofs the popliteal fossa. Deep pressure is then used to feel the pulse against the popliteal surface of the femur. This is not easy and requires plenty of practice. A slightly easier method for the beginner is to ask the subject to lie face down and bend his knee upwards to a right angle while deep pressure is applied over the popliteal surface of the femur.

The posterior tibial artery [**4**] is represented by a line commencing at the lower angle of the popliteal fossa and running to a point midway between the medial border of the tendo-calcaneus and the posterior border of the medial malleolus. Here it lies between the tendons of the flexor digitorum longus anteriorly and the flexor hallucis longus posteriorly.

The posterior tibial pulse is best felt by using the three middle fingers to compress the artery against the posterior aspect of the medial malleolus.

The posterior tibial artery divides into the medial and lateral plantar arteries before reaching the posterior border of the sustentaculum tali.

The medial plantar artery [**5**] passes directly forwards to the cleft between the first and second toes.

The lateral plantar artery [**6**] runs towards the medial side of the base of the fifth metatarsal and then swings round and arches towards the base of the first interosseous space to anastomose with the dorsalis pedis artery which dips down between the two heads of the first dorsal interosseous muscle.

The anterior tibial artery [**7**] runs from a point midway between the tibial tuberosity and the neck of the fibula, to a point in front of the ankle midway between the two malleoli, where it lies between the tendons of the extensor digitorum longus and the extensor hallucis muscles. From this point it continues as the dorsalis pedis artery.

The dorsalis pedis artery [**8**] commences in front of the ankle midway between the two malleoli and to the lateral side of the extensor hallucis longus muscle. It terminates by diving through the proximal part of the first intermetatarsal space to anastomose with the termination of the lateral plantar artery.

The dorsalis pedis pulse is best felt by using the middle three fingers and exerting very gentle pressure on the proximal part of the dorsal surface of the foot over the tarsal bones immediately lateral to the tendon of the extensor hallucis longus. There is a tendency to feel too far distally so always remember that the artery disappears through the proximal part of the first intermetatarsal space.

The peroneal artery [**9**]

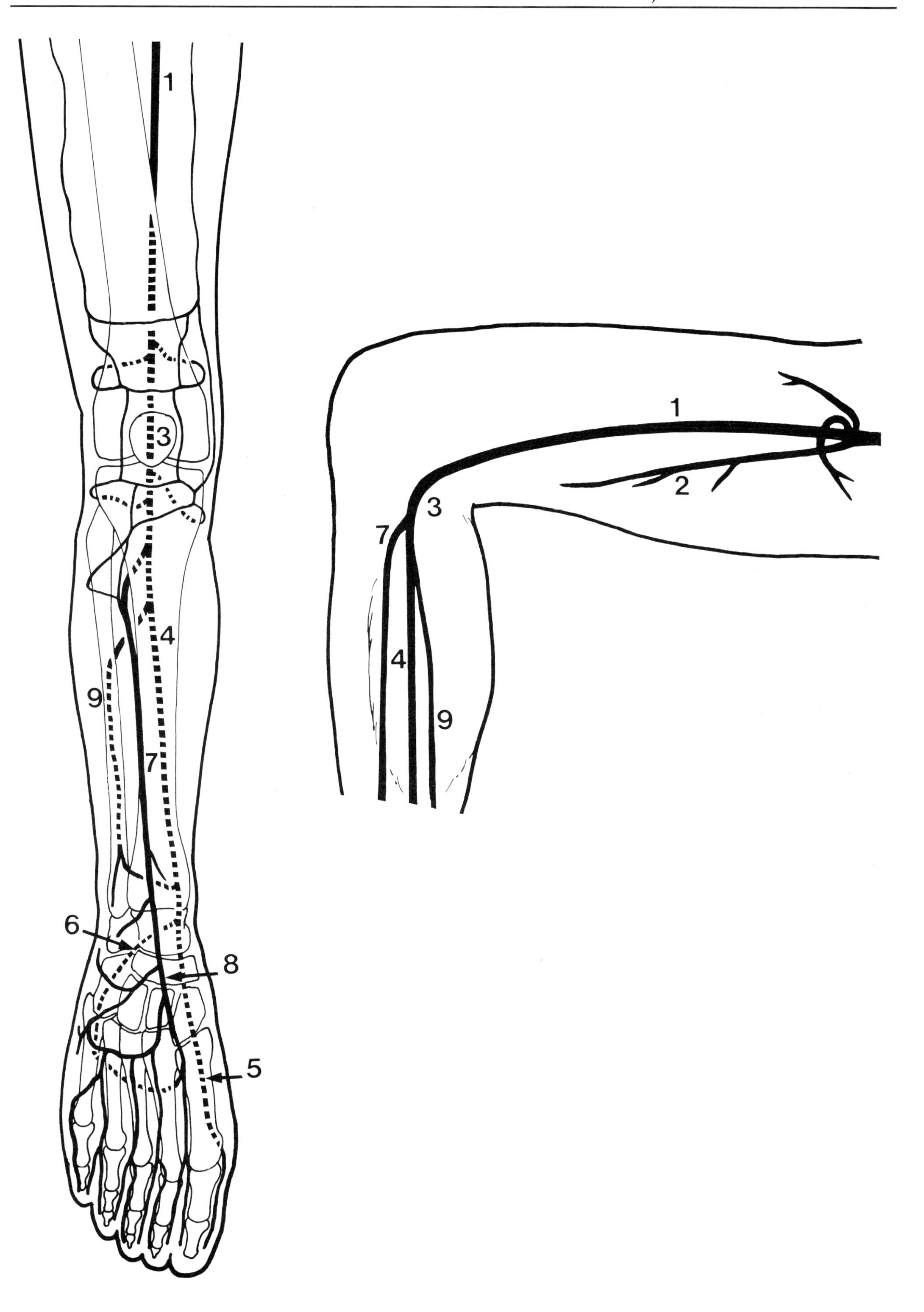

1
3
4
9
7
6
8
5
1
2
3
7
4
9

THE VEINS OF THE LOWER LIMB

The saphenous opening [**A**] lies at a point 3 cm below and lateral to the pubic tubercle.

The dorsal venous arch [**1**] is convex forwards with its ends overlying the bases of the first and fifth metatarsals.

The lateral marginal vein [**2**] of the foot begins at the lateral end of the dorsal venous arch and passes backwards over the upper surface of the cuboid to the lateral surface of the calcaneus where it turns upwards to become the short saphenous vein.

The short saphenous vein [**3**] begins behind the lateral malleolus as the continuation of the lateral marginal vein of the foot. Initially lateral to the Achilles tendon, it runs upwards in the midline of the calf to pierce the deep fascia between the two heads of the gastrocnemius at the lower angle of the popliteal fossa. It ends by joining the popliteal vein about 5 cm above the knee joint.

The long saphenous vein [**4**] commences at the medial end of the dorsal venous arch and passes backwards 2.5 cm in front of the medial malleolus. It crosses the distal third of the subcutaneous surface of the tibia obliquely backwards and runs up to the knee about one finger's breadth behind the medial border of the tibia. It passes immediately behind the medial condyle of the tibia, one hand's breadth from the medial border of the patella. Thereafter, it runs along the posterior border of the sartorius to pierce the cribriform fascia at the saphenous opening. On the thigh it is represented by a line between the adductor tubercle and the saphenous opening.

The position of the vein in front of the medial malleolus is constant and knowledge of this point may be life saving when an urgent transfusion is needed in obese or collapsed patients when other veins cannot be located.

Tributaries of the long saphenous vein in the thigh:
Lateral accessory vein [**5**]
Superficial circumflex iliac vein [**6**]
Superficial epigastric vein [**7**]
Superficial external pudendal vein [**8**]

During operations in the groin for varicose veins it is important that all of these tributaries are ligated as well as the main saphenous trunk.

The popliteal vein [**9**] is formed from the venae comitantes of the branches of the popliteal artery. Its main tributary is the short saphenous vein. Its surface marking is almost the same as for the artery, except that initially it lies on the medial side of the artery but as it ascends it crosses behind the artery to lie on its postero-lateral aspect in the adductor magnus hiatus where it becomes continuous with the femoral vessels. The hiatus in the adductor magnus lies level with the junction of the distal and middle thirds of the femur.

The femoral vein lies postero-lateral to the artery in the adductor magnus hiatus. It crosses behind the artery as it ascends to lie on the medial side of the artery behind the inguinal ligament.

Tributaries of the femoral vein:
Profunda femoris vein
Long saphenous vein
Venae comitantes of the descending genicular artery

Perforating Veins

The superficial veins of the calf are connected with the deep venous plexus by perforating veins which pierce the deep fascia. These veins contain valves which allow flow from the superficial to deep veins only.

On the medial side of the calf there are some four to six perforating veins connecting the long saphenous system with the deep veins. Their position is variable but there is usually one at mid-calf level, another one hand's breadth below the knee and a 'Hunterian perforator' one hand's breadth above the knee. The short saphenous system is similarly connected to the deep system by perforating veins.

Incompetence of the valves in these veins leads to an increase in venous pressure in the superficial veins which may give rise to varicosities.

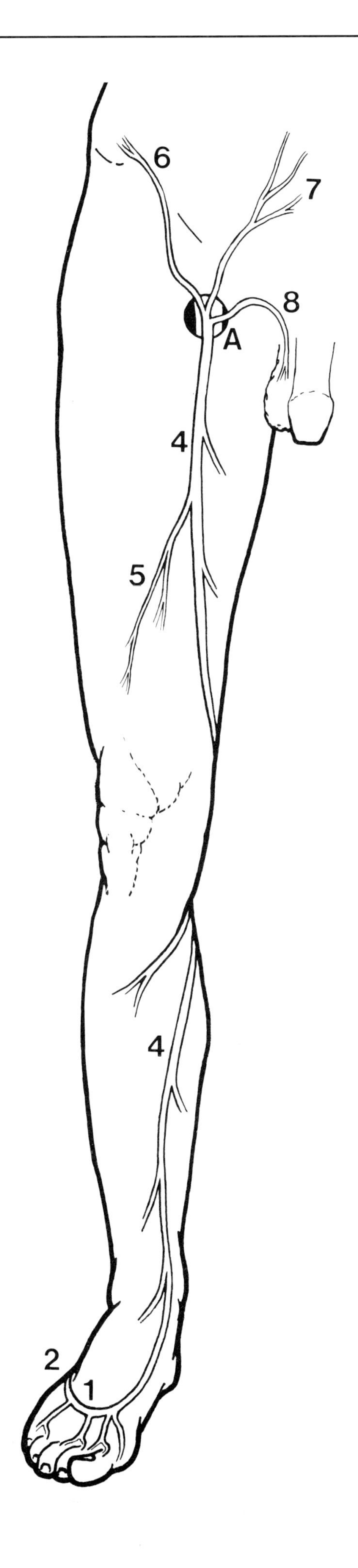
6
7
8
A
4
5
4
2
1

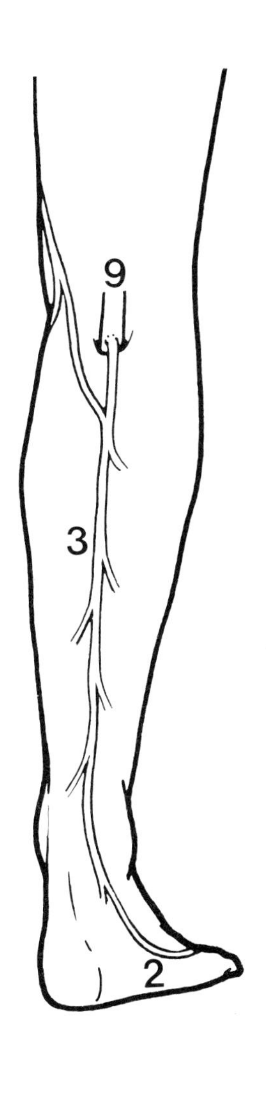
9
3
2

THE FEMORAL ARTERIOGRAM

This procedure is usually performed by injecting a bolus of radiopaque contrast medium into the lower aorta via a catheter inserted through the femoral artery. This allows demonstration of the aortic bifurcation, the iliac and femoral vessels as well as run off into the popliteal and lower leg vessels.

The major clinical indication for a study of this nature is suspected peripheral vascular disease, in particular atheroma and arteriosclerosis. Atheroma manifests itself as irregular plaques on the intima and is associated with narrowing of the vessel lumen, in some cases to the point of complete occlusion.

Femoral artery [**1**]
Profunda femoris artery [**2**]
Popliteal artery [**3**]
Descending genicular artery [**4**]
Superior medial and lateral genicular arteries [**5**]
Inferior medial and lateral genicular arteries [**6**]
Anterior tibial artery [**7**]
Posterior tibial artery [**8**]
Peroneal artery [**9**]

VENOGRAM OF THE LOWER LIMB

The study of veins in the lower limb is important for the differential diagnosis of pain in the calf, and in the search for the origin of pulmonary emboli. Occlusions due to thrombosis of the deep veins of the leg may easily be demonstrated by this technique. It is also useful in the pre-operative assessment of some patients with varicose veins to show the sites and number of incompetent perforating veins.

The procedure is performed by compressing the superficial veins around the ankle with a tourniquet and then injecting contrast medium into one of the veins on the dorsum of the foot. The passage of dye through the deep veins and of the leg is then monitored on a television screen using an image intensifier and appropriate films are taken.

Femoral vein [**A**]
Popliteal vein [**B**]
Soleal plexus of veins [**C**]
Valve [**D**]

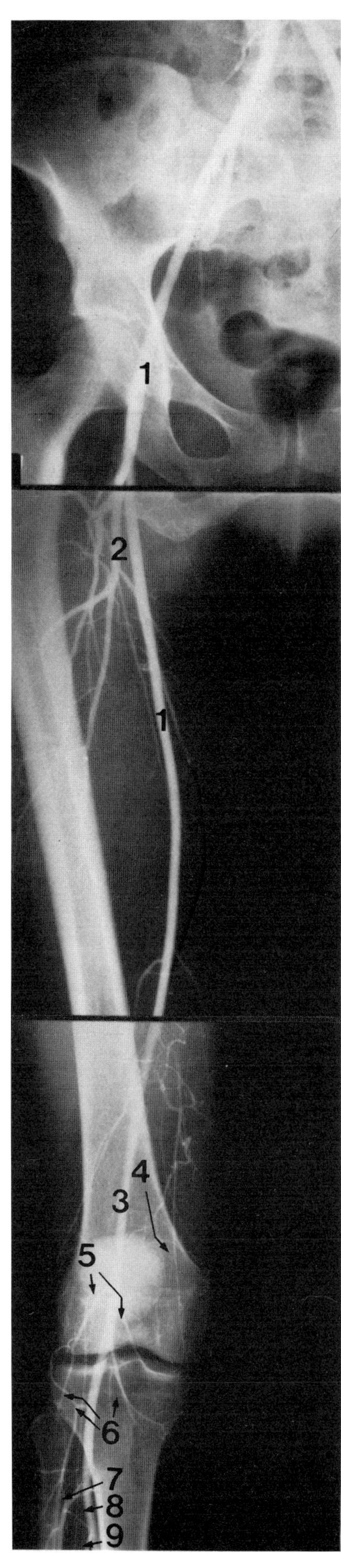

Arteriogram

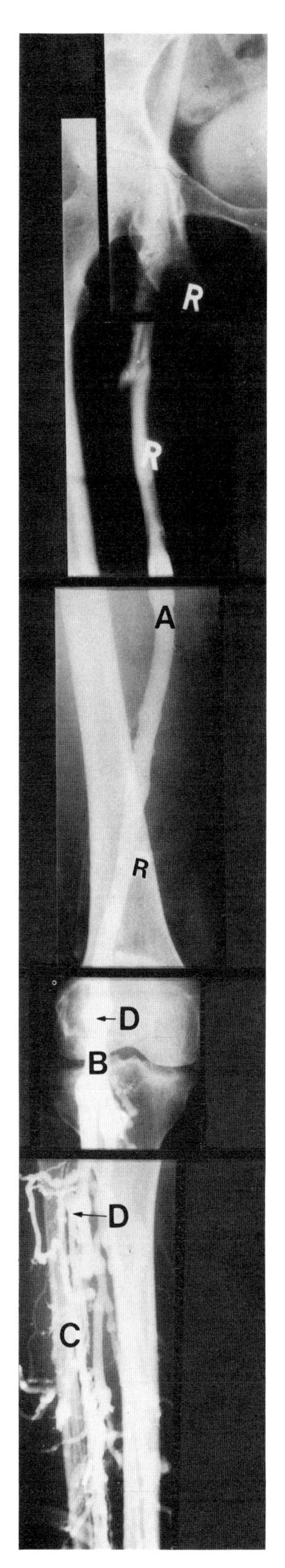

Venogram

LYMPHATICS OF THE LOWER LIMB

Clusters of lymph nodes are found in three areas only: the inguinal nodes in the groin, the popliteal nodes in the popliteal fossa and a small inconstant group close to the anterior tibial vessels.

Most of the lymph from the superficial tissues of the lower limb drains into the inguinal nodes via vessels accompanying the long saphenous vein. In addition, many large vessels spiral around the lateral side of the thigh to end directly in the inguinal nodes. Only a small area of skin and subcutaneous tissue over the heel and lateral side of the foot drains to the popliteal nodes via vessels accompanying the short saphenous vein. Therefore, even distal infections of the foot may result in inguinal lymphadenitis without involving the popliteal nodes.

The Inguinal Nodes

The inguinal nodes are divided into deep and superficial groups. The large and important superficial group drains into the deep group through the cribriform fascia.

The superficial inguinal nodes lie superficial to the deep fascia and are therefore readily palpable when enlarged. They are subdivided into three groups—vertical, medial and lateral.

The vertical group [**1a**] lies alongside the proximal 5 cm of the long saphenous vein. It receives all the lymph from the deep fascia and superficial tissues of the lower limb except from that small area of the heel and lateral side of the foot which drains to the popliteal nodes.

The lateral group [**1b**] receives lymph from the buttock, flank and back below the level of the umbilicus.

The medial group [**1c**] receives lymph from the abdominal wall below the level of the umbilicus and medial to a vertical line passing through the anterior superior iliac spine. It also receives lymph from the perineum and anal canal below Hilton's white line. Thus, tumours of the anal canal may present with enlarged inguinal lymph nodes. This group does *not* drain the testes.

The efferents from all three groups of the superficial inguinal nodes pass through the cribriform fascia into the deep inguinal nodes and thence to the external iliac nodes.

The deep inguinal nodes are relatively few in number and lie medial to the proximal part of the femoral vein. They receive lymph from the superficial inguinal nodes, the popliteal nodes and the vessels draining the deep part of the thigh which accompany the tributaries of the femoral vein. One often lies in the femoral canal.

The Popliteal Nodes [2]

This group receives lymph from the deep structures of the foot and leg and from the superficial tissues of the lateral part of the heel and foot as far as the little toe.

The Anterior Tibial Nodes [3]

This small and inconstant group lies in front of the interosseous membrane 5 cm below the level of the head of the fibula. The nodes receive lymph from the deep tissues of the front of the leg and their efferents pass to the popliteal nodes.

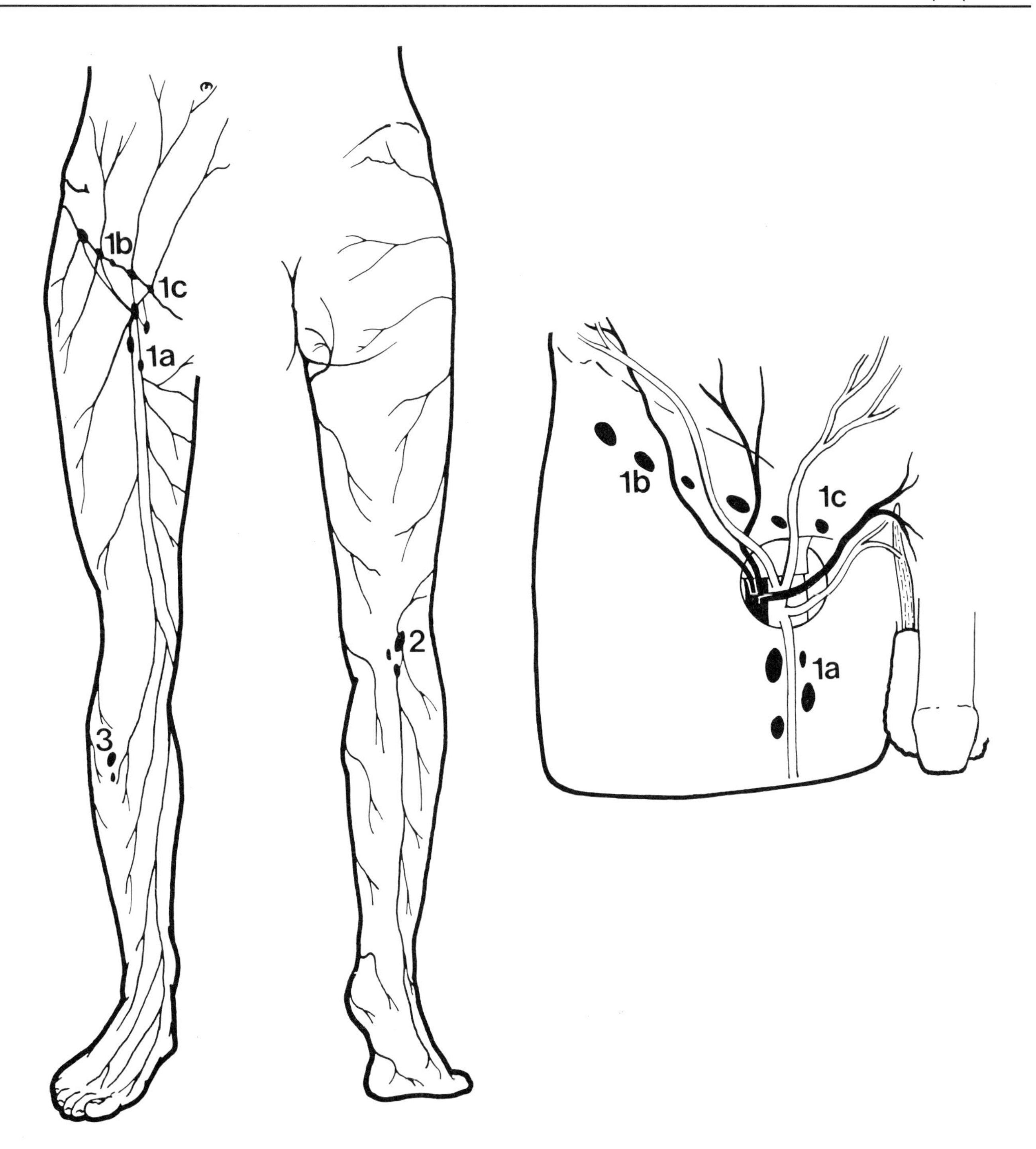
1b
1c
1a
2
3
1b
1c
1a

NERVES AND DERMATOMES OF THE LOWER LIMB

The femoral nerve leaves the abdomen behind the inguinal ligament 2 cm lateral to the femoral point. It runs into the thigh for about 3 cm before breaking up into branches which include three cutaneous branches—the saphenous nerve, and the medial and lateral anterior cutaneous nerves of the thigh.

The saphenous nerve may be represented by a line beginning at the point of division of the femoral nerve and drawn to the posterior part of the medial side of the knee, and then distally along the medial margin of the tibia as far as the lower third. Here the line of the nerve crosses the subcutaneous distal third of the tibia obliquely forwards and runs in front of the medial malleolus to the medial side of the foot as far as the ball of the great toe. The nerve pierces the deep fascia at the medial side of the knee and supplies sensation to the medial side of the leg and foot.

Cutaneous Nerves of the Lower Limb

Anterior aspect

Lateral cutaneous nerve of the thigh [**1**] (*See* The Nerves of the Abdominal Wall, p. 54)
Lateral anterior cutaneous nerve of the thigh [**2**], a branch of the femoral nerve
Medial anterior cutaneous nerve of the thigh [**3**], a branch of the femoral nerve
Lateral cutaneous nerve of the calf [**4**]
Infrapatellar branch of the saphenous nerve [**5**]
Superficial peroneal nerve [**6**]
Sural nerve, a branch of the tibial nerve [**7**]
Deep peroneal nerve [**8**]
Saphenous nerve, a branch of the femoral nerve [**9**]
Obturator nerve [**10**] (*See* The Nerves of the Abdominal Wall, p. 54)
Ilioinguinal nerve [**11**] (*See* The Nerves of the Abdominal Wall, p. 54)

Posterior aspect

Iliohypogastric [**1**] (*See* The Nerves of the Abdominal Wall, p. 54)
Dorsal rami of L1,2,3 [**2**]
Dorsal rami of S1,2,3 [**3**]
Perforating cutaneous nerve [**4**]
Posterior cutaneous nerve of the thigh [**5**]
Medial cutaneous nerve of the thigh [**6**]
Obturator nerve [**7**] (*See* The Nerves of the Abdominal Wall, p. 54)
Saphenous nerve [**8**]
Sural nerve [**9**]
Medial calcaneal branches of the tibial nerve [**10**]
Medial plantar nerve, a branch of the tibial nerve [**11**]
Lateral plantar nerve, a branch of the tibial nerve [**12**]
Superficial peroneal nerve [**13**]
Lateral cutaneous nerve of the calf [**14**]
Lateral cutaneous nerve of the thigh [**15**]

Dermatomes

The dermatomes of the lower limb are less clearly defined than those of the upper limb. One of the currently accepted patterns is given on the facing page.

Pre-axial border [**A**]
Ventral axial line [**B**]
Dorsal axial line [**C**]
Post-axial border [**D**]
Extension from dorsal axial line [**E**]

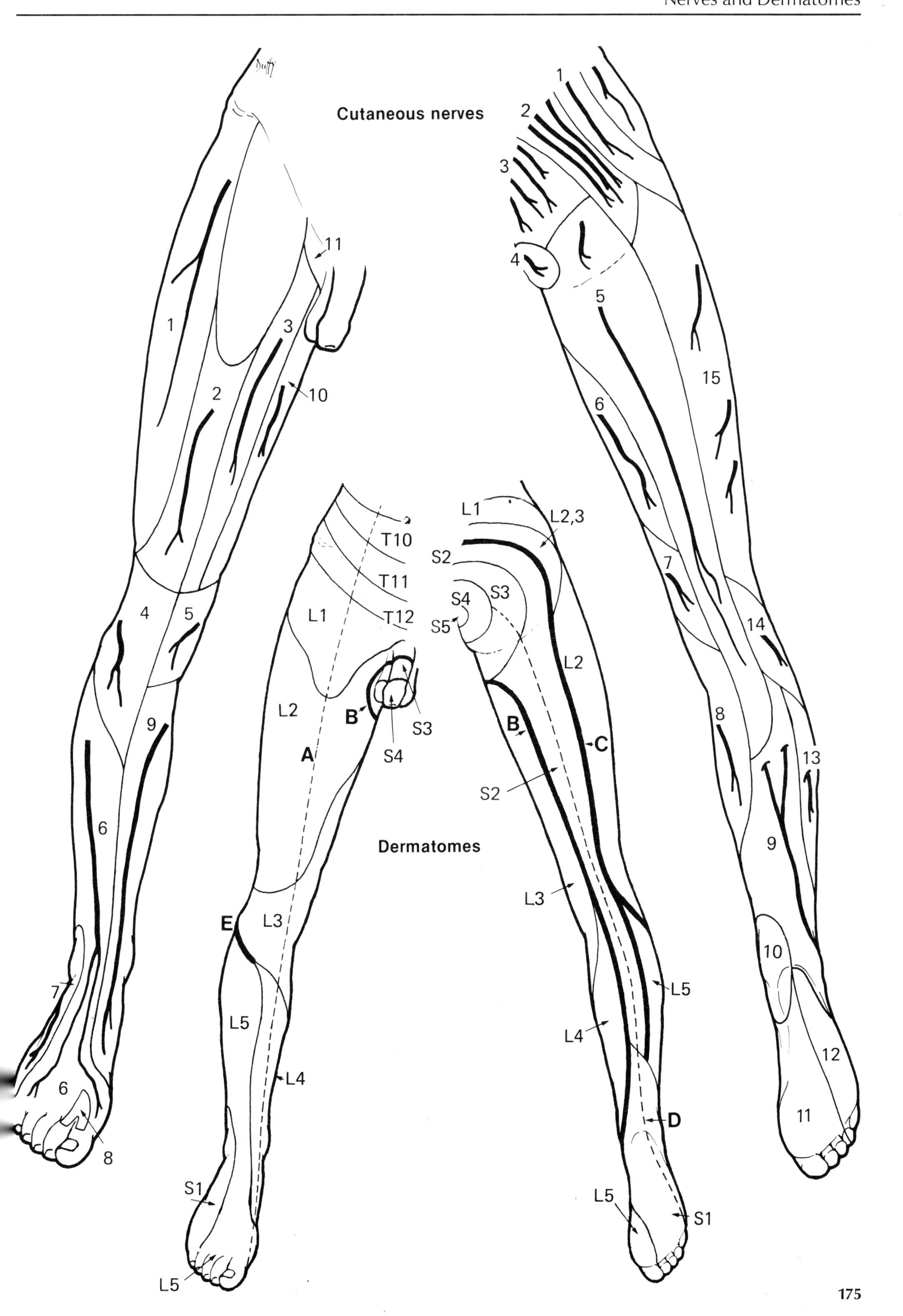
Cutaneous nerves
1
2
3
4
5
6
7
8
9
10
11
12
13
14
15
Dermatomes
T10
T11
T12
L1
L2
L2,3
L3
L4
L5
S1
S2
S3
S4
S5
A
B
C
D
E

THE SCIATIC NERVE

Root Value

L4, 5, S1, 2, 3

Course and Surface Markings

The sciatic nerve leaves the pelvis through the greater sciatic foramen and emerges from beneath the piriformis at a point opposite the junction of the upper and middle third of a line drawn between the greater trochanter and the ischial tuberosity. To mark its course, draw a line from this point to a point midway between the greater trochanter and the ischial tuberosity and thereafter down the middle of the back of the thigh to the apex of the popliteal fossa. Here the nerve divides into the posterior tibial and common peroneal nerves. The point of division is by no means constant and may occur anywhere along the line described, although division at the mid-thigh level is commonest.

The nerve is initially deep to the gluteus medius in the buttock and in the thigh it lies deep to the long head of the biceps muscle.

Muscles Supplied

Hamstrings
Ischial part of adductor magnus.

THE TIBIAL NERVE

Root Value

L4, 5, S1, 2, 3

Course and Surface Marking

This is the larger division of the sciatic nerve and it continues in the line of the sciatic nerve through the midline of the popliteal fossa. It then runs deep to the soleus along a line drawn from the inferior angle of the popliteal fossa to a point midway between the medial malleolus and the tendocalcaneus where it lies behind the tibial artery and its venae commitantes. From this point it then curves forwards to a point midway between the medial malleolus and the medial tubercle of the calcaneus where it divides into the medial and lateral plantar nerves.

The medial plantar nerve runs from the point of division of the tibial nerve towards the third toe until at the level of the middle cuneiform bone it divides into its digital branches. The digital branches may be represented by lines running from this point to the medial side of the big toe and the three medial clefts.

The lateral plantar nerve runs from the point of division of the tibial nerve to a point one thumb's breadth medial to the tuberosity of the fifth metatarsal bone. It then turns medially and runs to the base of the second metatarsal bone.

Muscles Supplied

Tibial nerve
Soleus
Gastrocnemius
Popliteus and plantaris
Tibialis posterior
Flexor digitorum longus
Flexor hallucis longus

Medial plantar nerve
Abductor hallucis
Flexor hallucis brevis
Flexor digitorum brevis
First lumbrical

Lateral plantar nerve
Flexor accessorius
Abductor digiti minimi
Flexor digiti minimi
Oblique head of adductor hallucis
All interossei
Lateral three lumbricals

Area of Cutaneous Supply

(*See* The Sole of the Foot, p. 180.)

The sural nerve is a cutaneous branch of the tibial nerve. It branches off in the popliteal fossa and runs down the midline of the calf from the inferior angle of the popliteal fossa to the back of the lateral malleolus. It lies at first lateral to the short saphenous vein and then behind it at the lateral malleolus. From the lateral malleolus it runs forwards along the lateral border of the foot to the little toe.

One hand's breadth above the heel it is joined by a communicating branch from the common peroneal nerve.

Medial calcaneal branch of tibial nerve. (*See* The Sole of the Foot, p. 180.)

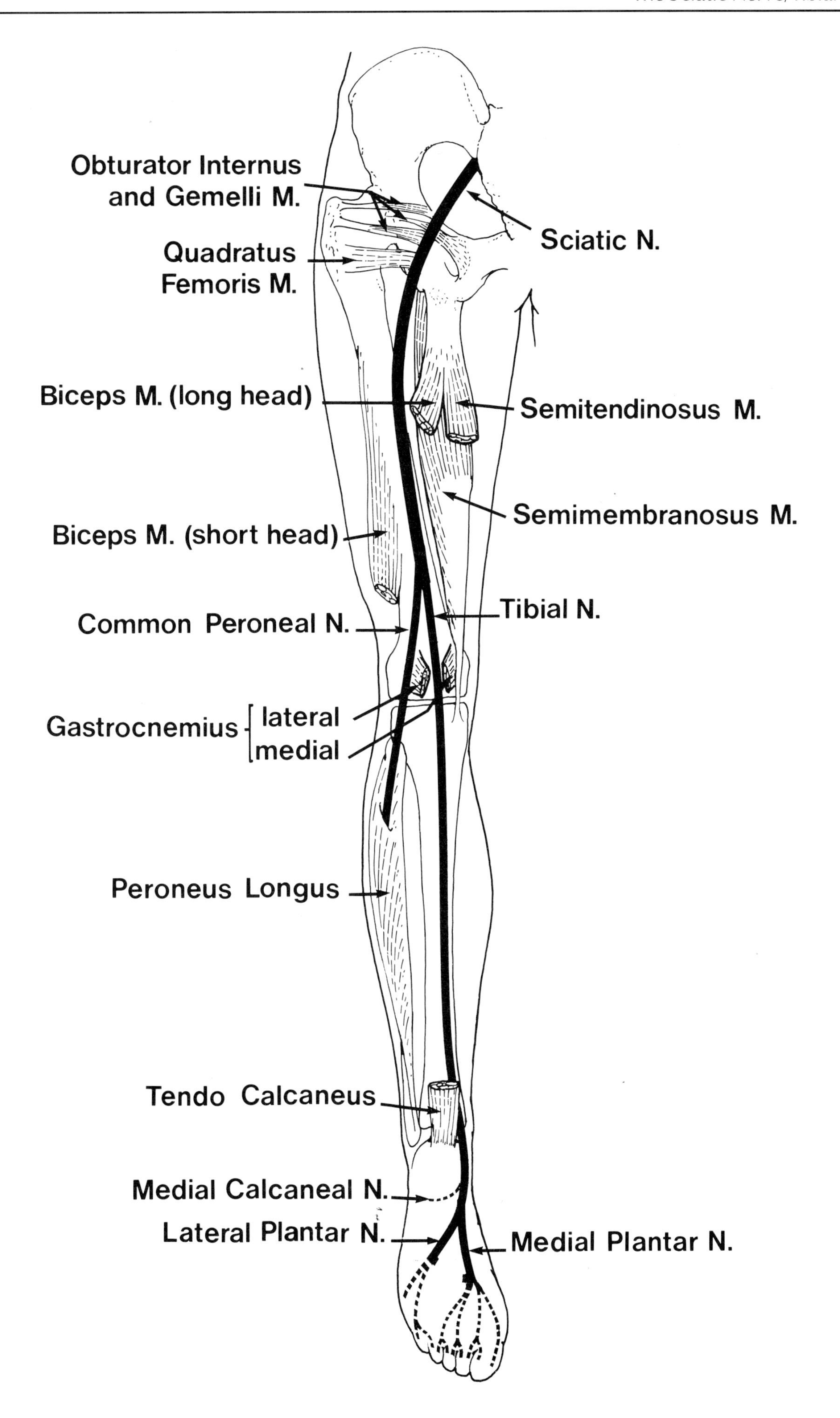
Obturator Internus
and Gemelli M.
Sciatic N.
Quadratus
Femoris M.
Biceps M. (long head)
Semitendinosus M.
Semimembranosus M.
Biceps M. (short head)
Tibial N.
Common Peroneal N.
Gastrocnemius
lateral
medial
Peroneus Longus
Tendo Calcaneus
Medial Calcaneal N.
Lateral Plantar N.
Medial Plantar N.

THE COMMON PERONEAL NERVE

Root Value

L4, 5, S1, 2

Course and Surface Marking

The common peroneal nerve is the smaller terminal division of the sciatic nerve. It is represented by a line commencing at the superior angle of the popliteal fossa which runs along the medial aspect of the biceps tendon to the back of the head of the fibula and then curves laterally around the neck of the fibula where the nerve ends, in the substance of the peroneus longus muscle, by dividing into the deep and superficial peroneal nerves.

The nerve is easily palpated through the skin at the back of the femoral condyle on the medial side of the biceps tendon. It may also be felt over the neck of the fibula where it is most susceptible to damage, particularly from tight plaster casts and 'bumper' fractures of the neck of the fibula.

Effects of Section

Damage to the nerve causes foot drop due to paralysis of the ankle and foot extensors. There is also a degree of inversion due to paralysis of the peroneal muscles with unopposed action of the foot flexors and invertors.

Anaesthesia occurs over the anterior and lateral aspects of the leg and foot as shown in the diagram.

Branches

Lateral cutaneous nerve of the calf
Sural communicating branch
Deep and superficial peroneal nerves

THE DEEP PERONEAL NERVE

This nerve is represented by a line beginning on the lateral aspect of the neck of the fibula which runs downwards and medially for 2.5 cm and then directly to a point midway between the malleoli. From here it runs forwards for 2.5 cm and divides into medial and lateral terminal branches.

The medial terminal branch is cutaneous and is marked by a line running directly to the web between the first and second toes.

The lateral terminal branch runs in a lateral direction from the point of division as far as the lateral margin of the extensor digitorum longus tendons. It supplies the extensor digitorum brevis and the tarsal and metatarsal joints.

Muscles Supplied

The muscles of dorsiflexion
- Extensor digitorum longus
- Peroneus tertius
- Extensor hallucis longus
- Tibialis anterior

THE SUPERFICIAL PERONEAL NERVE

This may be marked by a line beginning on the lateral aspect of the neck of the fibula and drawn to a point at the junction of the distal and middle thirds of the leg immediately in front of the fibula. Here the nerve divides into medial and lateral terminal branches, the final divisions of which are shown in the diagram. The nerve, which initially lies deep, becomes subcutaneous between the peronei and the extensor digitorum longus muscles halfway down the leg and pierces the fascia immediately prior to its point of division.

Muscles Supplied

Peroneus longus
Peroneus brevis

Area of Cutaneous Supply

The skin of the distal two-thirds of the lateral aspect of the leg and the dorsum of the foot, apart from the small area of skin between the first and second toes supplied by the deep peroneal nerve.

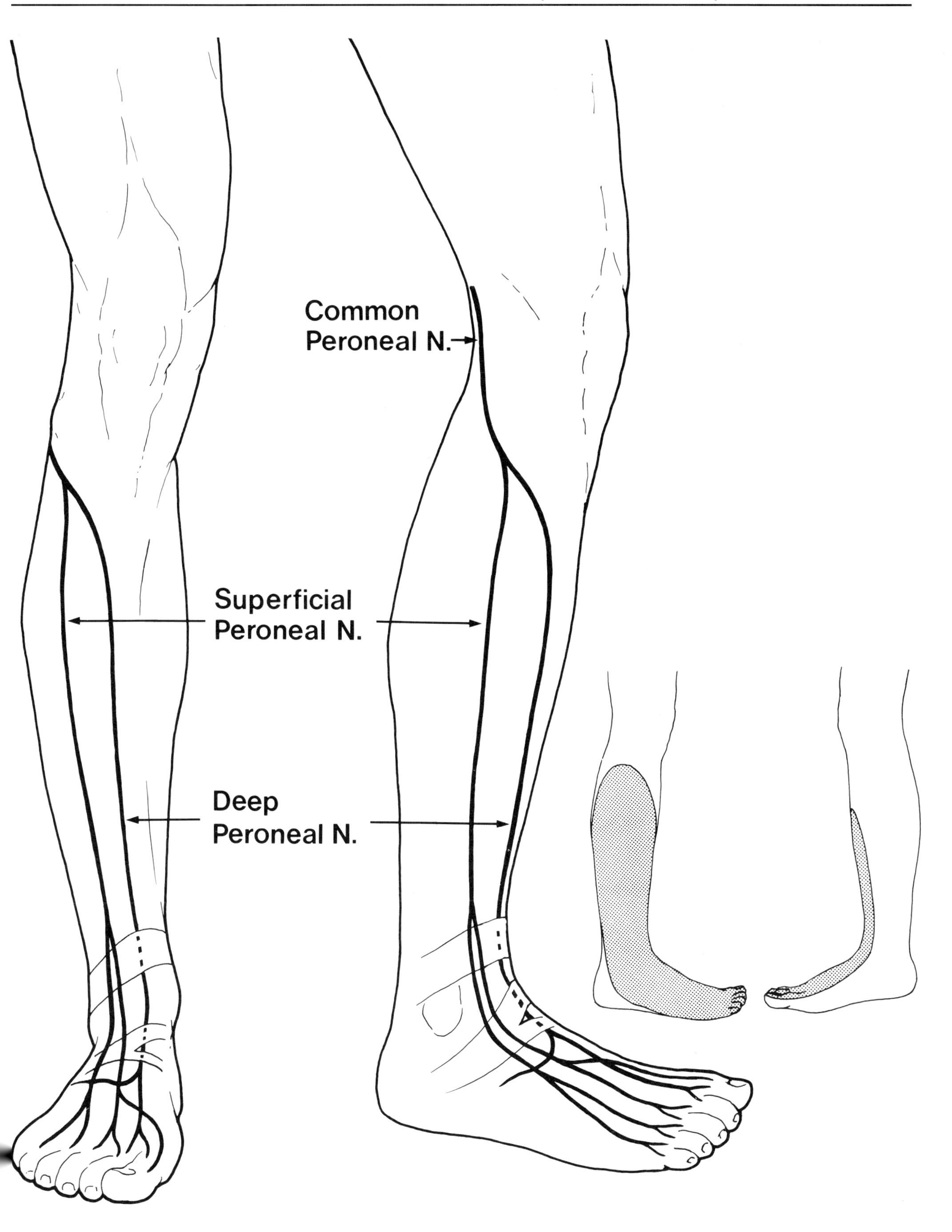
Common
Peroneal N.
Superficial
Peroneal N.
Deep
Peroneal N.

THE SOLE OF THE FOOT

The Skin

The skin of the sole of the foot is thick and hairless and contains many sweat glands. It is firmly anchored to the plantar aponeurosis by numerous fibrous septa which divide the subcutaneous fat into multiple small locules. These locules are filled with a rather fluid fat so that they act as shock absorbers during walking.

Flexure creases are seen at the sites of skin movements and individual dermatoglyphic patterns may be identified on the toes, the ball of the foot and the heel. The epidermis is thicker than elsewhere on the body, but in people who do not wear shoes it is particularly thick; indeed, it may become so thick that transverse cracks appear in the epidermis over the heel and ball of the foot and the dermatoglyphic pattern may then become obliterated.

Cutaneous Nerve Supply

Lateral plantar nerve [**A**]
Medial plantar nerve [**B**]
Medial calcaneal branch of tibial nerve [**C**]
Sural nerve [**D**]
Saphenous nerve [**E**]

Nerves and Arteries

Medial plantar nerve [**1**]
Muscles supplied
Abductor hallucis
Flexor hallucis brevis
Flexor digitorum brevis
First lumbrical

Medial plantar artery lying medial to nerve distally [**2**]

Medial calcaneal branches of posterior tibial nerve [**3**]

Lateral plantar artery lying lateral to nerve [**4**]

Lateral plantar nerve
Main trunk [**5**]
Muscles supplied
Flexor accessorius
Abductor digiti minimi

Superficial terminal branch [**6**]
Muscles supplied
Flexor digiti minimi
Interossei of fourth space

Deep terminal branch [**7**]
Muscles supplied
Abductor hallucis
Remaining interossei
Second, third and fourth lumbricals

Plantar arch lying distal to nerve [**8**]

Notice how the arteries are always lying 'outside' the nerves

Digital neurovascular bundle [**9**]. On the side of the toe the nerve is plantar to the artery.

Tendons in the Sole

Peroneus longus and distal sheath [**A**]

Insertion of peroneus longus into the base of the first metatarsal bone [B]

Tendons and sheath of flexor digitorum longus [**C**]

Tendons and sheath of flexor hallucis longus [**D**]

Plantar reflex. This is a reflex passing through the first and second sacral nerves.

A stroke with a fine blunt instrument such as a key or pencil forwards along the lateral aspect of the sole elicits a flexor movement of the big toe. However, in certain circumstances, such as an infant, deep sleep and upper motor neurone lesions, this flexor response is replaced with an extensor movement, otherwise known as the Babinski sign.

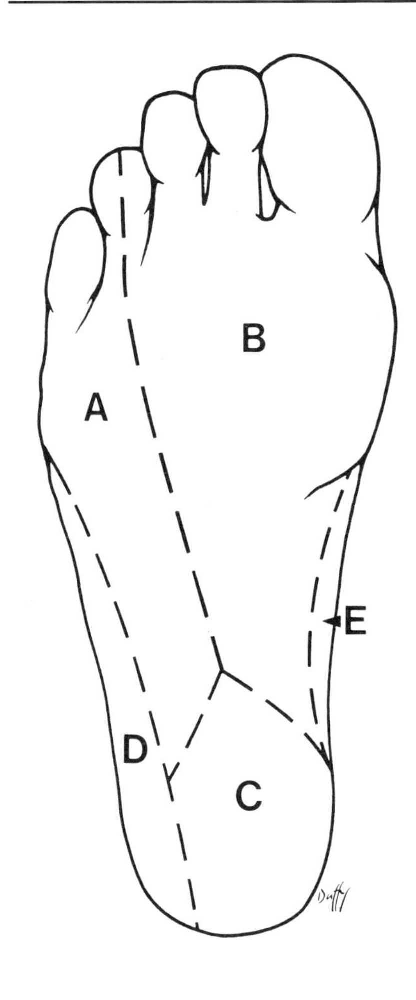

Cutaneous Nerves

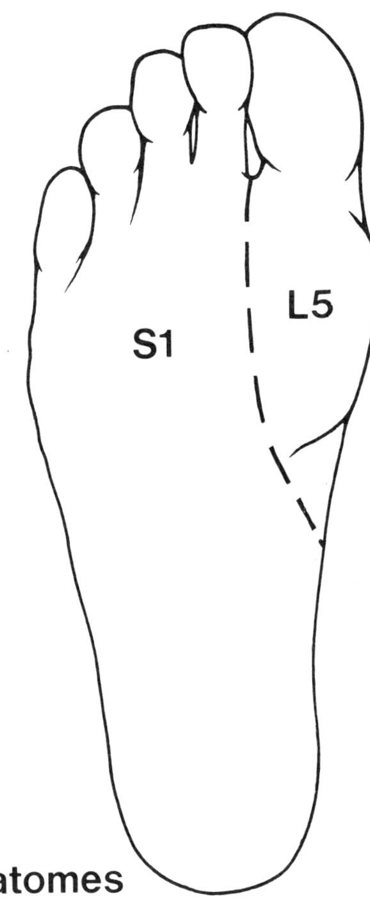

Dermatomes

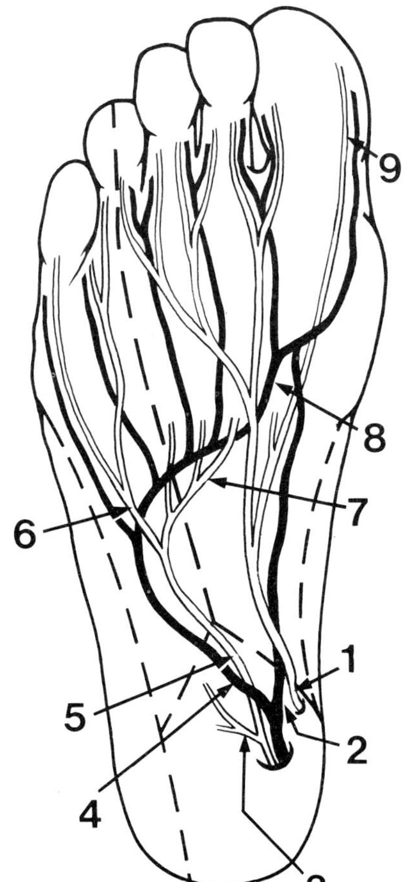

Arteries and Nerves

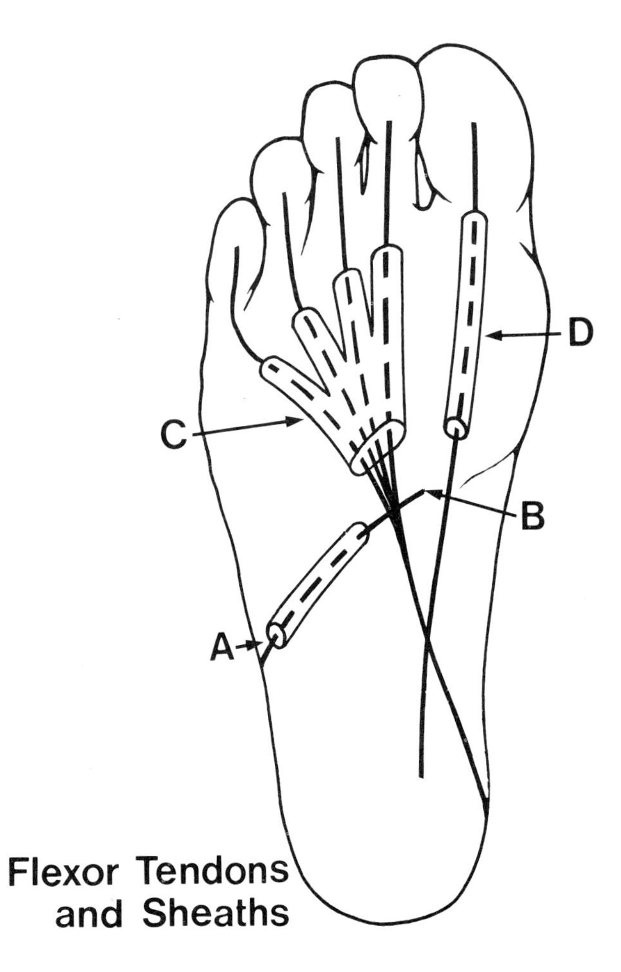

Flexor Tendons and Sheaths

10. The Head and Neck

BONY LANDMARKS OF THE HEAD

The frontal eminence [**1**] is the area of maximum convexity on either side of the forehead. It overlies part of the middle frontal gyrus.

The nasion [**2**] is the centre of the depression at the root of the nose where the frontal bones meet the nasal bones. It lies about 2 cm below the glabella.

The glabella [**3**] is the slight transverse elevation above the root of the nose. It lies in the midline opposite the origin of the superior sagittal sinus.

The superciliary arch [**4**] begins at the glabella and curves upwards and laterally above the medial half of the eyebrow. The medial part overlies the frontal air sinus on either side.

The supraorbital margin [**5**] lies below the eyebrow. It is indistinct medially but sharper laterally.

The supraorbital notch [**6**] may be palpated two finger breadths from the median plane along the supraorbital margin. It is more easily felt by applying pressure from below than from in front. It transmits the supraorbital nerve and vessels.

The zygomatic bone [**7**] forms the bony prominence of the cheek.

The maxilla [**8**] is the bone felt between the alar of the nose and the prominence of the cheek.

The infraorbital foramen [**9**] lies one finger's breadth from the side of the nose in the same vertical plane as the supraorbital notch and 0.5 cm below the infraorbital margin. It transmits the infraorbital nerve and vessels.

The mental foramen [**10**] may be marked in the same vertical plane as the supraorbital notch and infraorbital foramen about two finger breadths from the median plane and one finger's breadth above the lower border of the mandible.

The anterior nasal spine [**11**] may be felt indistinctly high in the philtrum between the nostrils.

The articulation between the frontal process of the zygomatic bone and the zygomatic process of the frontal bone [**12**] may be felt in the orbital margin just above the lateral end of the palpebral fissure.

Apertura piriformis [**13**]

The line of a vertical plane [**14**] through the supraorbital notch, the infraorbital foramen and the mental foramen.

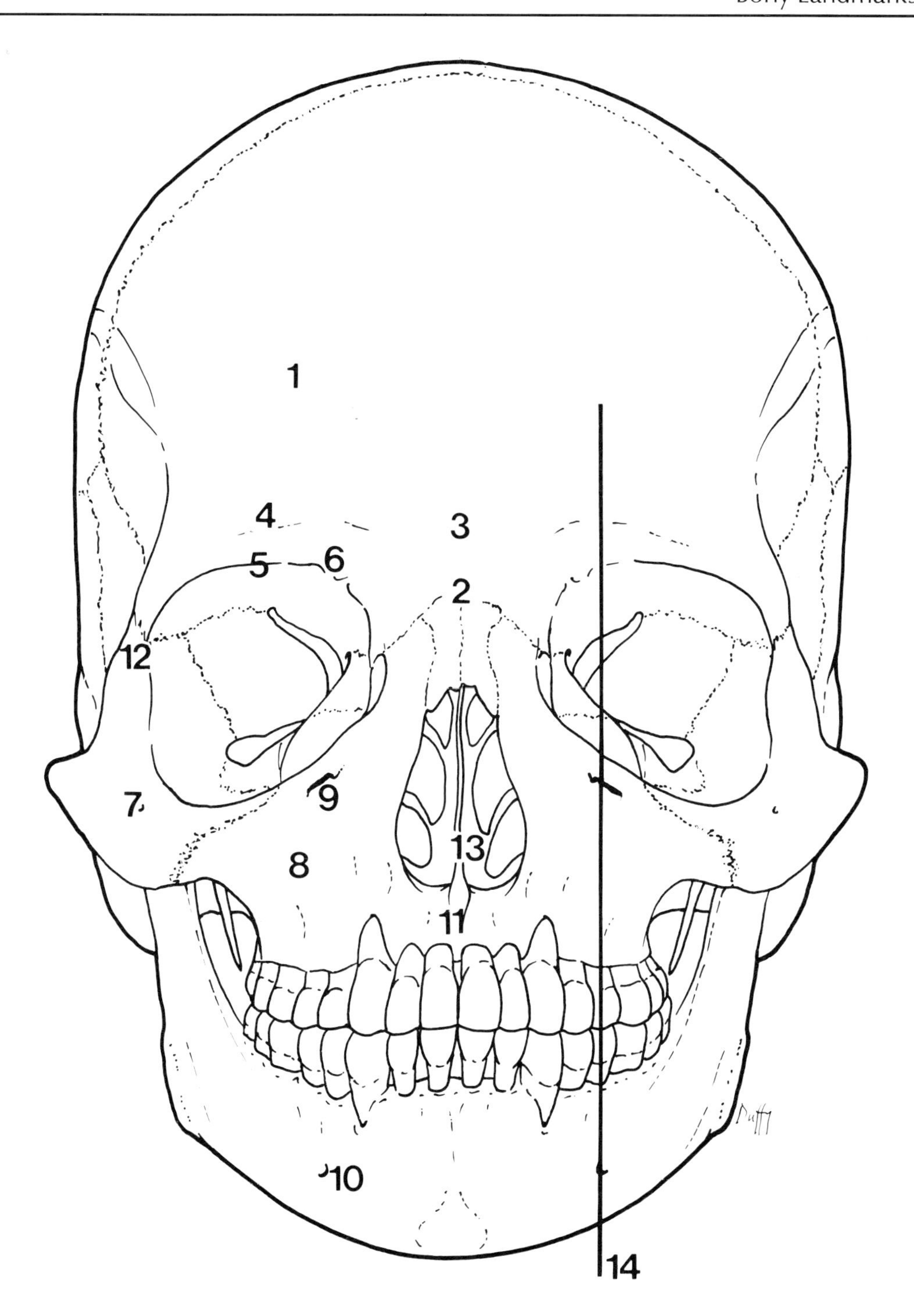
1
4
3
5
6
2
12
7
9
13
8
11
10
14

Lateral Aspect

The zygomatic arch [**1**] is formed by the zygomatic bone and the zygomatic process of the temporal bone. It spans the interval between the tragus and the bony prominence of the cheek. The lower border is obscured by the masseter and the upper border is difficult to feel because of the firm attachment of the temporalis fascia.

The jugal point [**2**] is an easily palpated angularity between the zygomatic arch and the frontal process of the zygomatic bone.

The sphenopalatine fossa lies 5 cm deep to the zygomatic arch at a point a little below and behind the jugal point. The fossa contains part of the maxillary nerve, the sphenopalatine ganglion and its branches, and the terminal branches of the maxillary artery.

The zygomaticofacial foramen [**3**] lies approximately at the junction of the vertical and horizontal lines in the planes of the lateral and inferior orbital margins respectively.

The tubercle of the root of the zygoma [**4**] is not easily felt, but the position is easily estimated. When the mouth is closed it lies just in front of the head of the mandible; when the mouth is wide open it lies above the head of the mandible. It is the surface marking of the foramen ovale through which the mandibular nerve leaves the cranium.

The pre-auricular point [**5**] is the point on the posterior root of the zygoma immediately anterior to the attachment of the auricle. The superficial temporal pulse may be felt here.

The suprameatal triangle (of Macewen) [**6**] is a small depression above and behind the external acoustic meatus, below the anterior end of the supramastoid crest. It is bordered above by the posterior root of the zygoma, anteriorly by the postero-superior border of the external acoustic meatus, and posteriorly by a vertical tangent from the posterior border of the meatus. Although covered by the crus of the helix, if the auricle is pulled forwards the depression can be felt from behind. The triangle is surgically important because it overlies the mastoid antrum. The superior border is level with the floor of the middle cranial fossa, the antero-inferior border lies along the course of the descending part of the canal for the facial nerve, and the posterior vertical border lies just in front of the course of the sigmoid sinus.

The supramastoid crest [**7**] is the continuation of the posterior root of the zygoma as a faint ridge above the acoustic meatus which curves upwards and backwards to meet the superior temporal line.

The superior temporal line [**8**] runs from the zygomatic process of the frontal bone upwards and backwards to a point a little below the parietal eminence. It then curves downwards and forwards to meet the supramastoid crest 3 cm above and behind the auditory meatus.

The parietal eminence [**9**] is best felt with the flat of the hand. It overlies the supramarginal gyrus which curves around the posterior limit of the lateral sulcus. It is from this point that the parietal bone begins to ossify centrifugally during the seventh week of intrauterine life.

The mastoid part of the temporal bone [**10**] contains air cells which develop after birth as diverticulae from the mastoid antrum.

The mastoid process [**11**] is an extension of the mastoid temporal bone which contains air cells. It is absent in infants, but is readily palpable in adults.

The inion [**12**] lies opposite the interval between the occipital poles of the cerebrum and marks the termination of the straight and superior sagittal sinuses and the commencement of the transverse venous sinuses.

The superior nuchal line [**13**] arches laterally from the inion on either side. It overlies part of the transverse sinus and the attachment of the tentorium cerebelli.

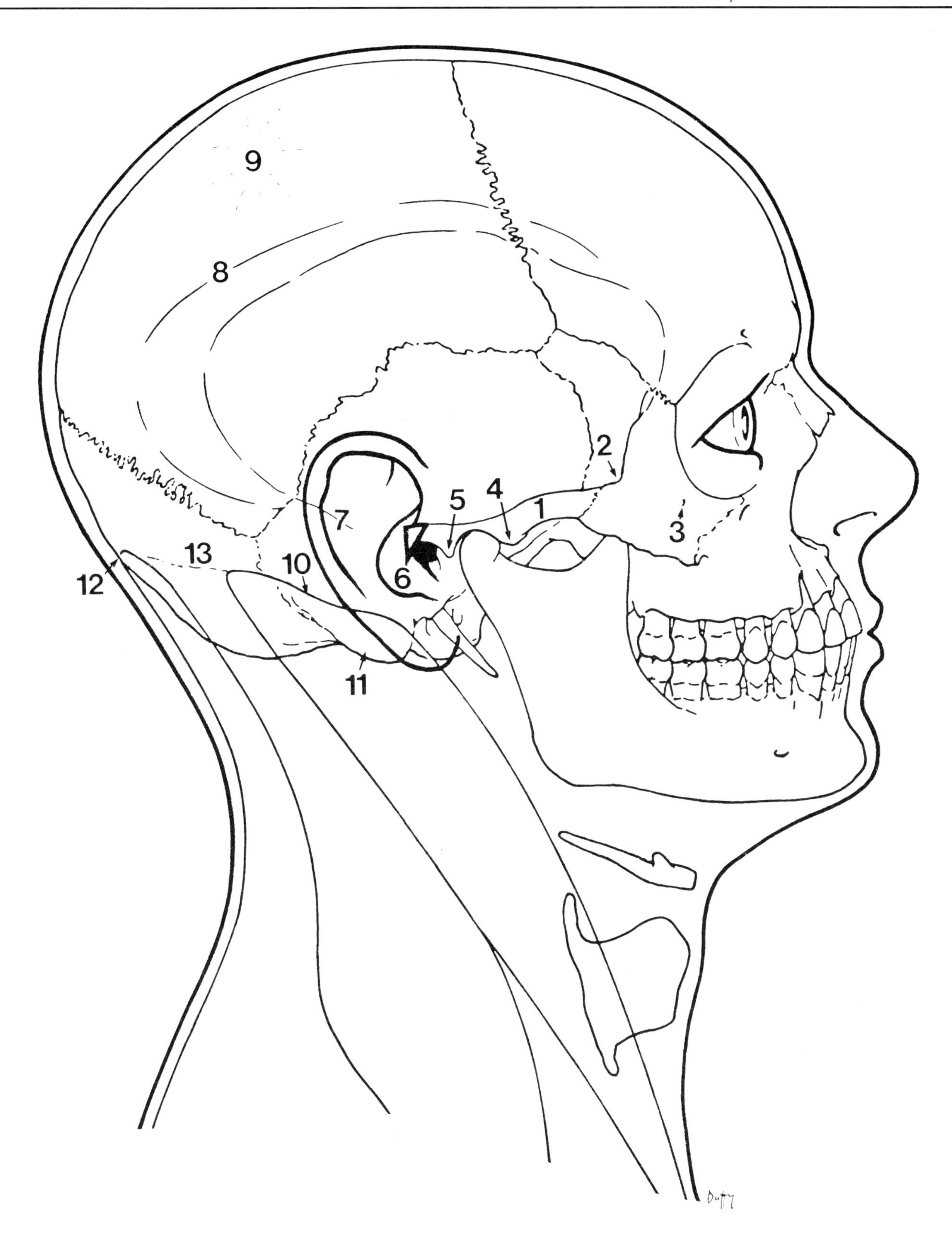
9
8
2
5
4
1
7
3
13
10
12
6
11
Duffy

RADIOLOGY OF THE SKULL

A series of skull x-rays is frequently requested in cases of head injury, for clinical or medico-legal reasons. There are several basic views of the skull, but usually only three views are routinely taken, unless further views are indicated on clinical grounds.

The three views are:
1. A lateral projection.
2. A postero-anterior (occipitofrontal) or anteroposterior (fronto-occipital) projection. The latter is usually reserved for situations when the patient cannot stand and it is easier or safer to slip the film casette beneath the patient's head while he remains recumbent.
3. A Towne's view, which is an antero-posterior view taken at 35° to the orbito-meatal line.

The orbitometal line is an imaginary line on the patient, drawn between the outer canthus of the eye and the external acoustic meatus.

Lateral View

This view is taken with the interpupillary line at right angles to the film and the beam centred on the mid-point of an imaginary line between the prominence of the forehead and the external occipital protuberance.

This is probably the most important of all the routine skull views especially in the casualty department where it is particularly useful for identifying fractures of the skull vault. Some radiology departments perform both right and left lateral views routinely, but if only one view is taken then it should be performed with the affected side against the film.

Key to the Lateral View

Skull tables [**1**]—the outer and inner cortical layers appear as dense white lines. The diploë lies between the cortical layers. In the young, the tables are difficult to identify because the diploë has not yet developed; equally it may be obliterated in old age.

The lambdoid suture [**2**] is nearly always recognisable even in old age.

The coronal suture [**3**] is often very difficult to see, particularly in the elderly when it may be united.

The middle meningeal groove [**4**]—when present appears as a dark line ascending behind the zigzag line of the coronal suture. It is caused by pressure on the inner skull table from the *anterior branches* of the *middle meningeal vessels*. It varies in width from 1–2 mm.

Diploic veins [**5**] are dark bands, 1–3 mm wide, which are usually broader and less well defined than the meningeal grooves and vary in width throughout their course. A converging group is frequently conspicuous in the parietal bone. They are sometimes mistaken for fractures but their irregularity and poor definition should distinguish them from the crisp, fine, well demarcated and often straight line formed by a vault fracture.

The position of the pineal body [**6**]. The pineal body may calcify in early adulthood. On the radiograph it varies in size from a small spot to a mottled opacity almost 10 mm in diameter.

The pituitary fossa [**7**] is lined by a thin layer of cortical bone, the lamina dura, which appears as a thin white line. This line may be lost or broken in certain pathological conditions.

The sphenoidal air sinus [**8**]. A fluid level due to blood may be seen in this sinus in some fractures of the skull.

The frontal air sinus [**9**]

The cribriform plate of the ethmoid bone [**10**]

The orbital roof [**11**]

Lateral orbital rim [**12**]

The zygomatic process of the maxilla [**13**]. The cortical bone of this strong buttress casts a 'U'-shaped shadow. The apex of the maxillary sinus projects laterally between the two limbs of the 'U'.

Hard palate [**14**]

Posterior wall of the maxillary sinus [**15**]

The superimposed pterygoid plates [**16**] are commonly seen only as a hazy white shadow posterior to the maxillary sinus.

The pterygopalatine fossa [**17**]

Mandibular ramus [**18**]

External acoustic meatus [**19**]

The pinnna [**20**]. The thick helix of the pinna often casts a circular shadow over the thin squamous temporal bone.

The petrous ridge [**21**]

The atlanto-occipital joint [**22**]

Anterior arch of the atlas [**23**]

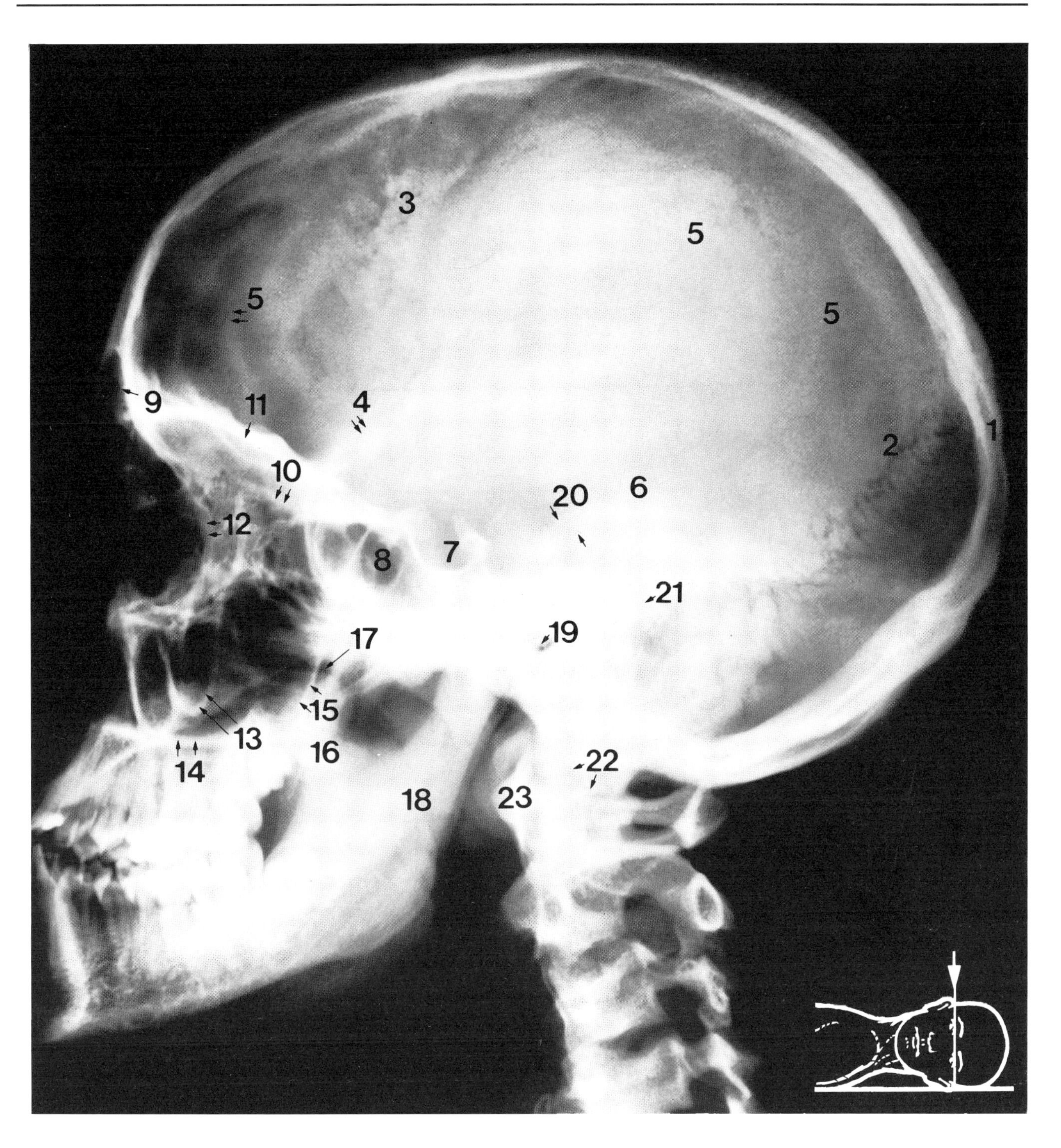

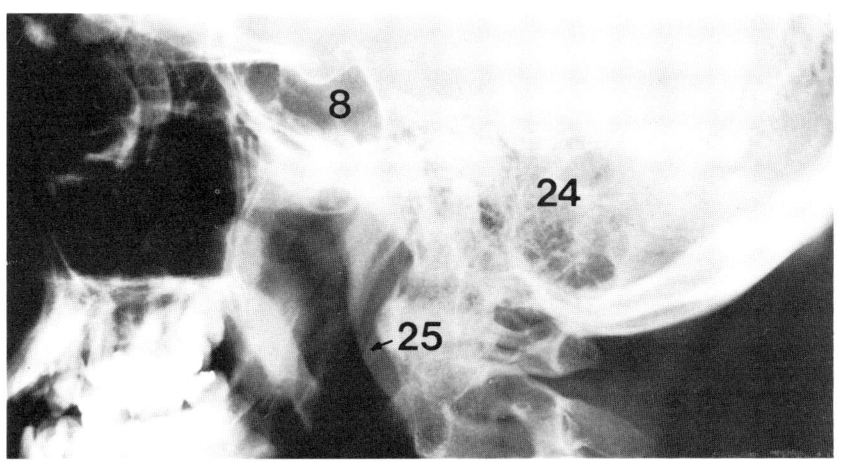

Inset

This has been included to show the mottled appearance of the mastoid air cells [24], and the large thin walled sphenoidal air sinus [8], and the styloid process [25] projecting in front of anterior arch of the atlas.

20° Occipitofrontal or Posteroanterior View

The view is taken with the subject's face against the film and the x-ray beam passing at a 20° angle to the orbito-meatal line, which is a line between the outer canthus of the eye and the tragus of the ear.

It is a useful view for assessing the overall symmetry of the skull and it provides a good view of the anterior and posterior aspects of the calvarium which is useful when looking for fractures in cases of head trauma. In addition, if the pineal gland is calcified it should lie in the midline and a shift of greater than 3 mm to one side implies intracranial midline shift.

This projection also provides a reasonable view of the orbital margins, the nasal cavity and the maxillary antra.

Key to the Posteroanterior View

Sagittal suture [**1**]

Lambdoid suture [**2**]

Wall of the frontal sinus [**3**]

Calcified midline septum between the frontal sinuses [**4**]

Superior orbital margin [**5**]

Lesser wing of sphenoid [**6**]

The orbital canal [**7**] in the lesser wing of sphenoid at the apex of the orbit.

The superior orbital fissure [**8**] between the greater and lesser wings of the sphenoid. Beneath the fissure the greater wing of the sphenoid is thin and casts only a light shadow. Here it forms the posterior part of the lateral wall of the orbit, and separates the orbit from the middle cranial fossa.

The medial wall of the temporal fossa [**9**] formed by the greater wing of the sphenoid.

Nasal septum [**10**]

Inferior concha [**11**]

Odontoid peg [**12**]

Atlantoaxial joint [**13**]—Notice the plane surfaces.

Upper central incisor [**14**]

Lower central incisor [**15**]

Lower third molar [**16**]

The outline of the maxillary sinus [**17**]

Foramen rotundum [**18**]

The petrous ridge [**19**] formed by the superior border of the petrous part of the temporal bone which obscures the lower part of the orbit.

The jugum sphenoidale [**20**] in continuity with the lesser wings of the sphenoid.

Crista galli [**21**]

Mastoid air cells [**22**]

The floor of the posterior cranial fossa [**23**] seen through the maxillary sinus.

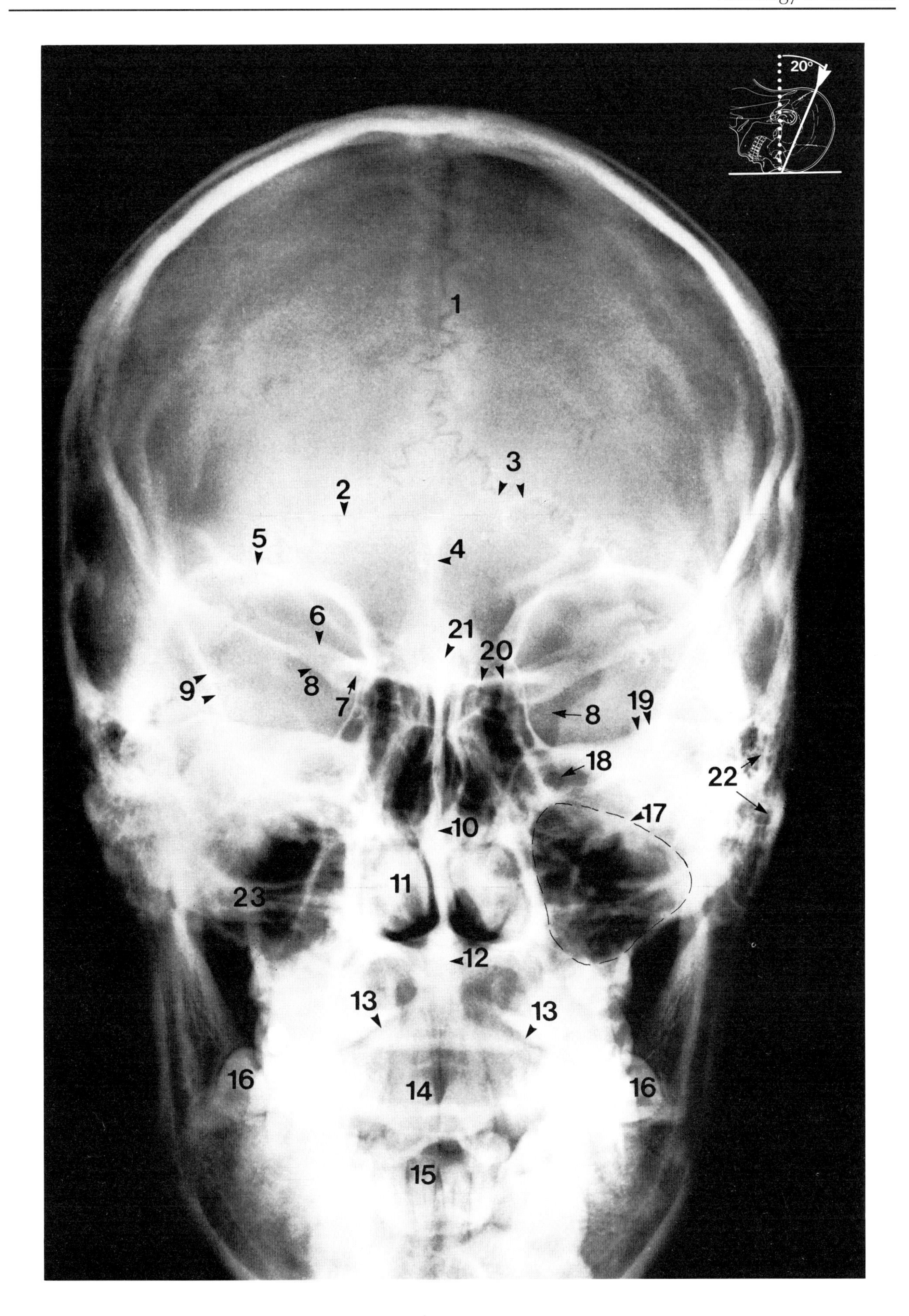
20°
1
2
3
4
5
6
7
8
9
10
11
12
13
13
14
15
16
16
17
18
19
20
21
22
23

35° Fronto-Occipital or Towne's View

This view is particularly useful for looking for fractures of the calvarium. It is also helpful in assessing symmetry and midline shift. It should project the dorsum sellae into the foramen magnum and is therefore often used in assessing the pituitary fossa.

This view is an inclined projection with the x-ray beam at 35° to the orbito-meatal line with the film against the back of the patient's head. To help understand this projection take a dry skull with the cap removed and hold it with its base towards you and the face pointing downwards. Now look through the foramen magnum and incline the skull so that the posterior clinoid processes are seen in the centre of the foramen magnum. A two-dimensional projection of this view is what is seen on the radiograph. Remember, however, that in the radiograph of the living person the upper cervical spine and face will be superimposed.

Key to the Towne's View

Lambda [1]

Lambdoid suture [2] separating the parietal and occipital bones

Groove for the sigmoid sinus [3]

Petrous ridge [4] formed by the posterior border of the petrous temporal bone

Arcuate eminence [5] overlying the superior semi-circular canal

Mastoid air cells [6] producing a mottled appearance

Foramen magnum [7]

Jugular tubercle [8] on the supero-lateral surface of the basiocciput

Arch of the atlas [9]

Arch of the axis [10]

The lateral wall of the orbit [11] formed by the greater wing of the sphenoid

The posterior wall of the maxillary sinus [12]

The inferior orbital fissure [13]

Nasal septum [14]

The articular processes of the superior cervical vertebrae [15] superimposed on the facial bones

The upper third molars [16]

The zygomatic process of the temporal bone [17] forming the posterior part of the zygomatic arch.

Mandibular ramus [18]

Head of mandible [19]

The temporomandibular joint space [20]

The region of the vestibule and cochlea [21]

The occipitomastoid suture [22] which is sometimes mistaken for a fracture

Sulcus for the transverse sinus [23]

Venous lakes [24]

Inset

The inset is part of a radiograph taken of a dried skull in the Towne's position and has been included to show more clearly the projection of the *dorsum sellae* [25] into the outline of the foramen magnum.

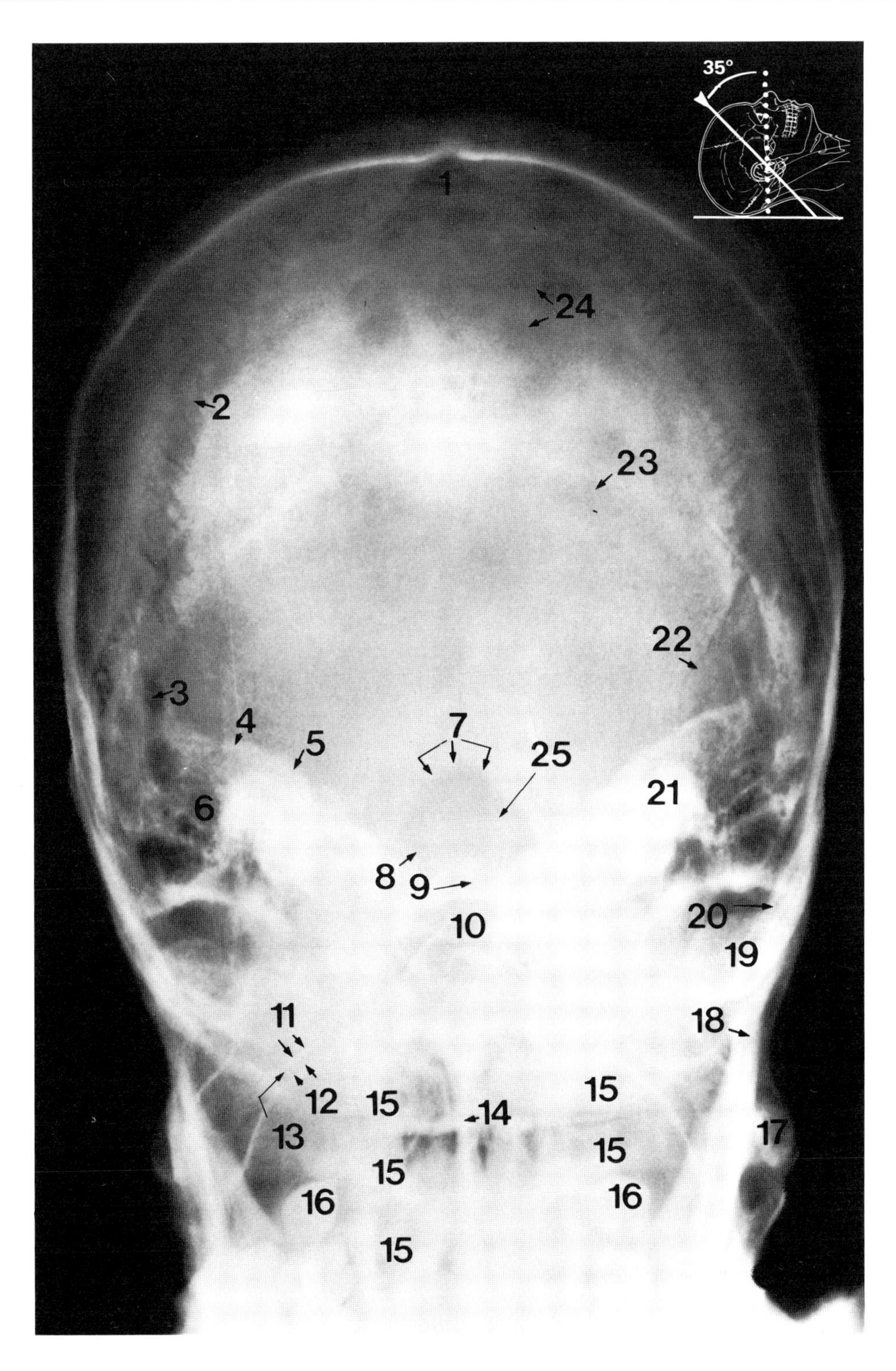

35°
1
24
2
23
22
3
4
5
7
25
21
6
8
9
20
10
19
11
18
12
15
14
15
13
17
15
15
16
16
15

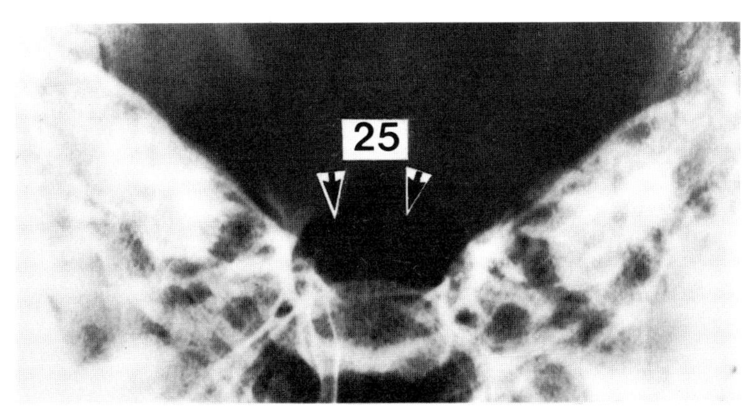

25

Submentovertical, Axial or Base View

This is a good view for looking at the base of the skull. It is taken at 90° to the orbito-meatal line with the x-ray beam entering through the neck and the film behind the patient's head.

This view will be best understood if it is studied in conjunction with the base of a dried skull, otherwise it is difficult to appreciate many of the superimpositions on the two-dimensional projection.

Key

Vomer **[1]**

Mandibular arch **[2]**

Posterior wall of maxillary sinus **[3]**

The greater wing of the sphenoid **[4]** forming the lateral wall of the orbit. The gap between this line and the line formed by the thin posterior wall of the maxillary sinus represents the inferior orbital fissure. It is not seen clearly in this particular radiograph.

The posterior limit of the ethmoid air cells **[5]**

The sphenoidal air sinuses **[6]**

Apex of the petrous temporal bone **[7]**

The foramen lacerum **[8]** which is bounded in front by the posterolateral part of the body of the sphenoid bone and the roots of the pterygoid processes and the greater wing, laterally by the apex of the petrous temporal bone and medially and posteriorly by the basiocciput.

The foramen ovale **[9]**

The foramen spinosum **[10]**

The carotid canal **[11]**

The anterior arch of the atlas **[12]**

The odontoid peg **[13]**

The posterior arch of the atlas **[14]**

Outline of the air-filled trachea **[15]** overlying the cervical vertebrae. Notice the narrowing at the larynx which is seen just above the posterior arch of the atlas.

Foramen transversarium in the lateral mass of the atlas **[16]**

Hazy mass of the atlanto-occipital joint **[17]**

Mottled mass of mastoid air cells **[18]**

External acoustic meatus **[19]**

The angle of the mandible **[20]** superimposed on the *head of the mandible*

The zygomatic arch **[21]**

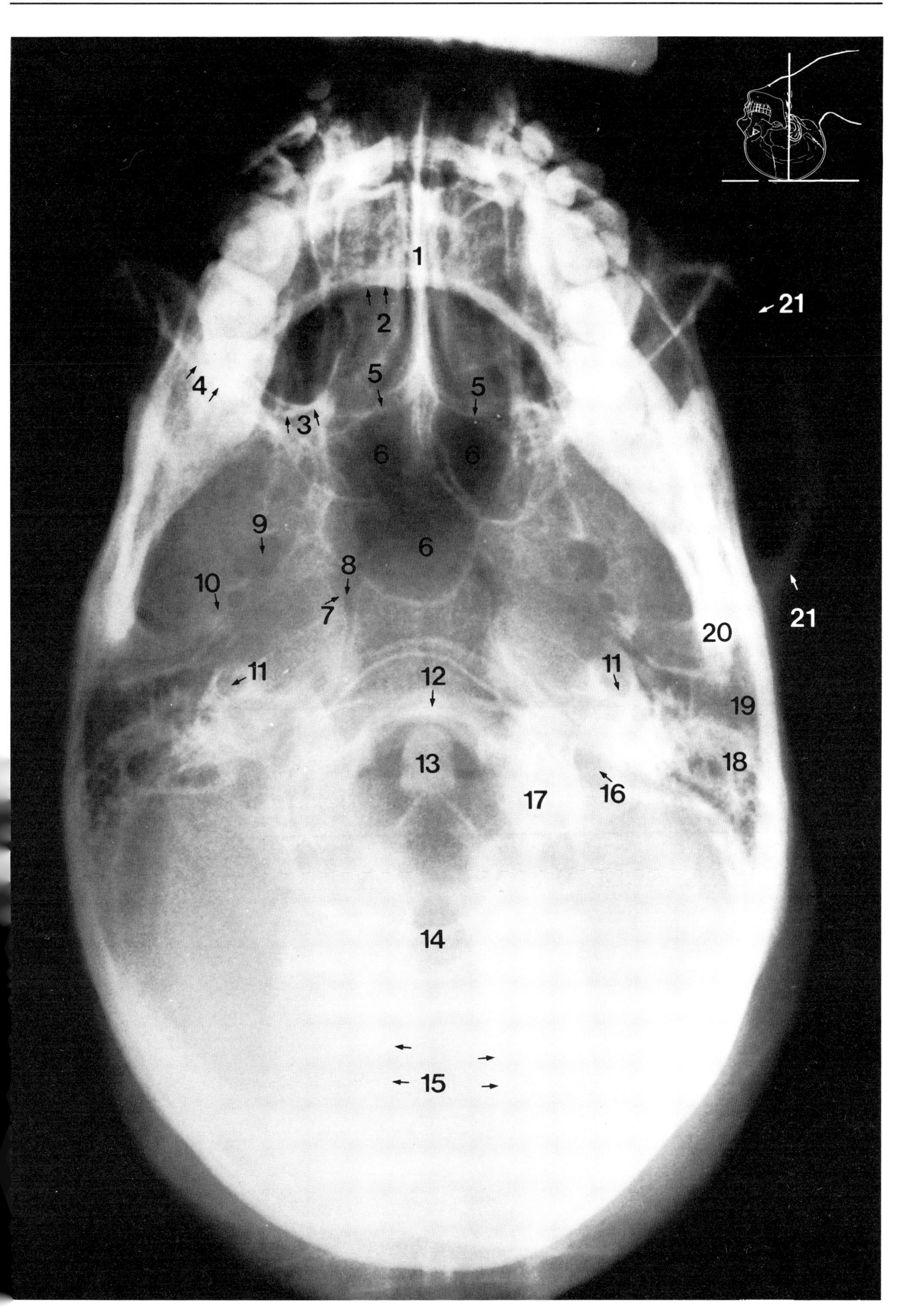
1
2
4
3
5
5
6
6
6
9
10
8
7
11
11
12
13
14
15
16
17
18
19
20
21
21

BONY LANDMARKS IN THE NECK

Spines of the Cervical Vertebrae

See The Back (p. 10).

The Cervical Transverse Processes

C1 The transverse process of the atlas lies just below the tip of the mastoid process. It is difficult to feel, but it may be palpated in the hollow below the ear. The deep palpation required may irritate the great auricular nerve which runs in the superficial fascia overlying the parotid gland with resultant pain or paraesthesia.

The positions of the transverse processes of the other cervical vertebrae can be estimated with less certainty. They form the bony resistance felt on palpation along a line between the tip of the mastoid process and the tip of the shoulder.

Approximate levels of the cervical transverse processes:

C2 Level with the angle of the jaw
C3 Just above the hyoid bone
C4 Level with the upper part of the thyroid cartilage
C5 Level with the middle part of the thyroid cartilage
C6 Level with the cricoid cartilage. It is level with the transition of larynx to trachea and pharynx to oesophagus

The Hyoid Bone [H]

This lies opposite the upper border of C4. With the head in the anatomical position, the hyoid may be moved from side to side between the thumb and index finger, about 1 cm below the level of the angle of the mandible. The tips of the greater horns may be felt below the angle of the mandible close to the anterior border of the sternocleidomastoid muscle and immediately in front of the vertebral column.

The Thyroid Cartilage [T]

The laryngeal prominence (Adam's apple) is usually visible in the male, but may be less evident in women and children because of the small antero-posterior diameter of the larynx. It is easily palpated. The vocal cords lie at the level of the junction of the upper and middle thirds of the symphysis.

The Cricoid Cartilage [C]

This may be felt as a narrow horizontal bar just beneath the lower border of the thyroid cartilage. It is best felt by an up and down movement of the finger.

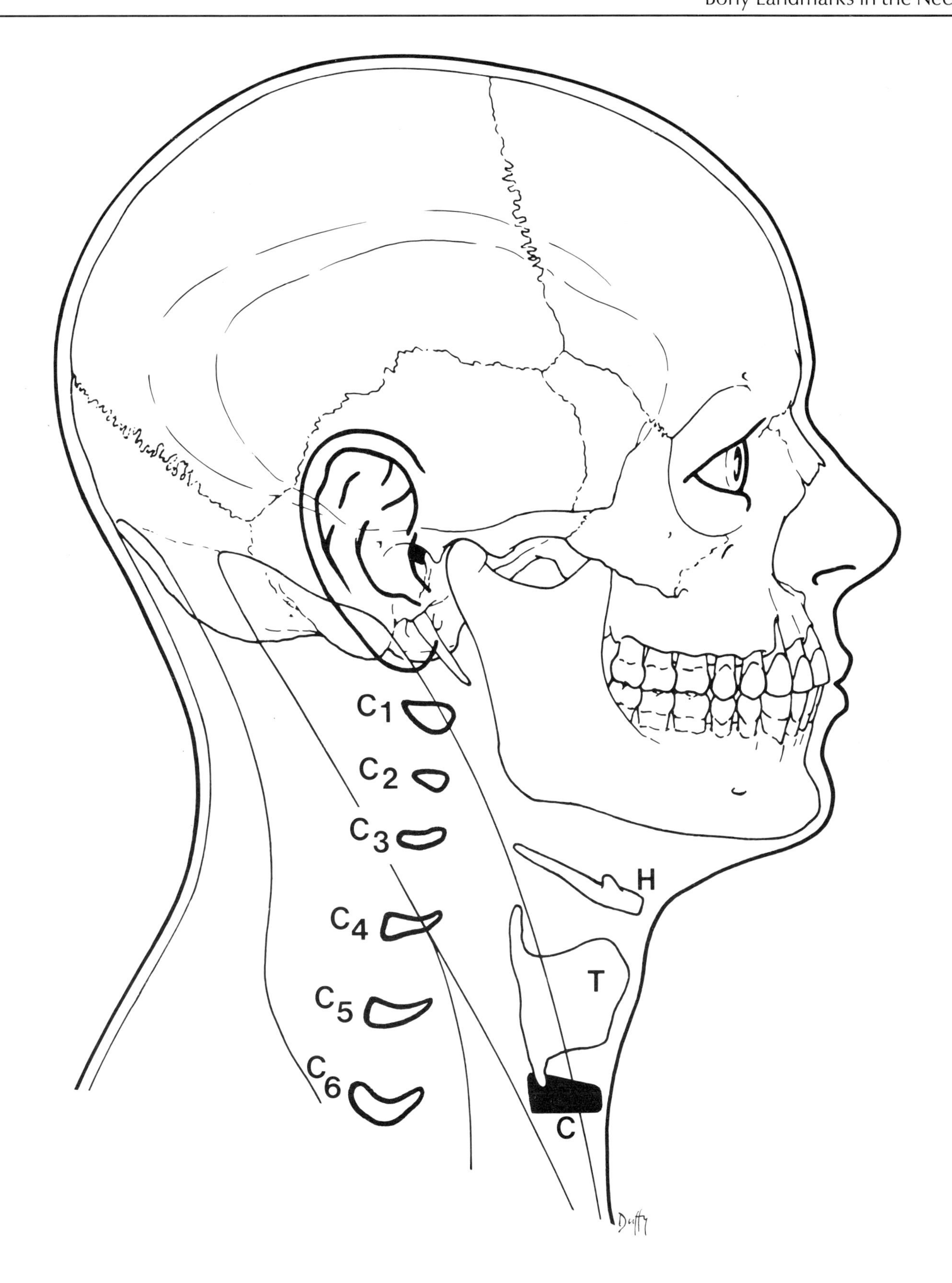
C1
C2
C3
C4
C5
C6
H
T
C
Duffy

RADIOLOGY OF THE CERVICAL SPINE

Routine radiographs of the cervical spine include antero-posterior and lateral projections plus two oblique views to show the intervertebral foramina of each side.

Antero-posterior View

This is taken with the head and cervical spine fixed in the neutral position and the jaw moving to prevent the mandibular shadow obscuring the atlas and axis.

In suspected fractures of the odontoid peg or suspected subluxation of the atlanto-axial joint, the view of the atlas and axis is clearest if taken through the open mouth, or with the jaw moving to blur the mandible.

Air in the trachea and larynx appears as a dark shadow overlying the lower cervical spines.

When studying a film consider the possibility of a cervical rib projecting from the seventh cervical vertebra.

The thyroid cartilage frequently shows signs of calcification or ossification in the older subject.

Odontoid peg [**1**]
Anterior arch of atlas [**2**]
Transverse process of atlas [**3**]
Posterior arch of atlas [**4**]
Atlanto-axial joint—plane synovial [**5**]
Spinous process of axis [**6**]
Cervical spinous processes [**7**]
Cervical transverse processes [**8**]
Joint of Luschka [**9**] a plane synovial joint between the bodies of adjacent vertebrae.
Transverse process of C7 [**10**]. In some radiographs it is possible to see the foramen transversarium in this transverse process
Transverse process of T1 [**11**]
First rib [**12**]

Lateral View

This should *always* include the whole spine from the base of the occiput to the body of the first thoracic vertebra. Horror stories abound of casualty officers missing fractures because the upper or lower extremities of the cervical spine were not visible on the radiograph.

The spinal canal is easily visible and the posterior borders of the cervical vertebral bodies which form the anterior wall of the canal should form a smooth continuous line. Similarly, the posterior limit of the canal should form a continuous line. Any irregularity in either line may indicate a fracture or subluxation. The antero-posterior diameter of the spinal canal varies between 14 and 21 mm. Cord compression is likely if the measurement is less than 10 mm. In patients with rheumatoid arthritis lateral views in flexion and extension may be useful to expose otherwise covert subluxation of the atlanto-axial joint.

The normal radiological joint space between the posterior border of the anterior arch of the atlas and the odontoid peg is 2 mm.

Atlanto-occipital joint [**1**]
Anterior arch of atlas [**2**]
Odontoid peg [**3**]
Posterior tubercle and arch of atlas [**4**]
Transverse process of axis [**5**]
Inferior articular process [**6**]
Superior articular process [**7**]
Spinous process of C4 [**8**]. The upper spinous processes are often bifid
Lamina of C5 [**9**]
Pedicle of C7 [**10**]
Intervertebral disc between C6–C7 [**11**]
First rib [**12**]
Clavicle [**13**]

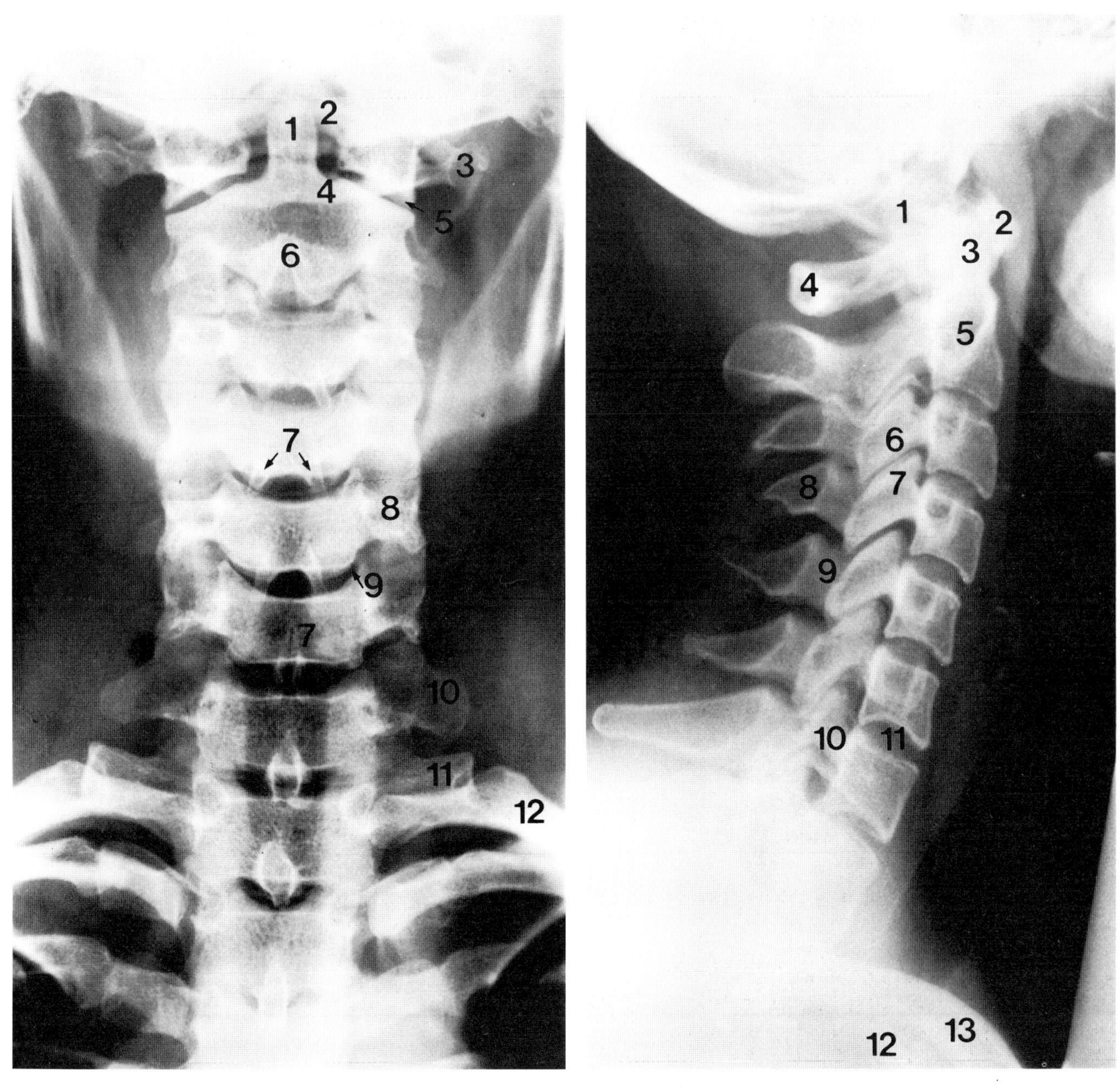

Anteroposterior view

Lateral view

Oblique Views

These views are taken with the subject turned at 45° from the antero-posterior position, i.e. halfway between the lateral and antero-posterior positions. The views are particularly useful for examining the intervertebral foramina. When the patient faces *right*, the *left* foramina come into view and vice versa. Likewise, the joints between the articular process are well visualised and fractures of the lamina may become apparent. If the diagram of the cervical vertebra on the facing page is studied, it can be seen that when the vertebra is rotated to the left the view of the left intervertebral foramen and left articular processes is obscured by the shadow of the vertebral body. However, the right intervertebral foramen and articular process cast an uninterrupted shadow onto the film.

Degenerative disease of the spine may result in the formation of osteophytes which may impinge on the cervical spinal nerve as it leaves the intervertebral foramen with resultant irritation or even compression. These changes may easily be detected in the oblique view.

Key to Oblique View

Occipital bone **[1]**
Tubercle of posterior arch of atlas **[2]**
Transverse process of atlas **[3]**
Anterior arch of atlas **[4]**
Odontoid peg **[5]**
Transverse process of axis **[6]**
Foramen transverium in the opposite transverse process of the axis **[7]**
Intervertebral foramina **[8]**
Spinous processes **[9]**, often bifid in the cervical spine
The laminae **[10]** are seen end on and therefore their cortical bone appears as dense ovals
Transverse process **[11]** lying beneath each intervertebral foramen
Transverse processes of the opposite side **[12]**
First rib **[13]**
Clavicle **[14]**
Hyoid bone **[15]**

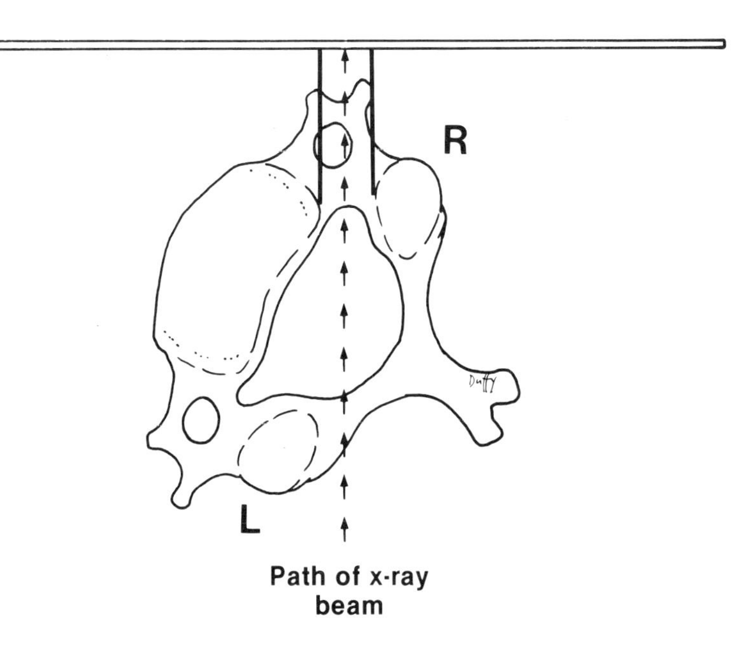
R
L
Path of x-ray beam

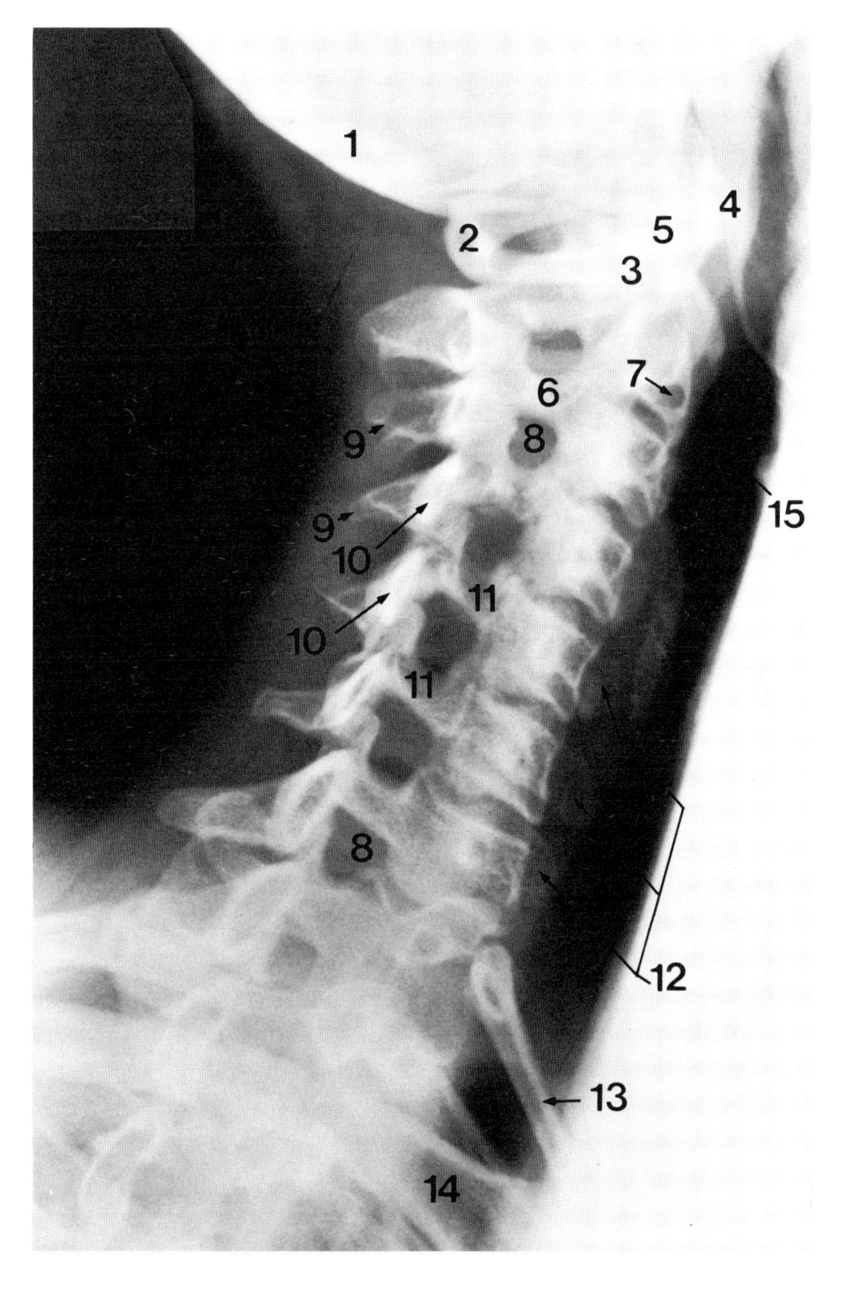
1
2
3
4
5
6
7
8
9
9
10
10
11
11
8
12
13
14
15

SUTURES AND FONTANELLES

The Adult

The bregma [1] is the meeting point of the parietal and frontal bones, and thus the sagittal and coronal sutures. It lies superiorly in the midline, 2.5 cm anterior to a line drawn in the perpendicular plane between one external acoustic meatus and the other.

The pterion [2] is an area about 2 cm in diamter where the greater wing of the sphenoid meets the parietal bone, separating the frontal bone and the squamous temporal bone as an H. It lies two finger breadths above the zygomatic arch and one finger's breadth behind the frontal process of the zygomatic bone. It is an important surface marking.

The coronal suture [3] runs between the bregma and the pterion of each side.

The lambda [4] is, as its name implies, an inverted V-shaped joint between the parietal bones and the apex of the occipital bone. It is about 5 cm above the inion.

The sagittal suture [5] runs in the midline between the bregma and the lambda.

The asterion [6] is where the parietal, occipital and mastoid temporal bones meet. It lies 1 cm above the midpoint of a line between the tragus and inion.

The lambdoid suture [7] slants down from the lambda to the asterion on either side.

The Newborn Infant

The metopic suture [8] runs from the nasion to the bregma or anterior fontanelle between the two frontal bones. It usually disappears during early childhood, but it may persist.

The anterior fontanelle [9] is an unossified diamond-shaped area in the region of the future bregma. Ossification spreads into it and is complete by the end of the second year in the normal child. The pulsation of intracranial arteries may be felt over the fontanelle, and any increase in intracranial pressure may cause the soft tissue of the fontanelle to bulge.

The posterior fontanelle [10] is a triangular-shaped area at the site of the lambdoid suture. Ossification from the bony margins produces closure before the end of the first year.

The different shapes of the two fontanelles allow the obstetrician to judge the position of a presenting fetal skull in the birth canal by vaginal examination.

Fontanelles are also present at the sites of the asterion and pterion and are known as the mastoid and sphenoid fontanelles respectively.

Mastoid fontanelle [11]

Sphenoid fontanelle [12]

Note the absence of the mastoid process in the newborn skull.

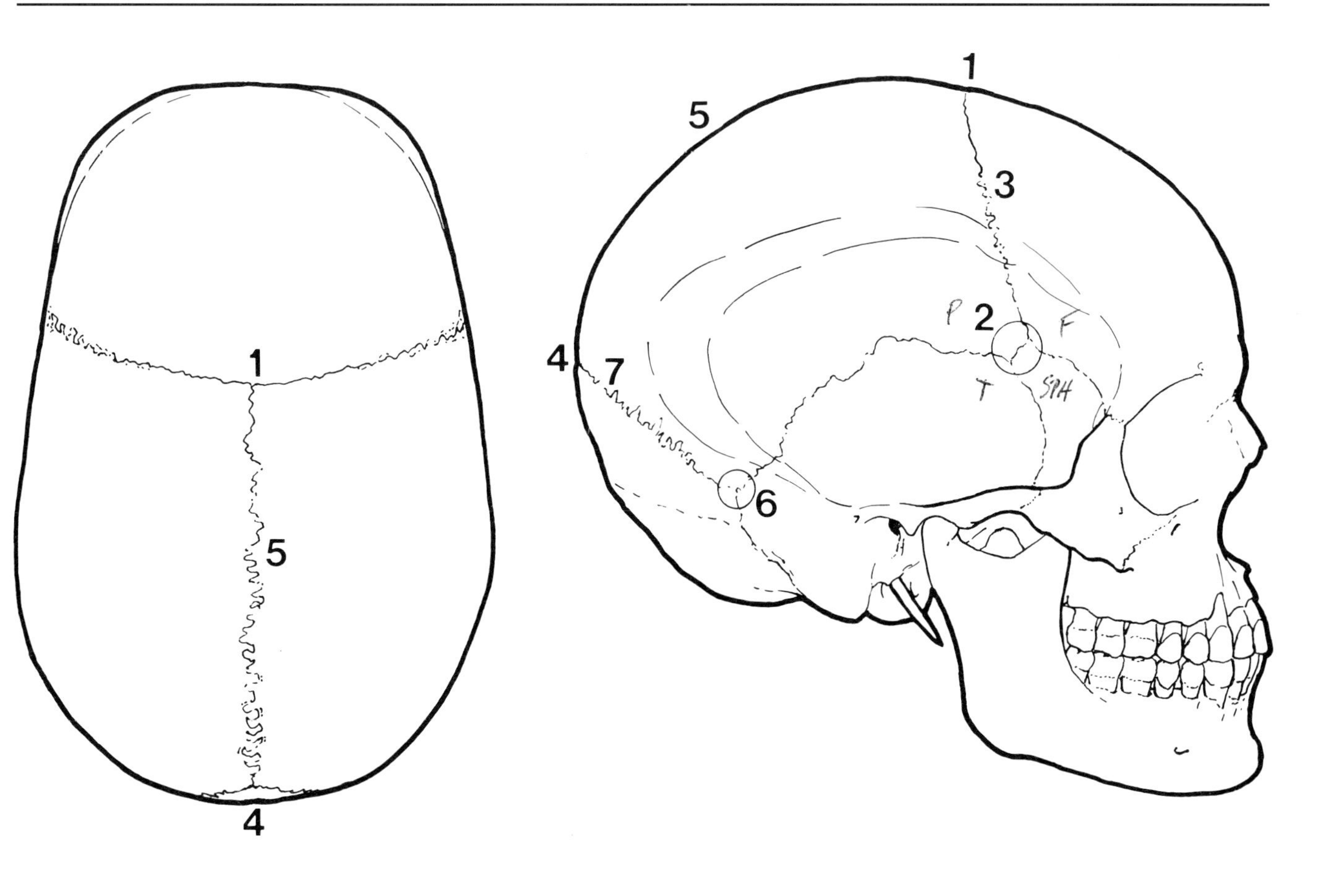

1
5
4
1
5
3
P
2
F
4
7
T
SPH
6

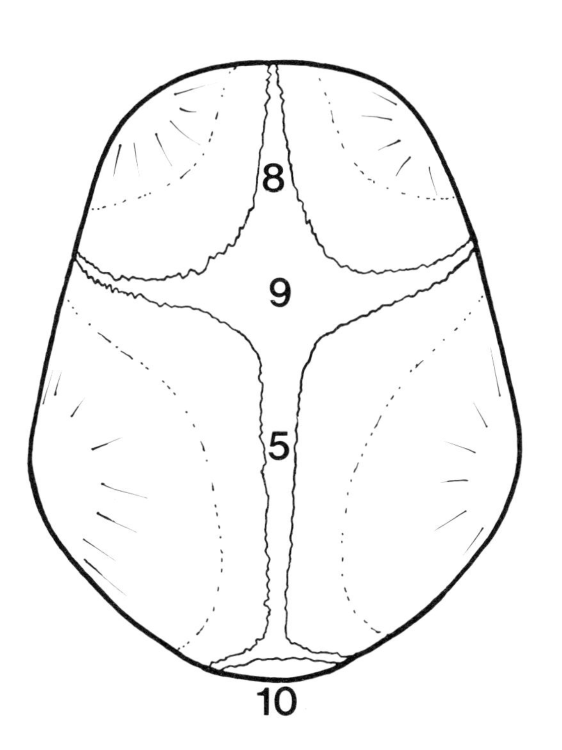

8
9
5
10

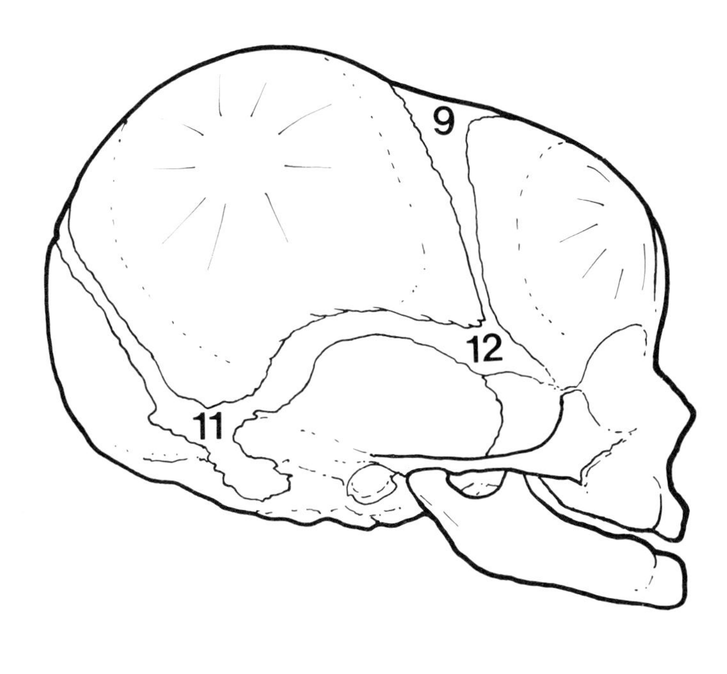

9
12
11

TEETH

The enthusiastic student is often intimidated by the formidable vocabulary of dental morphology which serves mainly to assist the dentist and the physical anthropologist. Medical practitioners need only familiarise themselves with sufficient terminology to ensure unambiguous identification of the teeth and their relations.

Man is *diphyodont* which means he has two sets of teeth in his lifespan—a deciduous or milk dentition of 20 teeth which gives way to a permanent dentition of 32 teeth. In addition, man is *heterodont* in that he possesses several different types of teeth namely, incisors, canines, premolars and molars, all with their own morphological characteristics.

Incisors are chisel-shaped with one root and are situated at the front of the mouth for biting.

Canines are single cusped teeth also with one root. They were large in our ancestors and were used for grasping, fighting and killing. In *Homo sapiens*, the canine is reduced to the same size as the other teeth but retains its primitive cusp.

Premolars have single roots except for the upper first premolar. They are all bicuspid, the buccal cusp usually larger than the lingual (see below).

Molars are multicusped and usually have three roots in the upper jaw, although two of these are sometimes fused, and two roots in the lower jaw. They are used mainly for grinding and crushing.

Tooth Surfaces and Cusps

Since teeth lie in an arc, the usual terms of anterior, posterior, medial and lateral are open to misinterpretation. Thus it is conventional to describe the tooth surfaces and its relations as shown in the diagram.

[**M**] Mesial surface
[**D**] Distal surface
[**A**] Oral surface. In the upper jaw this is known as the palatal surface and in the lower jaw as the lingual surface
[**B**] (i) the labial surface or (ii) the buccal surface depending on whether that surface of the tooth lies against the lip or the cheek
[**O**] Occlusal surface. The surface which contacts the tooth of the maxilla.

Dental Formula

Since the jaws are symmetrical, the teeth are counted and numbered from the midline. Thus, in the milk dentition, there are five teeth in each quarter of the mouth—two incisors, one canine, and two molars, represented as 2I, 1C, 2M, or 2102.

The permanent dentition, which consists of two incisors, one canine, two premolars and three molars in each quadrant, is similarly represented by the symbols

$$\frac{3212}{3212}\;\frac{2123}{2123}$$

The horizontal divisor between the two sets of figures represents the occlusive plane between the upper and lower jaws and the vertical divisor represents the midline.

To identify a tooth, the following shorthand is employed where the vertical line represents the midline, the horizontal line represents the occlusive plane and the number is the number of the tooth from the midline, thus:

$\lfloor 3$ identifies the third tooth in the *left upper* quadrant—the left upper canine
$4\rceil$ identifies the fourth tooth from the midline in the lower right quadrant—the first premolar

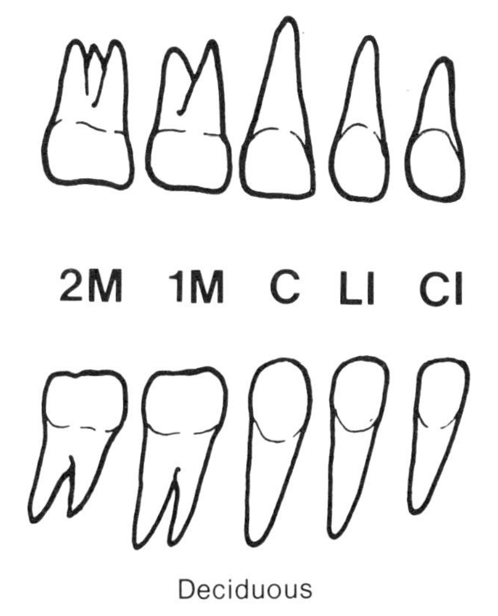
2M
1M
C
LI
CI
Deciduous

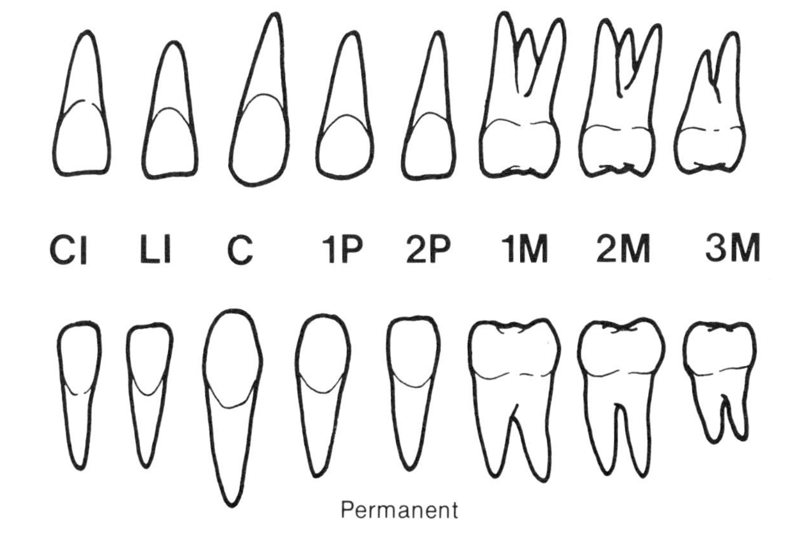
CI
LI
C
1P
2P
1M
2M
3M
Permanent

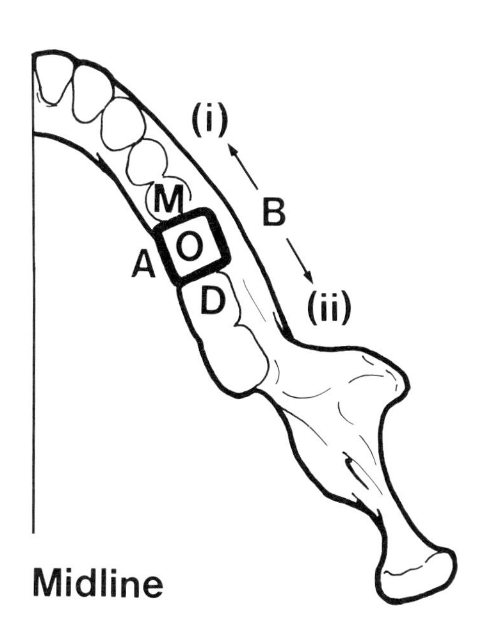
(i)
B
M
A
O
D
(ii)
Midline

Dental Occlusion

When the teeth are occluded optimally, so that there is maximum contact between the teeth of opposing arches, then the teeth are said to be in *centric occlusion*. In normal dentition it has the following features:

1. The lower canine is mesial to the upper canine.
2. The lower cheek teeth are slightly mesial to the upper cheek teeth.
3. The buccal cusps of the lower cheek teeth lie between the buccal and palatal cusps of the upper cheek teeth for maximum surface contact.

The combination of (2) and (3) means that the lower cheek teeth are a little lingual and mesial to the upper cheek teeth.

4. The upper incisors override the lower incisors by about one-third the crown length of the lower incisors. This is known as the *overbite*.
5. The upper incisors occlude in front of the lower incisors and the extent of the gap between the two is known as the overjet.

Normally the jaw is not held fully closed in the resting state. The space between the occlusal surfaces of the opposing teeth is known as the *freeway space* or the *interocclusal clearance*.

Dental Arches

Both the upper and lower dental arches approximate to a catenary curve which is the curve assumed by a chain fixed at both ends. The mandibular arch is normally narrower than the maxillary. This results in maximum intercuspation when the two arches are occluded.

The Curve of Spee

When a skull is viewed from the lateral aspect, it is seen that a line joining the buccal cusps follows a curve with its concavity facing upwards.

The Curve of Monson

The mandibular cheek teeth point slightly lingually, so that a line joining the buccal and lingual cusps of corresponding right and left teeth is curved with its concavity facing upwards. The upper cheek teeth have a corresponding tilt outwards.

The curves of Spee and Monson are of importance in the modelling of dentures.

Dental Eruption

Teeth erupt bloodlessly through the gums on the approximate dates given below:

	I		C	M		
Deciduous (upper)	7	9	18	12	24	
Deciduous (lower)	6	8	18	12	24	Months

	I		C	P		M			
Permanent (upper)	7	8	12	9	10	6	12	18+	
Permanent (lower)	7	8	12	9	10	6	12	18+	Ye

There is a simple pattern to these figures which is not difficult to learn. However, it should be appreciated that:

1. A mandibular tooth erupts earlier than its opposite number in the upper jaw.
2. The first permanent molar (the six-year molar) erupts before any deciduous teeth have been shed.
3. The first deciduous tooth is the lower central incisor. If it were an upper incisor the baby would have difficulty in suckling.

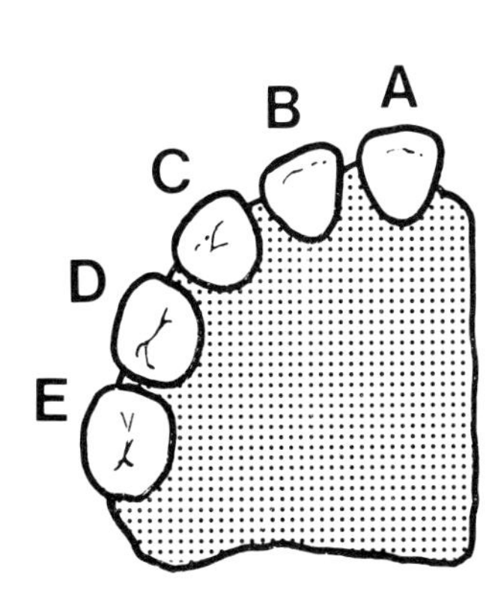

Deciduous

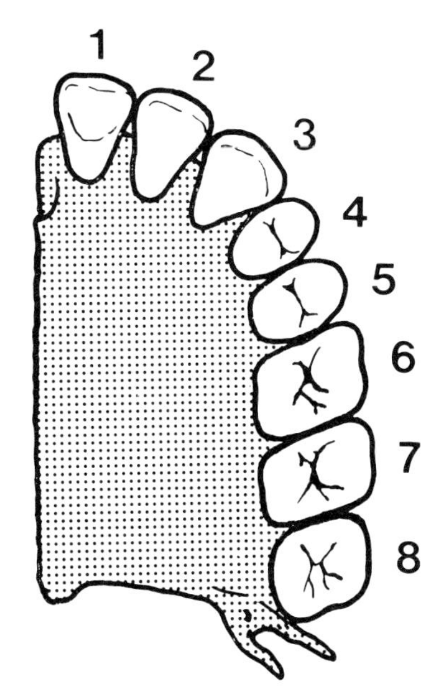

Permanent

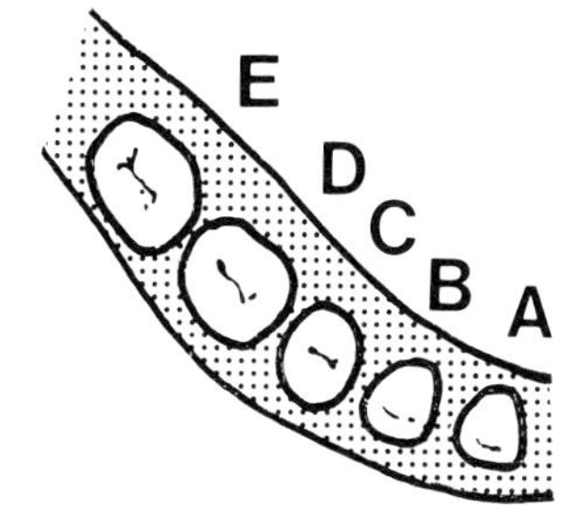

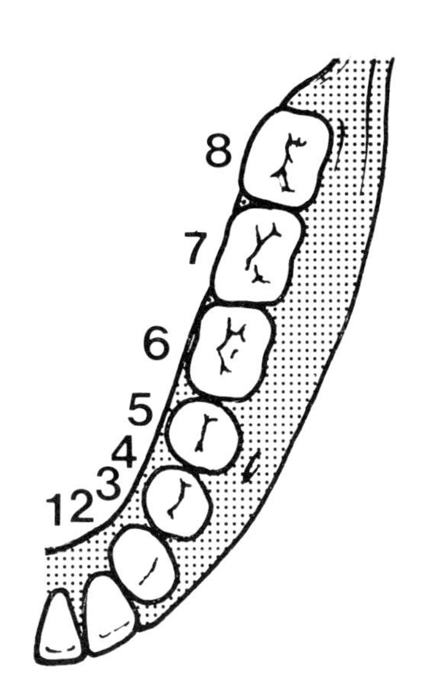

The Orthopantomogram

This is a tomogram designed to show the full set of upper and lower teeth in one sweep. Notice that in film [**C**] the extent of the sweep is such that it has taken in the cervical spine on both sides. Notice also that the front teeth are not usually as well seen as the cheek teeth.

The individual teeth are easily identified:
Central incisor [**1**]
Lateral incisor [**2**]
Canine [**3**]
First premolar [**4**]
Second premolar [**5**]
First molar [**6**]
Second molar [**7**]

The wisdom teeth (third molars) are the last teeth to arrive and the first teeth to be shed from the ageing mandible. There is a tendency to lose teeth in mammalian evolution and the wisdom tooth is a prime example. In some geographical areas 50% of the population will never develop a third molar tooth. The lower third molar is relatively variable and often has only one root, although it very occasionally has five! If one studies an articulated skull, it is seen that the lower third molar faces antero-superiorly and the upper third molar faces postero-inferiorly. The final direction of teeth reasonably represents their direction of eruption, and it is, therefore, easy to appreciate that during eruption the lower third molar may become impacted against the second molar giving rise to discomfort and predisposing to infection. Conversely, the upper third molar seldom becomes impacted.

This is well illustrated in film [**B**]. In film [**A**] the wisdom teeth are seen as developing tooth buds.

The structure of the teeth can also be examined. The highly mineralised enamel and dentine show up well. The enamel cap forms a more opaque cap over the slightly less opaque dentine and is seen particularly well in the unerupted permanent lower right second premolar which is beneath the second deciduous molar in [**A**], and also in the same premolar [**B**]. The tooth pulp is radiolucent and is surrounded by dentine except at the apex of the root where it appears to be in continuity with the radiolucent periodontal membrane which fastens the tooth into its socket. The socket itself is lined with dense, compact cortical bone which appears as a thin white line—*the lamina dura.* The periodontal membrane, therefore, is seen as the thin black line between the tooth and the lamina dura.

The laminae durae of adjacent sockets meet on the alveolar margin between the teeth to form the interdental crest which is seen as a sharp ridge between the anterior teeth, but is wider and flatter between the cheek teeth.

Several other structures may be seen:

Coronoid process of mandible [**8**]
Zygoma [**9**]
Mandibular canal [**10**]
Maxillary sinus beneath orbit [**11**]
Hard palate [**12**]
Nasal septum [**13**]

Notice the extensive root fillings in [**C**].

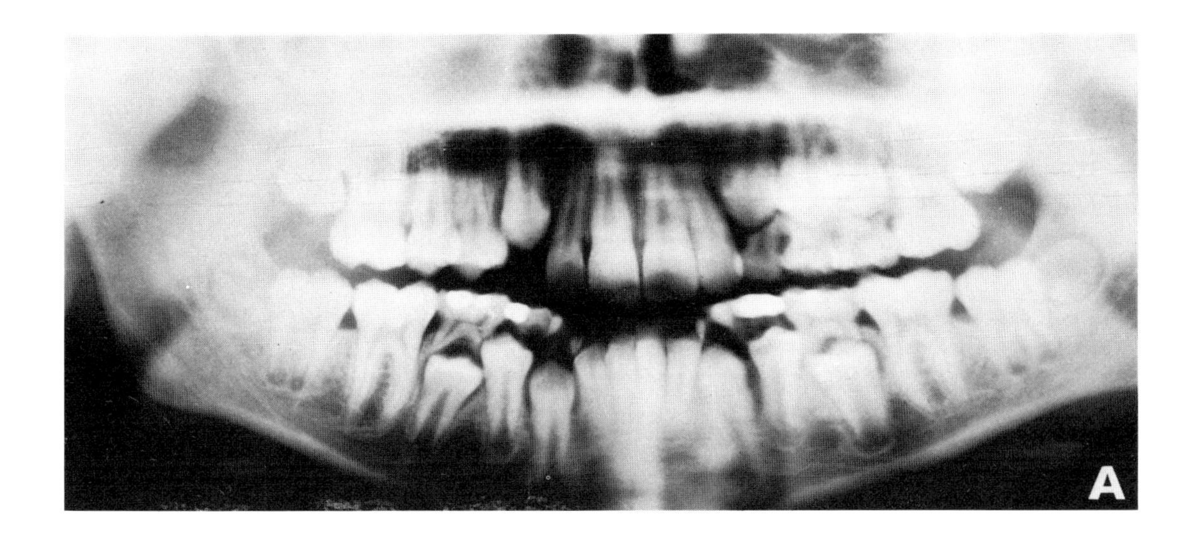
A

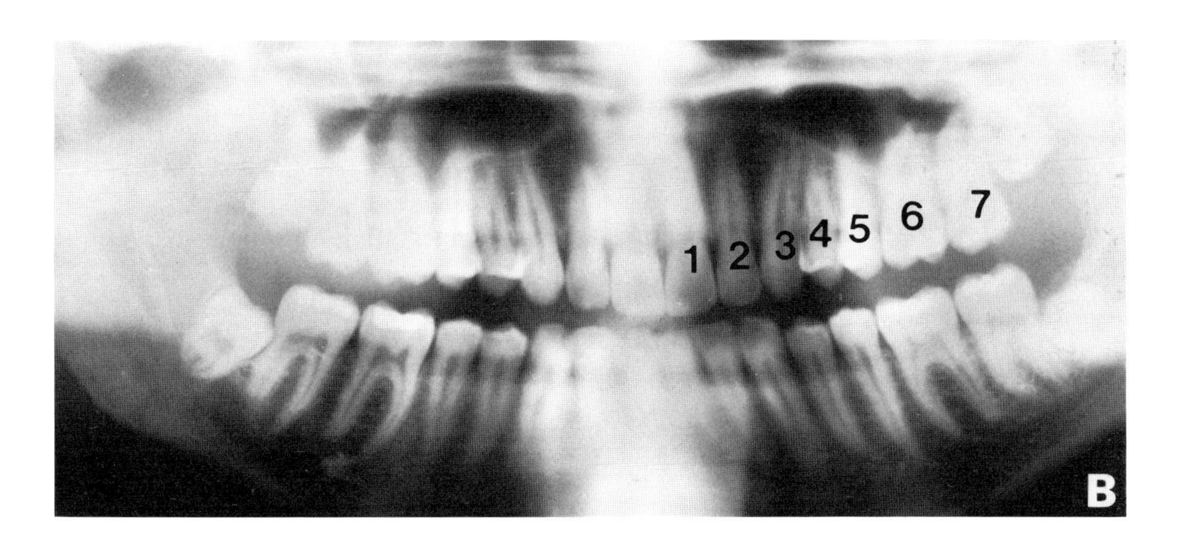
1
2
3
4
5
6
7
B

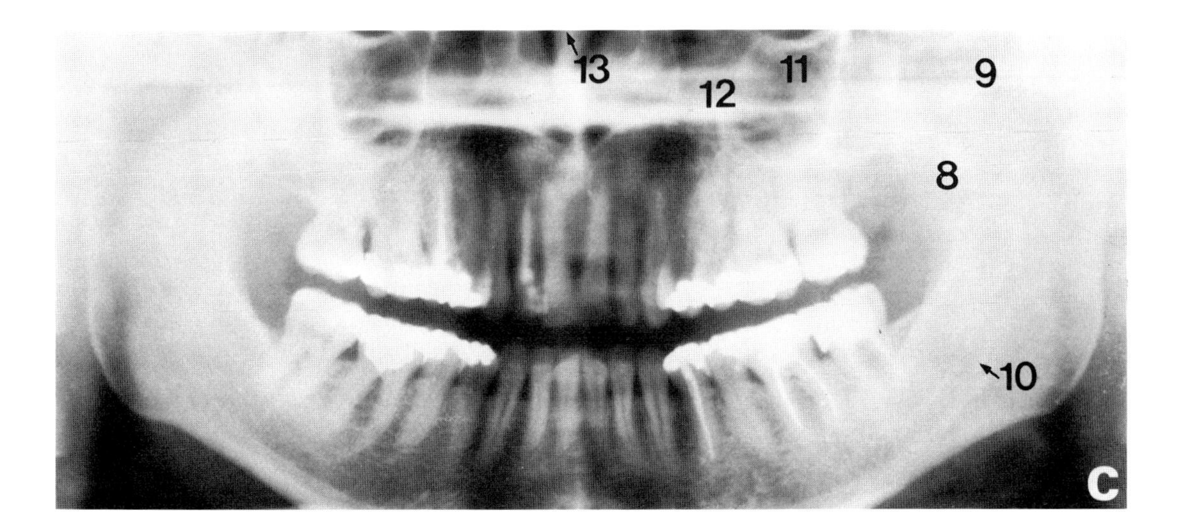
13
12
11
9
8
10
C

THE MUSCLES OF MASTICATION

Temporalis
Masseter
Medial pterygoid
Lateral pterygoid

Temporalis

This muscle may be felt to contract on the side of the skull beneath the temporal line during the act of chewing.

The remaining three muscles each have two borders, the surface projections of which may be represented by straight lines drawn on the side of the face.

Masseter

Anterior border. A line from 1 cm below the bony prominence of the cheek to the lower border of the mandible 3 cm from the angle. This border is easily palpable, and may even be seen in some gaunt individuals.

Posterior border. A line from immediately in front of the tubercle of the zygoma to a point on the posterior border of the ramus of the mandible 2 cm above the angle.

Medial Pterygoid [M]

Anterior border. A line from a point opposite the end of the upper alveolar margin to the lower border of the mandible 2 cm in front of the angle.

Posterior border. A line from a point one finger's breadth in front of the tubercle of the zygoma to the posterior border of the ramus of the mandible just below the level of the ear lobe.

Lateral Pterygoid [L]

Upper border. A line from the jugal point to the neck of the mandible.

Lower border. A line from a point opposite the end of the upper aveolar margin to the neck of the mandible.

The Jaw Jerk

All the muscles of mastication receive their motor supply from the trigeminal nerve. A sharp tap from a patella hammer to the index finger of the examiner lying on the mandible between the lower lip and the prominence of the chin with the mouth relaxed and half open will elicit a reflex contraction of the muscles of mastication.

NERVE SUPPLY OF THE TEETH AND DENTAL ANAESTHESIA

The Upper Teeth and Gums

The upper teeth are supplied by a plexus formed from branches of the anterior and posterior superior alveolar nerves. This plexus is occasionally supplemented by a further branch from the maxillary nerve—the middle superior alveolar nerve.

The buccal molar gingivae are supplied by the buccal nerve, whereas the rest of the buccal gingivae are supplied by branches from the infraorbital nerve which also supplies the adjacent lip. The palatal gingivae are supplied by the greater palatine nerve distally and the nasopalatine nerve mesially; these nerves communicate in the region of the canine tooth.

Anaesthesia. The molar teeth require a posterior superior alveolar nerve block, but the other teeth may be extracted using infiltration anaesthesia on the buccal and palatal surfaces.

The Lower Teeth and Gums

The lower teeth are all supplied by the inferior alveolar nerve. The buccal gingivae are supplied by the mental nerve distally, as far as the first premolar, thereafter they are supplied by the buccal nerve. The lingual gingivae are supplied by the lingual nerve.

Anaesthesia. All the lower teeth on one side of the jaw can be anaesthetised by inferior alveolar nerve block. The gums are anaesthetised by lingual nerve block for the lingual gingivae, buccal nerve block for the molar buccal gingivae and local infiltration or mental nerve block for the labial gingivae.

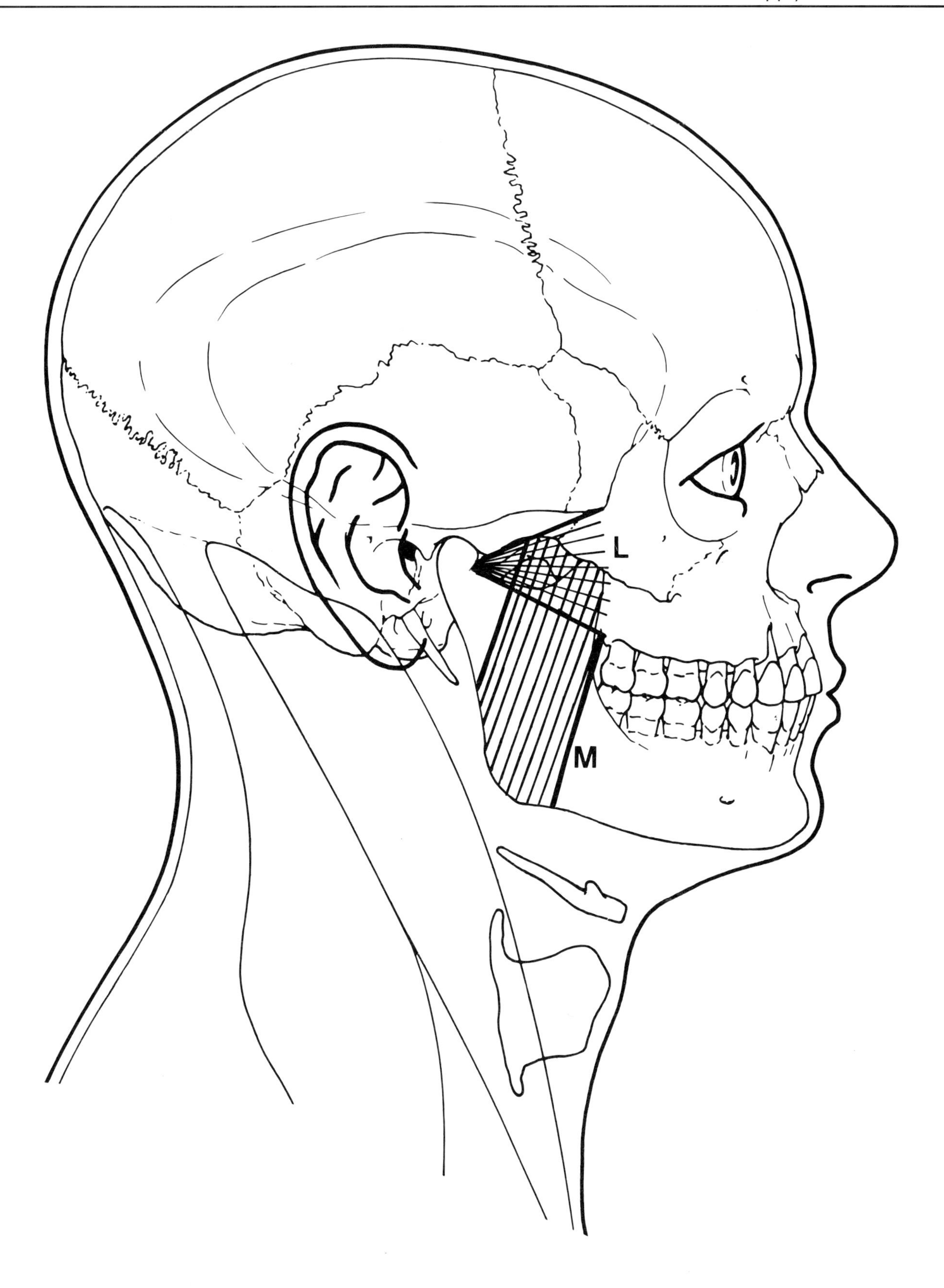
L
M

MUSCLES OF THE NECK

Sternocleidomastoid [1]

This important landmark is brought into relief when the muscle contracts to turn the face to the *opposite* side. The anterior border is then conspicuous and may be traced from the manubrium to the mastoid process. The posterior border is far less obvious, particularly in the lower third, where it extends from the junction of the medial and middle thirds of the clavicle to a point on the superior nuchal line, midway between the mastoid process and inion.

It divides the lateral side of the neck into two triangles, anterior and posterior.

The anterior triangle is bounded in front by the median line of the neck; above by the line of the inferior border of the mandible continuing to the anterior border of the sternocleidomastoid muscle; and behind by the anterior border of the sternocleidomastoid muscle.

Trapezius [2]

The anterior border runs from a point midway between the inion and the mastoid process to the junction of the middle and lateral thirds of the clavicle.

The posterior triangle is bounded in front by the posterior border of the sternocleidomastoid; behind by the anterior border of the trapezius; and below by the middle third of the clavicle.

Omohyoid

The intermediate tendon of the omohyoid passes through its fascial sling overlying a point on the internal jugular vein, beneath the sternocleidomastoid muscle, at approximately the level of the cricoid cartilage.

The anterior belly [**3**] may be represented by a line drawn from the point of the fascial sling to the hyoid bone one finger's breadth from the midline.

The posterior belly [**4**] is represented by a line from the same initial point to a point on the clavicle a little lateral to the junction of the middle and lateral thirds. It divides the posterior triangle into an upper *occipital triangle* [**5**] and a lower *supraclavicular triangle* [**6**].

Digastric

The intermediate tendon lies on the hyoid bone, two finger breadths from the midline.

The anterior belly [**7**] stretches from this point above the hyoid to the lower border of the mandible adjacent to the symphysis menti. It can be demonstrated by opening the mouth against resistance when it can be felt to harden.

The posterior belly [**8**] may be represented by a line from the intermediate tendon to the tip of the mastoid process.

Together, the digastric and omohyoid muscles serve to subdivide the anterior triangle into smaller triangles which form important surgical landmarks.

The digastric triangle [**9**] between the anterior and posterior bellies of the digastric and the line of the lower border of the mandible, and floored by the mylohyoid.

The carotid triangle [**10**] between the posterior belly of the digastric above, the anterior belly of the omohyoid in front, and the sternocleidomastoid behind.

The muscular triangle [**11**] between the median line of the neck from the hyoid to the sternum in front, the anterior belly of the omohyoid postero-superiorly, and the anterior border of the sternocleidomastoid postero-inferiorly.

The submental triangle [**12**] is a median triangle beneath the chin. It is bounded by the body of the hyoid bone below and the anterior bellies of the digastric muscles on either side.

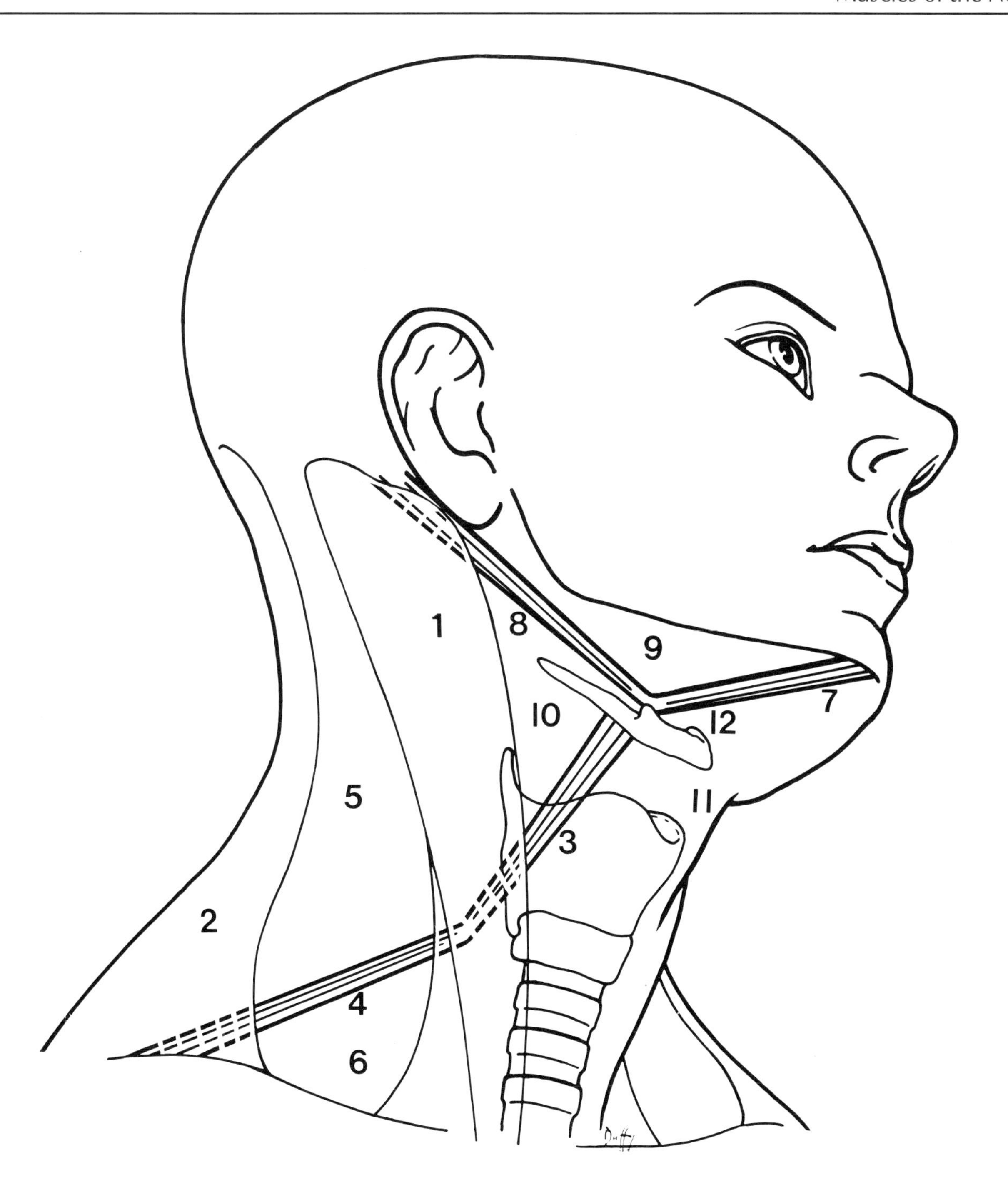
1
2
3
4
5
6
7
8
9
10
11
12

ARTERIES AND PULSES IN THE NECK

The two main arteries on either side in the root of the neck are the subclavian and common carotid arteries. On the right, they are formed by the division of the innominate artery behind the right sternoclavicular joint. However, on the left, they both arise independently in the thorax from the arch of the aorta.

Subclavian Artery

The right subclavian artery [**1**] may be represented by a broad line, convex upwards, commencing behind the right sternoclavicular joint and arching over the front of the cervical pleura reaching its highest point behind the scalenus anterior muscle, one thumb's breadth medial to the posterior border of the sternocleidomastoid and one thumb's breadth above the clavicle, and then dropping to the lower border of the middle of the clavicle, where it crosses the lateral border of the first rib to become the axillary artery.

The left subclavian artery [**2**] arises from the arch of the aorta behind the middle of the left manubrial border and ascends to behind the left sternoclavicular joint, after which its course is similar to that of the right subclavian artery.

The pulse. Each subclavian artery is divided into three parts by the scalenus anterior muscle which passes anterior to the artery at the highest point of its arch. The first two parts lie deep in the neck but, because the lateral border of the scalenus anterior corresponds closely to the posterior border of the sternocleidomastoid, the third part is superficial in the antero-inferior angle of the posterior triangle of the neck. Here the pulse can be felt and sometimes seen. The artery may be compressed against the first rib by pressure directed firmly backwards, downwards and medially.

Branches of the subclavian artery
Vertebral
Internal thoracic
Thyrocervical trunk
Costocervical trunk
Dorsal scapular

The vertebral artery [**3**] may be indicated by a line drawn, with the head facing forwards, from the sternoclavicular joint to the posterior border of the sternocleidomastoid at the level of the cricoid cartilage, where the artery enters the sixth foramen transversarium, and thereafter to the hollow behind the ear. From here the artery turns backwards for 3.5 cm around the lateral mass of the atlas, and then turns medially in the suboccipital triangle before piercing the posterior atlanto-occipital membrane to gain the foramen magnum. The part in the suboccipital triangle is opposite a point about one thumb's breadth postero-medial to the tip of the mastoid process.

The internal thoracic artery [**4**] *See* The Thorax (p. 34).

Thyrocervical trunk [**5**] arises at the medial border of the scalenus anterior behind the internal jugular vein. It lies opposite a point between the two heads of the sternocleidomastoid, one thumb's breadth above the clavicle. It divides almost immediately into: inferior thyroid artery; transverse cervical artery; and suprascapular artery.

The inferior thyroid artery [**6**] corresponds to a line drawn from the point for the thyrocervical trunk vertically upwards to the level of the cricoid cartilage and then medially towards the cartilage.

The transverse cervical artery [**7**] may be marked by a line from the point for the thyrocervical trunk to a point two finger breadths above the junction of the middle and lateral thirds of the clavicle.

The suprascapular artery [**8**] is indicated by a line beginning opposite the thyrocervical trunk and running obliquely downwards to the lower end of the posterior border of the sternocleidomastoid, and then along the clavicle to the junction of the middle and lateral thirds. The artery is first behind the sternocleidomastoid, then behind the middle third of the clavicle and finally it runs down under cover of the trapezius to reach the scapula.

The costocervical trunk arises from the back of the subclavian artery just lateral to the thyrocervical trunk. Although all the other branches of the subclavian artery arise from the first part, this artery may occasionally arise from the second part, especially on the right. It arches backwards over the cervical pleura to reach the neck of the first rib where it divides into the superior intercostal and deep cervical arteries.

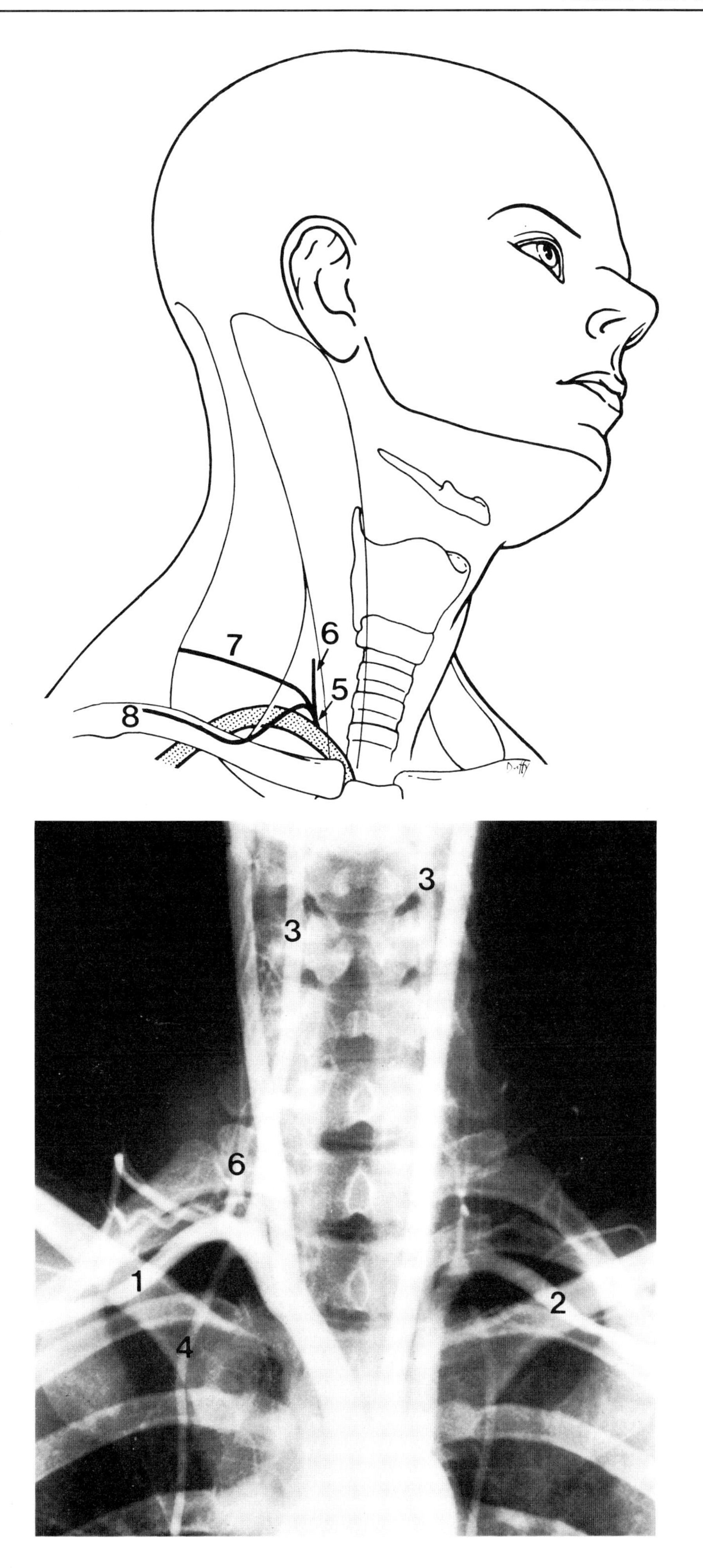
7
6
5
8
3
3
6
1
2
4

The Common Carotid Arteries [A]

The right common carotid artery commences as a terminal division of the innominate artery behind the right sternoclavicular joint. It is represented by a line from this point to a point just behind the anterior border of the sternocleidomastoid muscle at the upper border of the thyroid cartilage, where it divides into the internal and external carotid arteries.

The left common carotid artery arises from the aorta immediately to the left of the centre of the manubrium sterni, and enters the neck behind the left sternoclavicular joint. Thereafter, its course is similar to that of the right common carotid artery.

The carotid pulse may be felt by sliding the three middle fingers postero-laterally over the lateral surface of the thyroid cartilage under the anterior border of the sternocleidomastoid immediately above the level of the laryngeal prominence.

The internal carotid artery [**B**] runs up the side of the pharynx to the base of the skull. Its course is indicated by a line from a point on the sternocleidomastoid near its anterior border at the level of the upper border of the thyroid cartilage to a point just behind the *head* of the mandible.

The external carotid artery [**C**] has almost the same surface marking as the internal carotid artery except that the line lies a little anterior to that for the internal carotid and ends behind the *neck* of the mandible.

Branches of the external carotid
Ascending pharyngeal
Superior thyroid
Lingual
Facial
Occipital
Posterior auricular
Superficial temporal
Maxillary

Ascending pharyngeal branch [**1**] arises just above the origin of the external carotid artery and ascends up the pharynx to the base of the skull opposite the same line as the internal carotid artery.

The superior thyroid artery [**2**] arises in the carotid triangle just below the hyoid bone and runs towards the cricoid cartilage.

The lingual artery [**3**] arises *at* the level of the tip of the greater horn of the hyoid bone and follows a line along the upper border of the horn for 3 cm, then upwards and forward towards the angle of the mouth as far as the lower border of the mandible where it enters the tongue and becomes the arteria profunda linguae, which runs forward in the inferior surface of the tongue close to the frenulum.

The facial artery [**4**] arises in the carotid triangle just above the tip of the greater horn of the hyoid bone. It leaves the triangle by running upwards to a point opposite one finger's breadth above and in front of the angle of the mandible. From this point it bends downwards and forwards to enter the face by wrapping around the lower border of the mandible at the anterior border of the masseter, from where it takes a sinuous course towards the medial angle of the eye, passing 1 cm lateral to the angle of the mouth.

The facial pulse may be felt at the anterior border of the masseter where the artery winds round the lower border of the mandible.

The occipital artery [**5**] arises at the same level as the facial artery and may be marked by a line from that point on the external carotid artery passing upwards and backwards to a point one thumb's breadth lateral to the inion. Initially it lies superficially in the carotid triangle, before passing deep to the muscles attached to the mastoid process; but it becomes superficial again while passing through the apex of the posterior triangle of the neck. It pierces the trapezius to enter the scalp with the greater occipital nerve one thumb's breadth lateral to the inion.

The posterior auricular artery [**6**] is small and may be marked by a line beginning just above the angle of the mandible and extending upwards behind the root of the auricle.

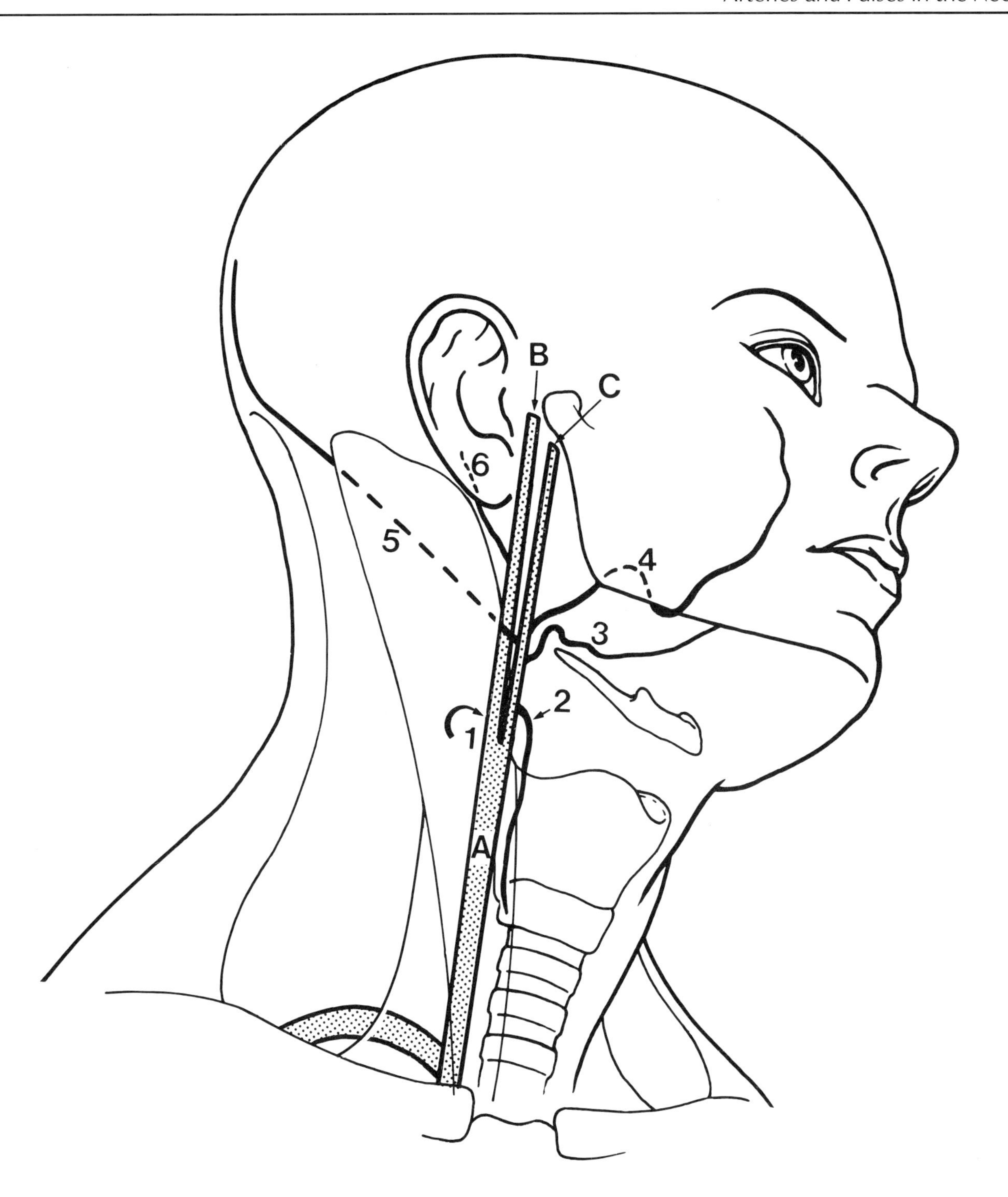
B
C
6
5
4
3
2
1
A

The superficial temporal artery [7] arises in the parotid gland as the continuation of the external carotid artery after its terminal division into the maxillary and superficial temporal arteries. It runs upwards across the zygomatic arch at the pre-auricular point. About 2.5 cm above the pre-auricular point it divides into the parietal and frontal branches passing to the parietal and frontal eminences respectively.

Branches
Transverse facial
Middle temporal

The transverse facial artery runs forward on the masseter between the zygomatic arch and the parotid duct.

The middle temporal artery arises immediately above the zygomatic arch and pierces the temporal fascia to anastamose with the deep temporal branches of the maxillary artery within the substance of the temporalis muscle.

The superficial temporal pulse may be felt at the pre-auricular point. It is of particular use to anaesthetists who do not always have access to the radial pulse. It is also a useful compression point to help to control bleeding from some scalp injuries. Paediatricians may also occasionally use this artery for obtaining samples of arterial blood for blood gas analysis.

The maxillary artery [8] commences behind the neck of the mandible within the substance of the parotid gland. It is represented by a line drawn from the back of the neck of the mandible to the upper border of the zygomatic arch just behind the jugal point; at this point the artery turns more medially and dives into the sphenopalatine fossa.

Branches
The maxillary artery has many branches but from a clinician's viewpoint the *middle meningeal artery* is singularly the most important (*see* Clinical Significance).

The course of the middle meningeal artery [**a**] may be divided conveniently into extracranial and intracranial parts.

The extracranial part is short and runs opposite a line drawn upwards from the root of the neck of the mandible to the anterior part of the temporomandibular joint, where it passes through the foramen spinosum about 3 cm deep to the surface.

The intracranial portion of the artery is more important. Shortly after entering the middle cranial fossa through the foramen spinosum, the artery divides at a variable point into an anterior and posterior branch.

The course of the anterior branch [**b**] may be marked by a line commencing at the upper border of the midpoint of the zygomatic arch and passing upwards towards the pterion and, thereafter, backwards to a point on the vertex midway between the glabella and the inion. Here it lies superficial to the precentral gyrus.

The course of the posterior branch [**c**] may be represented by a line beginning at the upper border of the midpoint of the zygomatic arch and drawn backwards towards the lambda. It overlies the middle temporal gyrus.

Clinical Significance
In its intracranial course, together with its companion vein, the middle meningeal artery lies in immediate juxtaposition to bone between the two layers of dura mater formed by the dura mater proper and the periosteum of the skull itself. The vein tends to lie between the artery and the bone and thus it is the vein which is responsible for the grooves seen on the prepared skull. Examine a skull and notice how thin the bone is in the region of 'the temple' and how deeply the vascular bundle grooves the bone. It is easy to understand why the vascular bundle is extremely susceptible to damage in certain fractures of the skull, particularly in the region of the pterion where the grooving is deep and may even be converted to a tunnel through the greater wing of the sphenoid.

Damage to either the artery or vein by tearing when the inner skull table is fractured results in serious bleeding and haematoma formation which may compress the cerebral cortex.

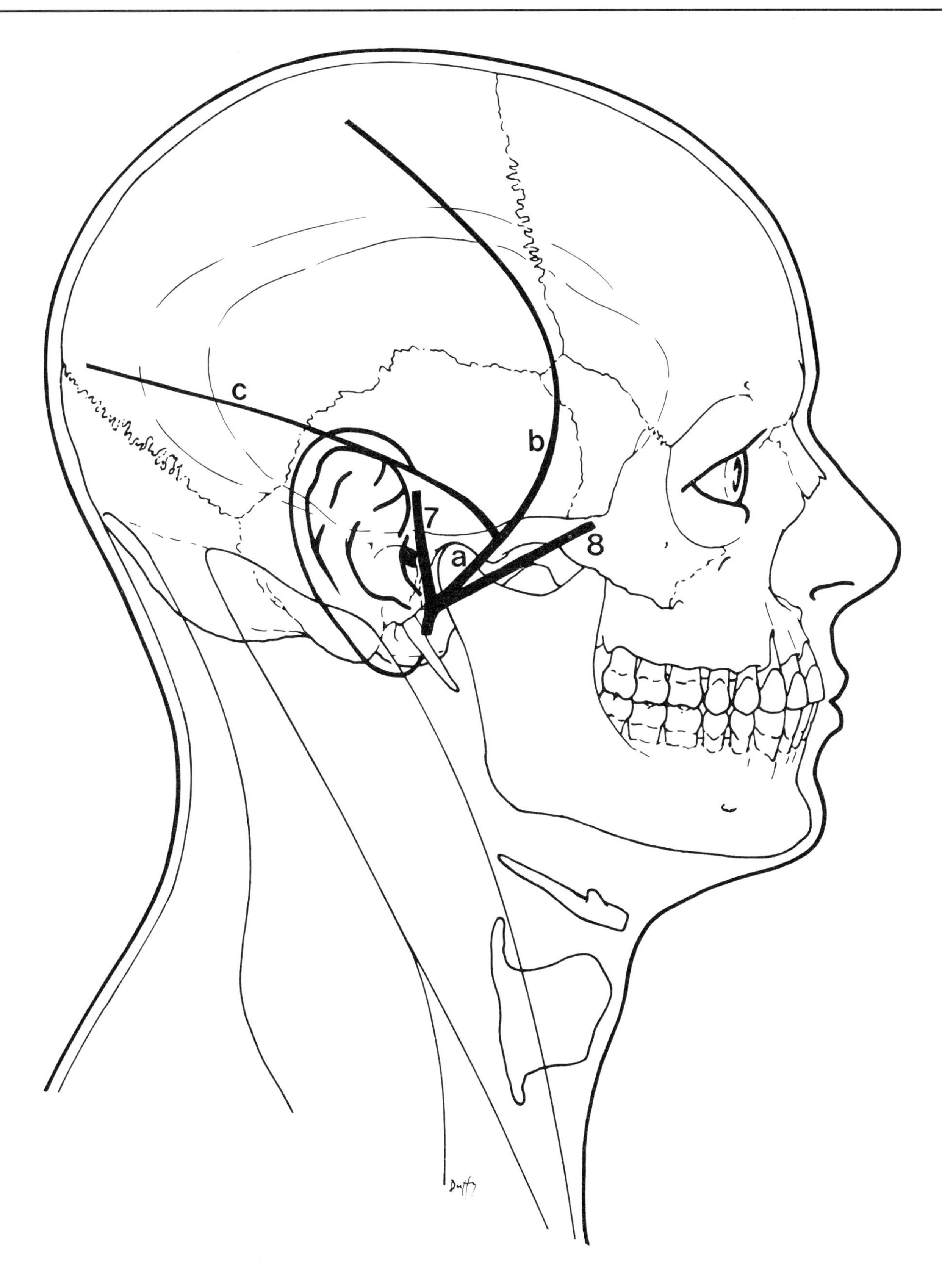
c
b
7
a
8

Internal Carotid Angiography

Carotid angiography is performed by introducing a bolus of a water soluble radiopaque dye into the carotid arteries by selective catheterisation via the femoral artery or by direct puncture. Sequential films are then taken as the dye is circulating through the cerebrovascular system.

Carotid angiography is divided into four phases:

1. The arterial phase lasting about 2 seconds
2. The capillary phase lasting about 1 second
3. The early venographic phase
4. The late venographic phase

Subtraction

This is a photographic technique employed in contrast radiography to eliminate unwanted images thus making diagnostically important images easier to see.

Three conditions must be fulfilled in order to perform a successful subtraction study:

(a) A *control* or *scout* radiograph is taken prior to the introduction of contrast medium.

(b) Immediately after the scout film has been taken contrast medium is introduced into the patient.

(c) It is imperative that there should be no movement between the scout film and the contrast radiographs.

Let us consider cerebral angiography. Once the patient has been fully prepared a scout film of the head is taken to show the structural details of the skull and soft tissues. Immediately after this exposure, with the patients head held absolutely still in the same position, the contrast medium is introduced into the cerebrovascular system while sequential films are taken.

Both the *scout film* and the *angiograms* are developed normally so that bone and contrast medium are white and air is black. In addition, a contact radiograph is made of the *scout film*. In this reversed film, known as the *mask film*, bone is dark and air is white.

If the mask film is then superimposed on the angiogram and a contact radiograph made, the mask film will cancel out the bony and soft tissue shadows of the angiogram leaving the contrast image unattenuated. The resultant radiograph shows a 'subtracted' image in which the previously white contrast medium appears black and the bony and soft tissue shadows have fused into an homogeneous grey background.

The commercial production of automatic developers has now made this process easy, fast and relatively inexpensive.

Lateral view

Cervical part of internal carotid artery [1]
Intrapetrous part of internal carotid artery [2]
Cavernous part of internal carotid artery; note the S-shape [3]
Ophthalmic artery [4]
Anterior cerebral arteries [5]—the opposite one has filled through the anterior communicating artery
Middle cerebral artery [6]
Position of posterior communicating artery [7]
Pericallosal artery arising from anterior cerebral artery [8]

Antero-posterior view

Cervical part of internal carotid artery [1]
Intrapetrous part of internal carotid artery [2]
Cavernous part of internal carotid artery [3]
Ophthalmic artery [4]
Anterior cerebral artery [5], passing medially to gain the longitudinal fissure. Once in the longitudinal fissure it lies close to the opposite anterior cerebral artery and occasionally they form a single vessel
Site of the anterior communicating artery [6] which is usually about 4 mm long
The middle cerebral artery [7] passing laterally to lie in the lateral fissure before reaching the lateral surface of the hemisphere. It is the largest branch of the internal carotid artery and is therefore the most susceptible to embolism

In both the lateral and antero-posterior views the branches of the middle cerebral artery are labelled [**MC**] and those of the anterior cerebral artery are labelled [**AC**].

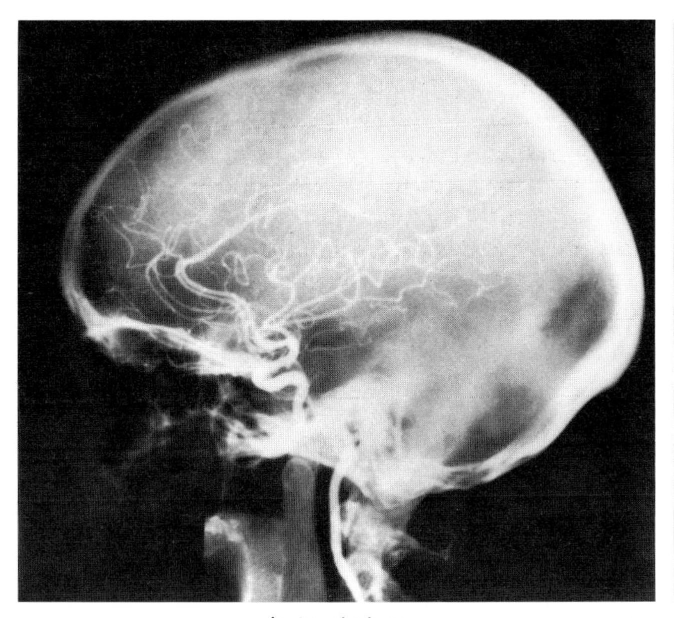

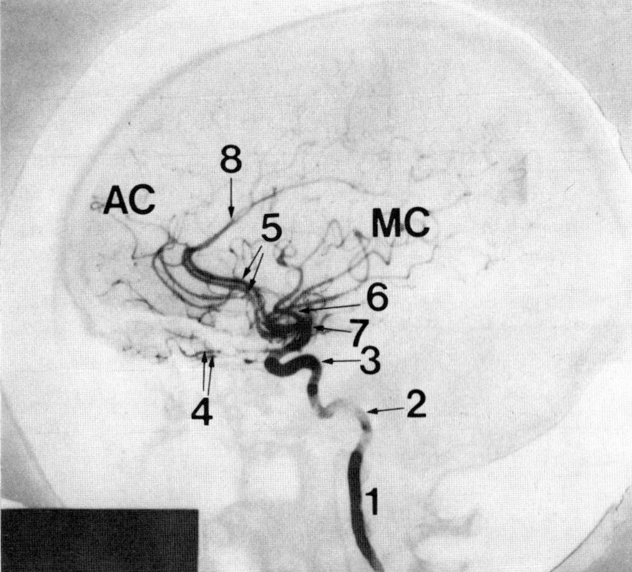

Lateral view

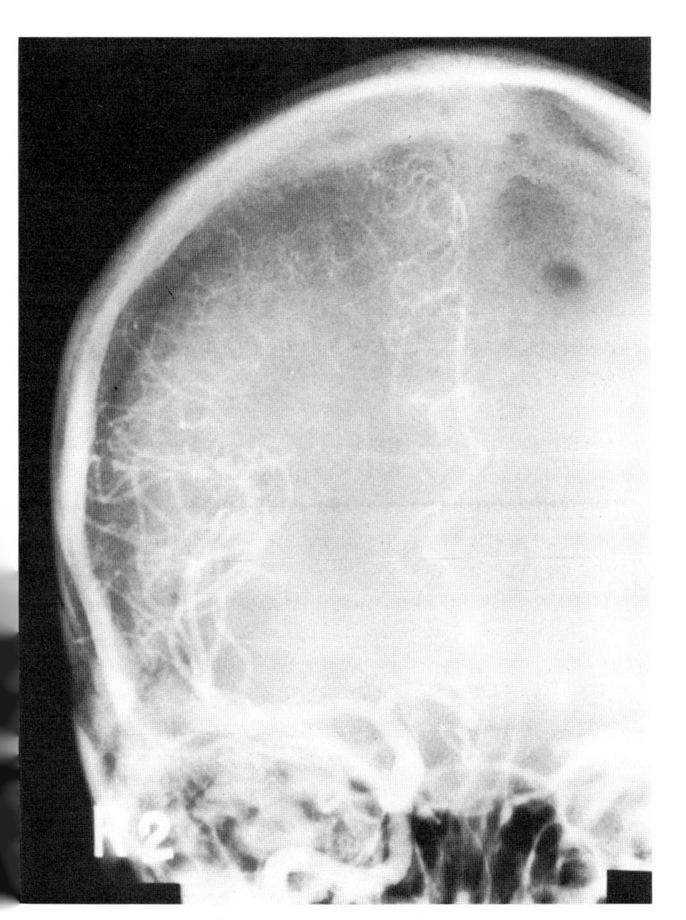

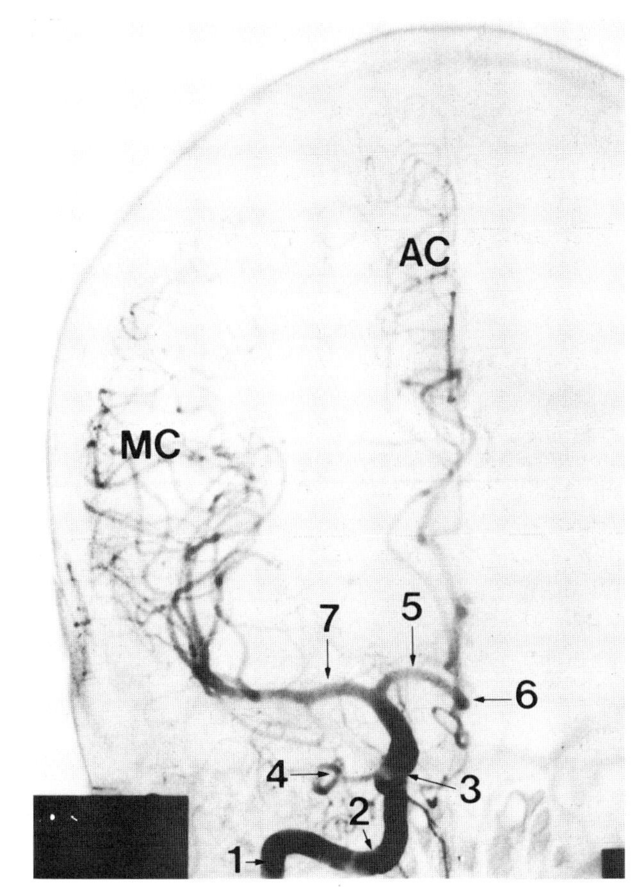

Antero-posterior view

THE VEINS OF THE BRAIN AND THE VENOUS PHASE OF THE CAROTID ANGIOGRAM

The Veins of the Brain

These consist of the cerebral and cerebellar veins and the veins of the brain stem.

The cerebral veins may be divided into the external veins which drain the surfaces of the hemispheres and the internal veins which drain the inner regions of the hemispheres. The external veins may themselves be divided into a superficial group which drains the convex surface of each hemisphere and a deep group which drains the medial surface.

The superficial external group consists of superior, middle and inferior cerebral veins, together with the superior and inferior anastomotic veins, which connect the middle superficial cerebral vein with the sagittal and transverse sinuses respectively.

The middle superficial cerebral vein lies between the lips of the posterior ramus of the lateral fissure and ends in the cavernous sinus. Its surface marking is therefore the same as that for the posterior ramus.

The deep external group consists of the anterior cerebral vein, the deep middle cerebral vein and the striate veins which unite to form the basal vein. This wraps around the cerebral peduncle to drain into the greater cerebral vein of Galen.

The internal cerebral veins are fairly constant in position. They are formed on each side near the interventricular foramen of Monro by the union of the thalamostriate and choroidal veins. They join the basal vein of Rosenthal to form the great cerebral vein of Galen, which drains into the straight sinus.

Venous Phase of the Carotid Arteriogram

The early venous phase tends to show the variably placed external cerebral veins.

The late venous phase highlights the deep external veins and the internal cerebral veins together with the venous sinuses. The deep veins are fairly constant in position and their displacement may be useful in localising space occupying lesions within the substance of the brain.

Lateral view

Superior sagittal sinus [**1**]
Superior cerebral veins [**2**]. There are usually 8–12 in number, and they tend to follow sulci. Notice their anterior course at the point of entry into the superior sagittal sinus
The superior anastomotic vein [**3**]. The middle superficial cerebral vein is not shown
Inferior sagittal sinus [**4**]
Straight sinus [**5**]
Transverse sinus [**6**]
Internal cerebral vein [**7**]
Basal vein [**8**]
Great cerebral vein (of Galen) [**9**] formed by the junction of [**7**] and [**8**]
Thalamostriate vein [**10**]
Septal vein [**11**]
A superficial cerebral vein [**12**] draining into the sphenoparietal sinus
Cavernous sinus [**13**]

The angle between the septal and thalamostriate veins [**A**] usually marks the position of the interventricular foramen of Monro.

Antero-posterior view

The superior sagittal sinus, inferior sagittal sinus and straight sinus are all single midline structures and are, therefore, superimposed [**1**]
Superior cerebral veins [**2**]
Transverse sinuses [**3**]—notice asymmetry
Origin of sigmoid sinus [**4**]
Sphenoparietal sinus [**5**]
Superior orbital margin [**6**]
Optic canal in lesser wing of the sphenoid [**7**]
Cavernous sinus [**8**]

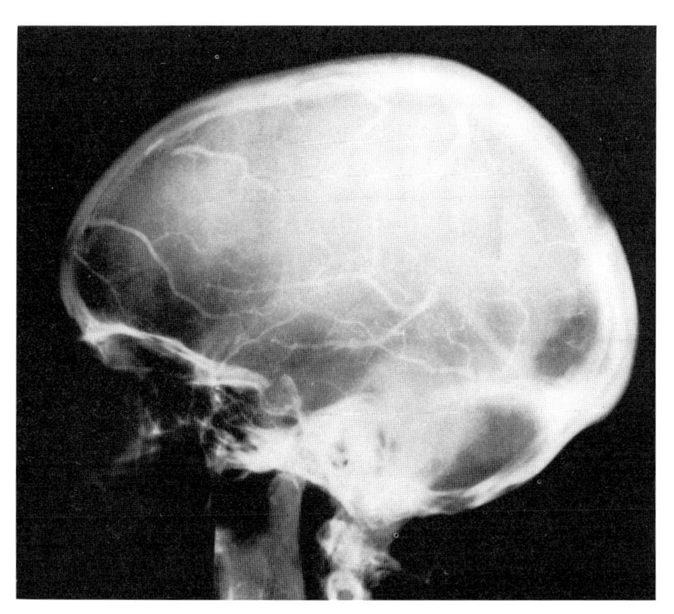

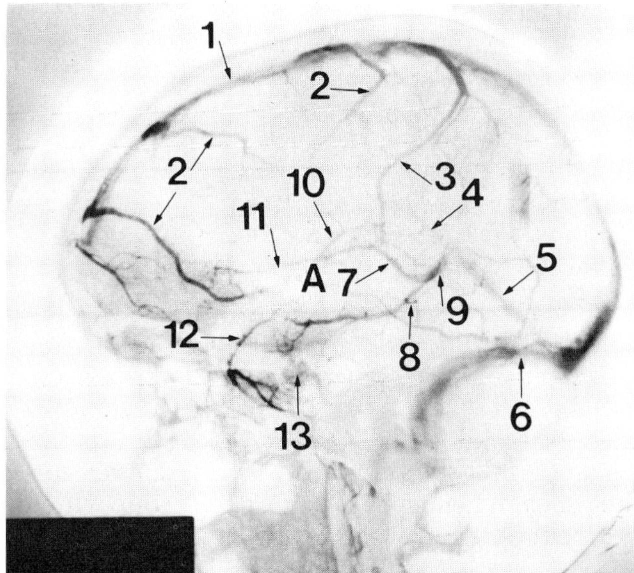

Lateral view

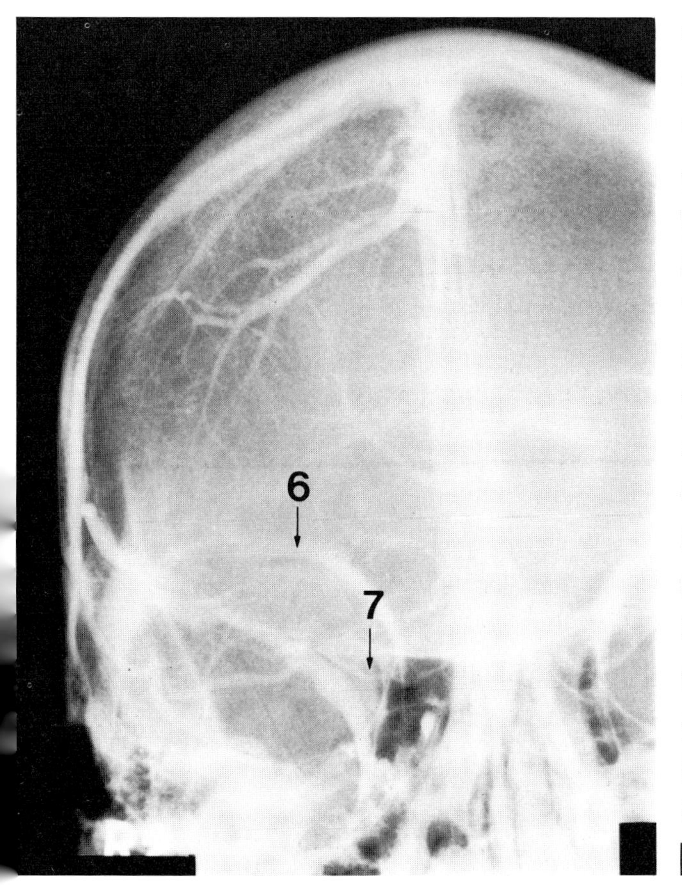

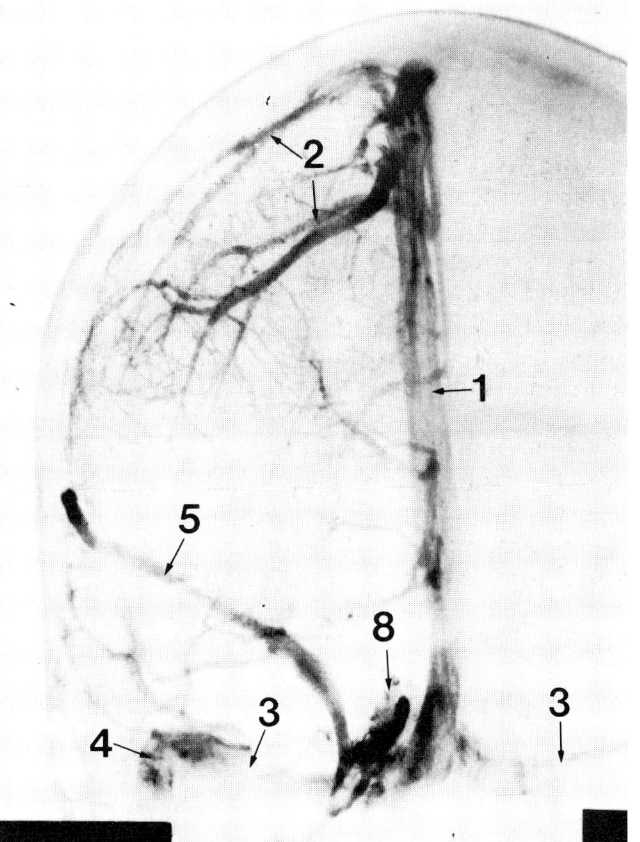

Antero-posterior view

VENOUS SINUSES OF THE HEAD

These are intracranial vascular spaces lying between the two layers of the dura mater. They are filled with venous blood, but have neither muscle nor valves in their walls. The cerebrospinal fluid drains from the subarachnoid space into the venous sinuses via the arachnoid granulations.

The superior sagittal sinus [**1**] runs in the upper margin of the falx cerebri, arching in the midline, from the glabella in front to the inion behind. It is usually continuous with the *right transverse sinus.*

Lacunae laterales [**2**] are short wide diverticulae draining into the superior sagittal sinus on either side. There is usually a large one about 2 cm in diameter overlying the apex of the motor cortex on either side.

Inferior sagittal sinus [**3**]

The straight sinus [**4**] is formed by the junction of the inferior sagittal sinus and the great cerebral vein (of Galen). It is represented by a line beginning 2 cm above the top of the auricle running downwards and backwards to the internal occipital protuberance which is represented externally by the inion. It is usually continuous with the left transverse sinus.

The transverse sinus [**5**] is indicated by a line commencing just lateral to the inion and running horizontally with a slight upward convexity to the back of the root of the auricle.

The sigmoid sinus [**6**] is the continuation of the transverse sinus and runs from the back of the root of the auricle downwards for 2 cm to the level of the lower margin of the external auditory meatus. This sinus is important clinically because of its close relation laterally to the mastoid air cells and anteriorly to the tympanic antrum, both frequent sites of infection.

The cavernous sinuses [**7**] lie on either side of the body of the sphenoid behind the apex of the orbit opposite two fingertip breadths above the tubercle of the zygoma. Infection in the sphenoidal air sinuses may spread to involve the cavernous sinus and maxillary nerve with serious consequences, and pituitary tumours may compress the sinus, injuring the related nerves and obstructing venous return from the orbit.

All of these sinuses, with the exception of the sigmoid sinus, can be clearly seen in the films of the venous phase of the carotid angiogram on the previous page.

Superior sagittal sinus puncture

The superior sagittal sinus is easily accessible through the anterior fontanelle and it is occasionally used as a site for venepuncture or drips. However, the absence of a muscular wall makes uncontrollable bleeding a real risk; this, combined with the risk of infection, has resulted in this site falling from favour and it is now only used as a last resort.

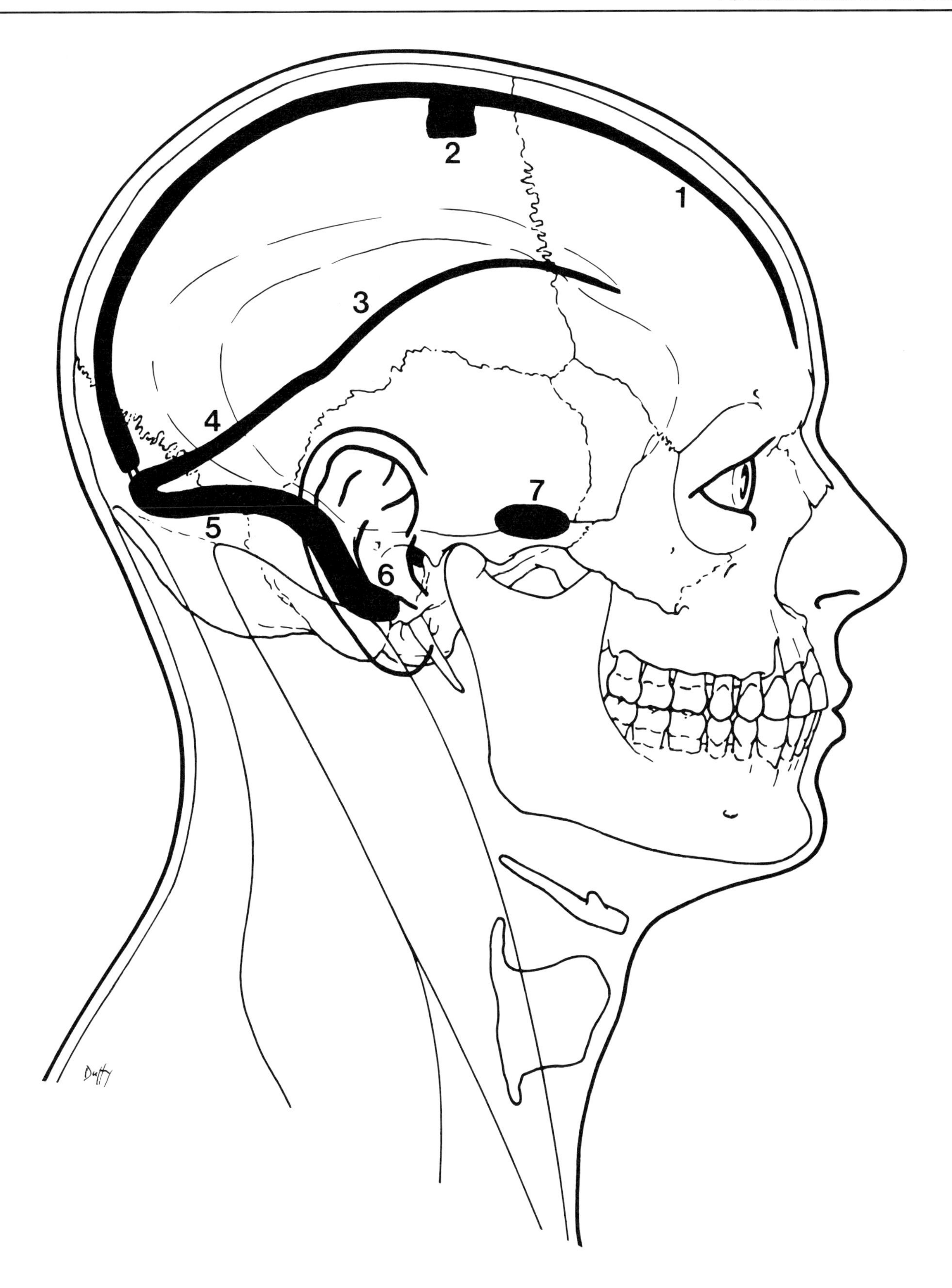
1
2
3
4
5
6
7
Duffy

SOME VEINS OF THE HEAD AND NECK

The internal jugular vein [**1**] begins at the jugular foramen as the continuation of the sigmoid sinus and ends by joining the subclavian vein to form the brachiocephalic vein. It runs opposite a broad band drawn from the lower margin of the external auditory meatus to the upper border of the clavicle between the two heads of the sternocleidomastoid. Its surface marking is important as it is often cannulated in order to administer total parenteral nutrition or to measure the central venous pressure.

The superficial temporal vein [**2**]

The junction of the superficial temporal and the maxillary veins behind the neck of the mandible, [**3**] forming the retromandibular vein.

The retromandibular vein [**4**] may be represented by a line running just behind the ramus of the mandible between the levels of the neck and the angle of the mandible.

The division of the retromandibular vein [**5**] into the anterior and posterior branches.

Posterior auricular vein [**6**]

The junction of the posterior auricular and the posterior branch of the retromandibular veins [**7**] forming the external jugular vein.

The external jugular vein [8] corresponds to a line beginning a little behind and below the angle of the mandible and extending down to the point at which the posterior border of the sternocleidomastoid meets the clavicle. It is often visible, particularly on a hot day, or when a patient is reclined. About 2.5 cm above the clavicle it pierces the investing layer of the deep cervical fascia. The vein at this point is held open by the fascia and cannot collapse when empty. If the vein is severed above this point, air may be sucked into the vein during inspiration, resulting in a fatal air embolus.

The facial vein [**9**] is marked by a line from the medial angle of the eye to the point at which the masseter meets the lower border of the mandible. At this point the vein pierces the deep fascia and continues in the same direction for a further 2–3 cm and then joins the anterior branch of the retromandibular vein to form the common facial vein.

The deep facial vein [**10**] leaves the facial vein midway between the prominence of the cheek and the lower border of the mandible and connects with the pterygoid plexus which, in turn, is connected to the cavernous sinus. Thus, retrograde infection may spread from the face to the intracranial cavernous sinus via the deep facial vein.

The common facial vein [**11**] arises from the confluence of the facial vein [**9**] and the interior branch of the retromandibular vein and drains into the internal jugular vein [**1**].

The anterior jugular vein [**12**] corresponds to a line drawn one finger's breadth from the midline from the chin down to 2.5 cm above the sternal end of the clavicle. From here it turns laterally and runs just above the clavicle behind the sternocleidomastoid muscle to join the external jugular vein. The anterior jugular veins of each side communicate with each other between the two layers of the investing deep cervical fascia one finger's breadth above the manubrium.

The occipital vein [**13**] arises in the scalp and pierces the cranial attachment of the trapezius 2.5 cm lateral to the inion to enter the suboccipital triangle.

The supratrochlear frontal vein [**14**] is the most medial vein on the forehead, about 1 cm from the midline. It is often visible when dilated on a hot day.

The supraorbital vein [**15**] arises near the zygomatic process of the frontal bone and runs medially along the upper margin of the orbit to join the supratrochlear vein and form the facial vein. It sends a branch through the supraorbital notch to join the superior ophthalmic vein, which drains into the cavernous sinus. This provides a potential route for the spread of infection from the face to the intra-cranial sinuses—a process facilitated by the fact that the facial vein has no valves.

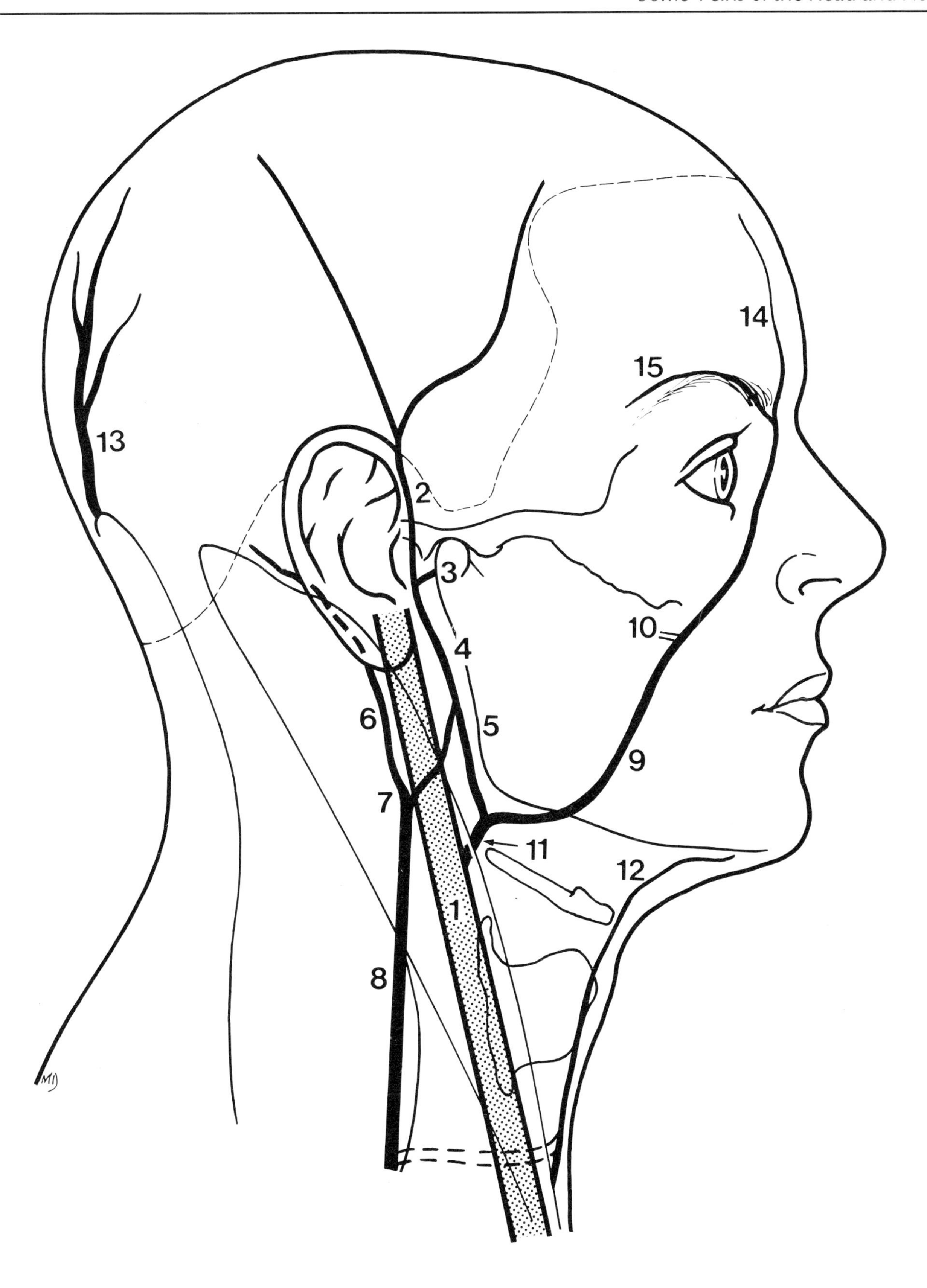
13
14
15
2
3
10
4
6
5
9
7
11
12
1
8

THE BRAIN

The frontal pole [**1**] is opposite the root of the nose close to the median plane.

The occipital pole [**2**] lies to the side of the inion.

The sylvian point [**3**] lies opposite the pterion, 4 cm vertically above the middle of the zygomatic arch and marks the beginning of the lateral sulcus.

The temporal pole [**4**] is represented by a curved line drawn, convex forward, from the sylvian point to one finger's breadth behind the jugal point.

The lateral sulcus (sylvian fissure) begins at the sylvian point and divides into three rami—anterior, ascending and posterior.

The anterior ramus of the lateral sulcus [**5**] is represented by a line, 2.5 cm long, drawn horizontally forward from the sylvian point.

The ascending ramus of the lateral sulcus [**6**] is represented by a line, 2.5 cm long, drawn vertically upwards from the sylvian point.

The posterior ramus of the lateral sulcus [**7**] is marked by a line from the sylvian point with a slight upward concavity to a point 1 cm below the parietal eminence.

The central sulcus [**8**] begins on the sagittal suture at a point one finger's breadth behind the midpoint between the glabella and inion. It passes sinuously downwards and forwards until it almost meets the posterior ramus of the lateral sulcus, 2.5 cm posterior to its origin.

The pre-central sulcus [**9**] runs parallel to the central sulcus, 1 cm anterior to it; between the two lies the *motor cortex*.

The post-central sulcus [**10**] runs parallel to the central sulcus, 1 cm posterior to it, between the two lies the *sensory cortex*.

The parieto-occipital sulcus [**11**] is situated mainly on the medial side of the occipital lobe and, therefore, only cuts into the supero-lateral surface to the extent of the depth of the sulcus, about 2.5 cm. It is opposite a line, 2.5 cm long, drawn laterally from the lambda. The lambda is usually recognisable by palpation.

The superior temporal sulcus [**12**] runs parallel to the posterior ramus of the lateral sulcus about 1 cm below it.

Inferior frontal sulcus [**13**]

The motor speech area of Broca [**14**] lies mainly in the inferior frontal gyrus between the ascending rami of the lateral sulcus and the pre-central sulcus. It also extends into the pars triangularis between the anterior and ascending rami of the lateral sulcus.

The motor speech area of Wernicke [**15**] is large. It lies partly on the superior surface of the temporal lobe within the lateral sulcus and curves around the upper end of the lateral sulcus to occupy a large area of the parietal and temporal lobes.

The acoustic area [**16**] is situated just below the middle of the lateral sulcus.

The visual cortex [**17**] lies mainly in the medial aspect of the occipital lobe on either side of the calcarine sulcus, but it becomes superficial as a small area just above the occipital pole.

Before the advent of CT scanning, air was sometimes used as a contrast medium to outline the ventricles of the brain by introducing it through a lumbar puncture needle into the spinal canal, or occasionally directly into the ventricle. A pneumo-encephalogram is shown on the facing page.

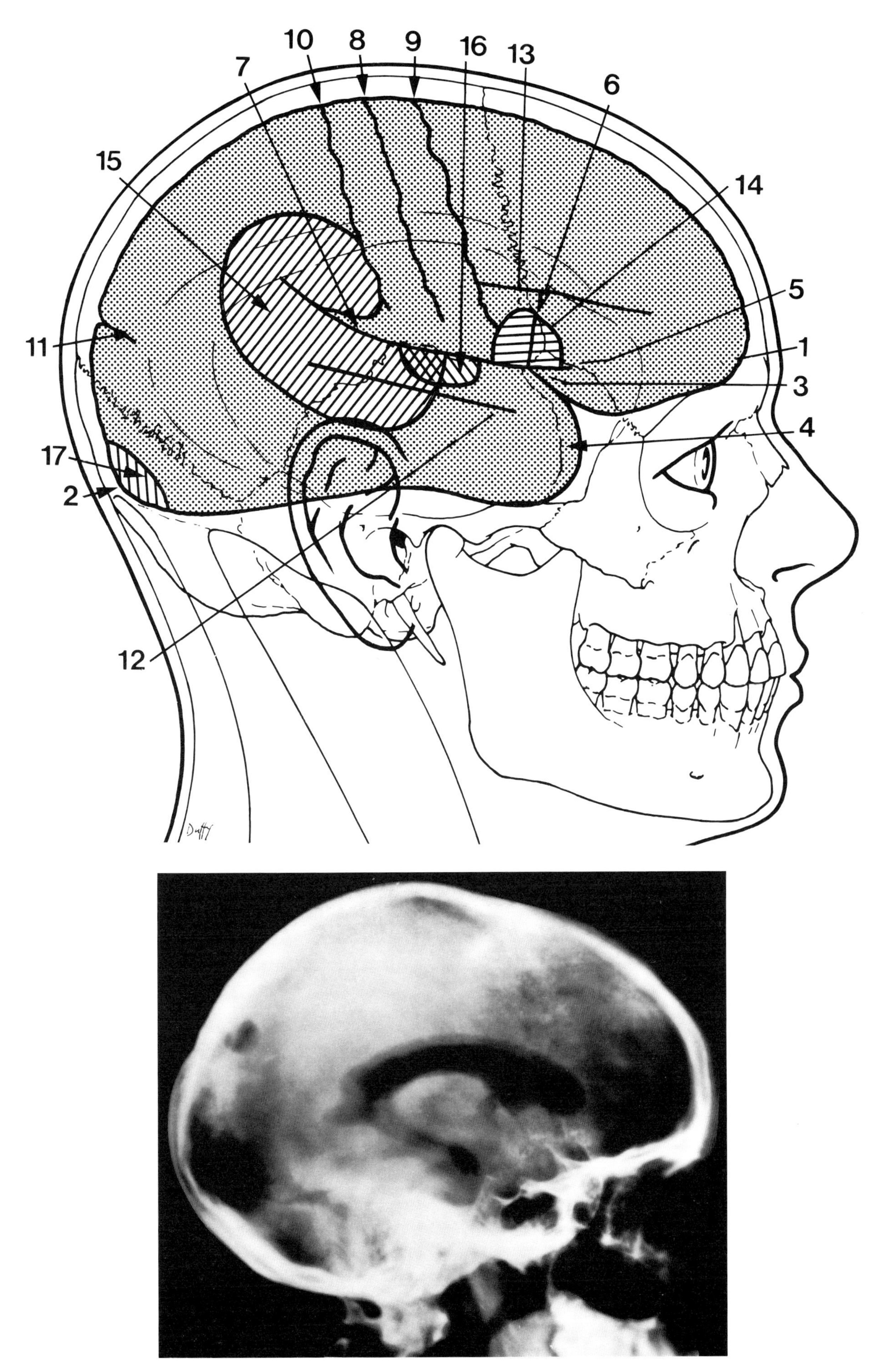

Pneumo-encephalogram

THE NERVES OF THE HEAD

The skin of the head and neck is supplied partly by the trigeminal nerve and partly by the anterior and posterior primary rami of the upper cervical nerves. In addition, the vagus, facial and glossopharyngeal nerves contribute to the supply of the auditory meatus.

The Trigeminal Nerve

This nerve divides at the semilunar ganglion into ophthalmic, maxillary and mandibular branches.

The semilunar ganglion lies about 5 cm from the surface of the head opposite the head of the mandible.

The ophthalmic nerve runs forward in the lateral wall of the cavernous sinus and divides into the lacrimal, frontal and nasociliary nerves before entering the orbit through the superior orbital fissure. It supplies sensation to the contents of the orbit, the mucosal lining of the frontal and ethmoidal air sinuses and part of the mucosal lining of the nostrils. Some of the branches reach the skin.

The cutaneous branches are:
The supraorbital and supratrochlear branches of the frontal nerve
The infratrochlear and external nasal branches of the nasociliary nerve
The palpebral branch of the lacrimal nerve.

Supraorbital nerve [1] emerges from the supraorbital notch, two finger breadths from the median plane and runs upwards to supply the skin as far as the vertex.

The supratrochlear nerve [2] emerges from the supraorbital margin, one finger's breadth from the median plane and runs up as far as the hair line.

The infratrochlear nerve [3] emerges just above the medial end of the palpebral fissure and runs halfway down the side of the nose.

The external nasal nerve [4] is the terminal branch of the nasociliary nerve and emerges between the nasal bone and cartilage to supply the distal half of the nose.

The palpebral branch [5] of the lacrimal nerve is a twig which supplies the lateral half of the upper eyelid.

The maxillary nerve runs beneath the ophthalmic nerve in the lateral wall of the cavernous sinus. It leaves the cranial cavity through the foramen rotundum and traverses the roof of the sphenopalatine fossa to enter the orbit through the infraorbital fissure. Once in the orbit, it is known as the *infraorbital nerve* which runs along the floor of the orbit before entering the infraorbital canal to emerge on the face through the infraorbital foramen.

The sphenopalatine ganglion hangs from the maxillary nerve in the sphenopalatine fossa. It lies deep to the zygomatic arch a little behind the jugal point.

The maxillary nerve supplies sensation to the teeth and gums of the upper jaw, the mucus membrane of the maxillary and sphenoidal air sinuses, part of the upper lip and some of the skin of the face and head.

The cutaneous branches are:
The infraorbital nerve
The zygomatic nerve which divides into the zygomaticotemporal and zygomaticofacial nerves.

The infraorbital nerve [6] lies two finger breadths from the median plane and sends branches to the side of the nose, the upper lip and the cheek below the eye.

The zygomaticotemporal nerve [7] emerges from the foramen of the same name and supplies the skin over the anterior part of the temple.

The zygomaticofacial nerve [8] similarly supplies the skin over the bony prominence of the cheek.

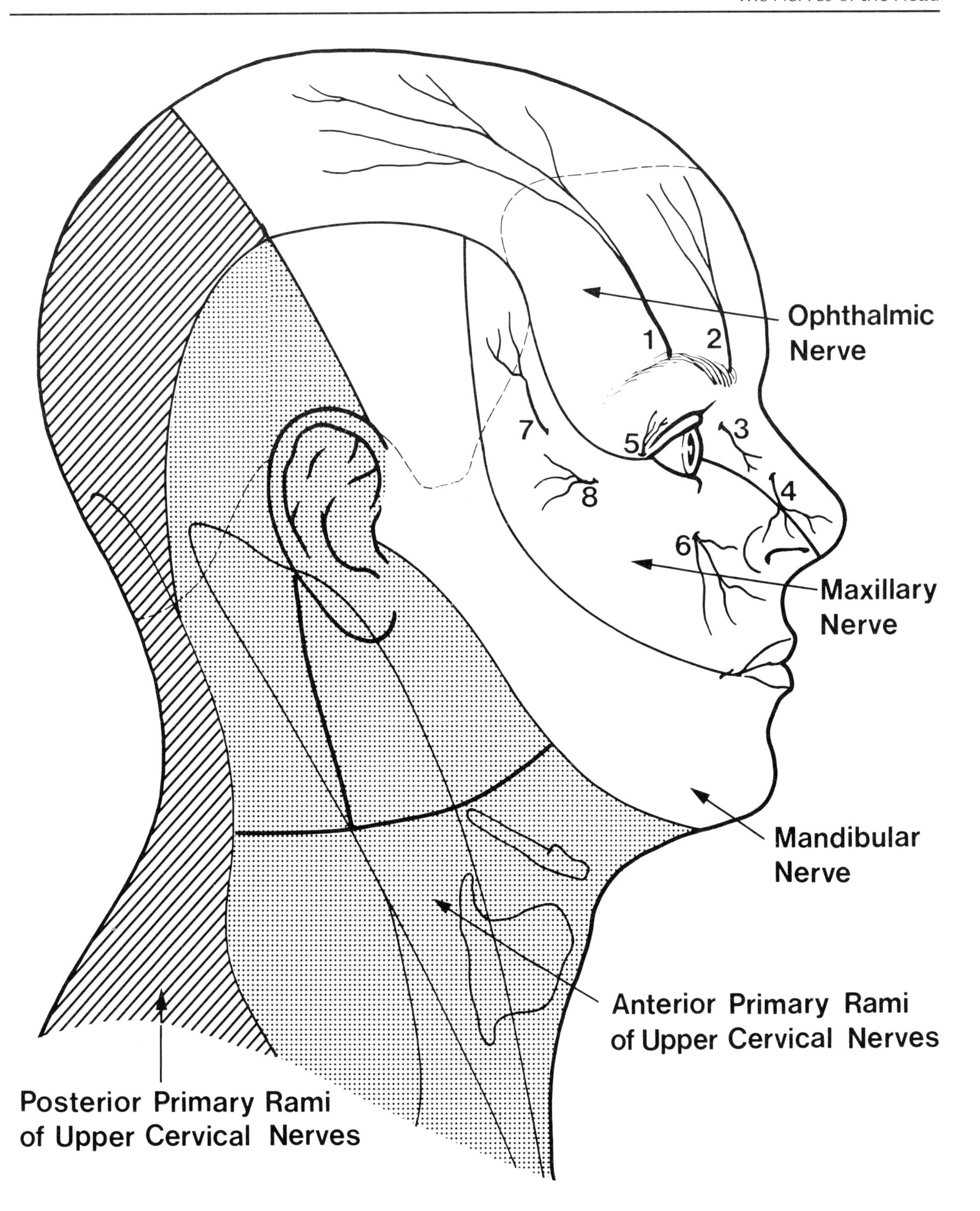
Ophthalmic Nerve
1
2
3
4
5
6
7
8
Maxillary Nerve
Mandibular Nerve
Anterior Primary Rami of Upper Cervical Nerves
Posterior Primary Rami of Upper Cervical Nerves

The mandibular nerve leaves the skull through the foramen ovale about 3.75 cm deep to the tubercle of the root of the zygoma. Just beneath this point it divides into the lingual nerve, the inferior alveolar nerve, the buccal nerve, the auriculotemporal nerve and the motor nerves to the muscles of mastication.

Thus, all these nerves begin opposite the tubercle of the root of the zygoma and the optic ganglion, which lies posterior to the main trunk of the mandibular nerve which also lies at about this level.

The lingual nerve [**1**] runs under the cover of the mandible along a line from the tubercle of the root of the zygoma to a point just above the lower border of the mandible, midway between the angle and the symphysis menti, and from there to the canine tooth.

The submandibular ganglion [**2**] lies under cover of the mandible and the submandibular gland at the lowest point of the lingual nerve.

The inferior alveolar nerve [**3**] runs opposite a line from the tubercle of the root of the zygoma to the centre of the ramus of the mandible, and then downwards and forwards midway between the two borders of the body of the mandible almost as far as the symphysis menti.

The mental nerve [**4**] is the continuation of the inferior alveolar nerve which emerges through the mental foramen and supplies the mucus membrane of the lower lip and the skin covering the mandible between the symphysis menti and the masseter.

The buccal nerve [**5**] is marked by a line from the tubercle of the root of the zygoma to the angle of the mouth. It supplies the skin and mucus membrane of the soft part of the cheek. It must not be confused with the buccal branch of the facial nerve which is a motor nerve to the buccinator muscle.

The auriculotemporal nerve [**6**] runs backwards from the tubercle of the root of the zygoma under cover of the lateral pterygoid muscle and the neck of the mandible. It then turns upwards in front of the auricle to supply the skin over the side of the head and over the upper part of the masseter. It also supplies the upper half of the lateral surface of the auricle and the upper part of the acoustic meatus and tympanic membrane.

The facial nerve leaves the skull through the stylomastoid foramen which lies deep to the posterior margin of the external acoustic meatus. It enters the deep surface of the parotid gland and passes forwards and laterally, crossing the styloid process and retromandibular vein, to lie in the substance of the parotid gland superficial to the carotid artery. Once in the superficial substance of the gland, the nerve divides into several branches which usually radiate from a point immediately anterior to the tragus and leave the gland by its upper and anterior borders to supply the muscles of the face and scalp including the platysma.

The terminal branches are:

Temporal—upwards and forwards towards the temple and forehead

Upper and lower zygomatic—forwards towards the nose and orbit

Buccal—towards the mouth

Marginal mandibular—along the lower border of the mandible before turning upwards and forwards over the body of the mandible to join the branches of the mental nerve

Cervical—passes 1 cm below the angle of the mandible to reach the neck.

Some of the branches may be rolled on bone as they cross the neck of the mandible or the zygomatic arch.

The hypoglossal nerve [**7**] lies superficially in the carotid triangle of the neck where it curves forward over the external carotid artery, about 2 cm above the bifurcation of the common carotid artery and less than 1.25 cm above the tip of the greater horn of the hyoid bone.

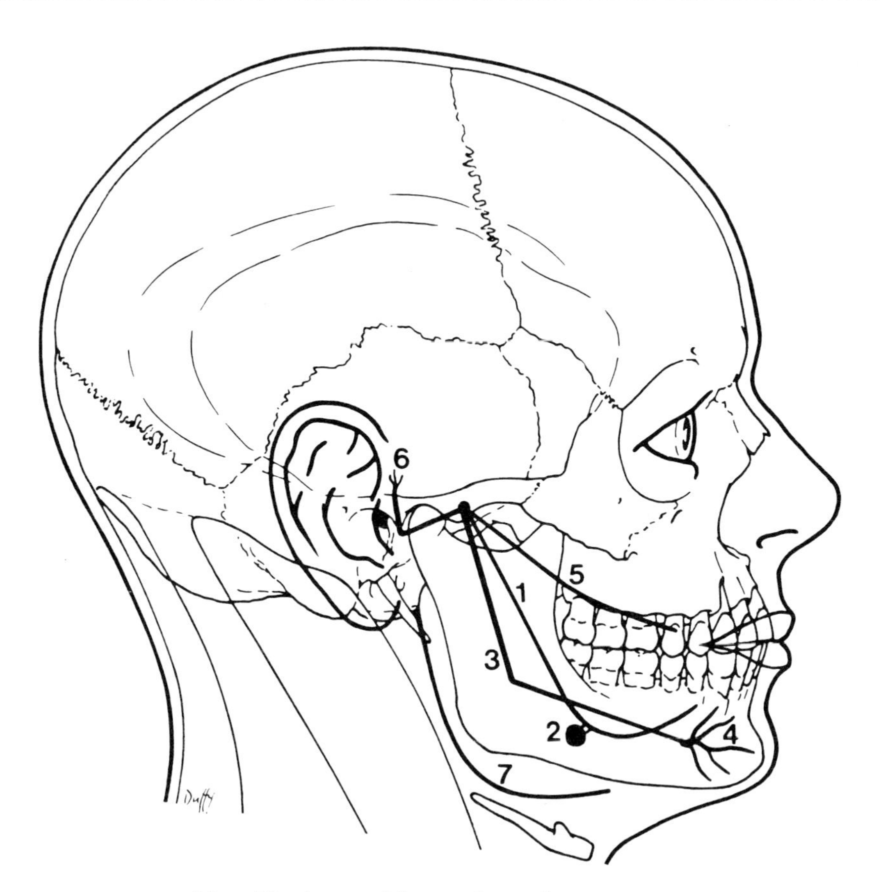

Mandibular and hypoglossal nerves

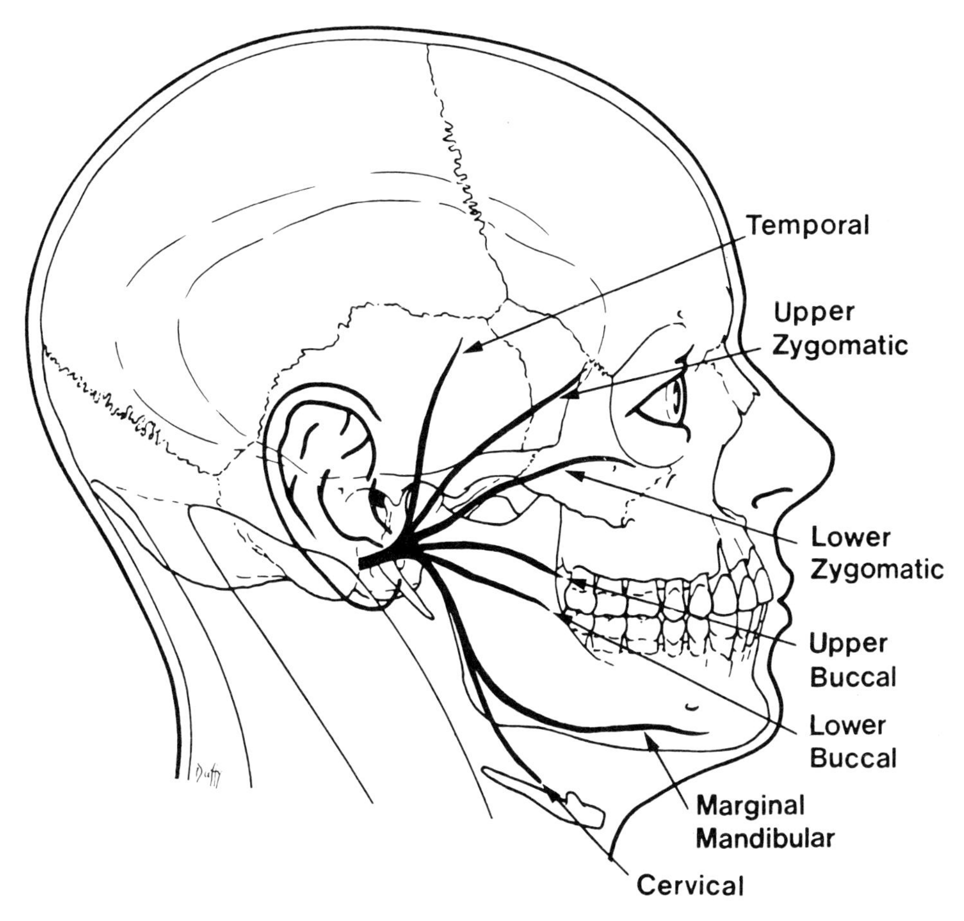

Facial nerve

THE NERVES OF THE NECK

The glossopharyngeal nerve [1] lies opposite a line drawn from the lower margin of the external acoustic meatus to the angle of the mandible. It runs down between the internal and external carotid arteries and breaks up into multiple branches in the submaxillary region to supply the palatine tonsil and the posterior third of the tongue.

The Cervical Nerves

There are eight cervical nerves. The upper seven emerge from the vertebral foramen above the corresponding vertebra, and the eighth emerges from between the eighth cervical and first thoracic vertebrae. Each divides into anterior and posterior primary rami.

The posterior primary rami supply the extensor musculature of the neck and the overlying skin. The first does not, and the seventh and eighth may not, have cutaneous branches. The cutaneous branch of the second posterior primary ramus forms the greater occipital nerve and the small cutaneous branch of the third posterior primary ramus forms the third occipital nerve.

The greater occipital nerve [2] is represented by a line beginning halfway between the transverse processes of the atlas and axis, to a point on the superior nuchal line, 2 cm from the inion. Thereafter, it breaks up into multiple branches supplying the skin as far as the vertex.

The third occipital nerve [3] supplies the skin over the inion and the nape of the neck.

The anterior primary rami form the cervical plexus from the first four units, and the lower four unite with the anterior primary ramus of the first thoracic nerve to form the brachial plexus.

The cervical plexus lies beneath the upper half of the sternocleidomastoid opposite a vertical line from the tip of the mastoid process to the level of the upper border of the thyroid cartilage.

The cervical plexus gives rise to four important cutaneous nerves, all of which emerge in a cluster from behind the middle of the posterior border of the sternocleidomastoid muscle to supply the areas shown on the diagram.

The lesser occipital nerve [4]—C2

The greater auricular nerve [5]—C2, 3

The transverse cutaneous nerve of the neck [6]—C2, 3

The supraclavicular nerve [7]—C3, 4

The supraclavicular nerve divides into medial, intermediate and lateral branches which supply skin over the respective thirds of the clavicle as far down as the second rib.

The phrenic nerve—*see* The Thorax p. 24.

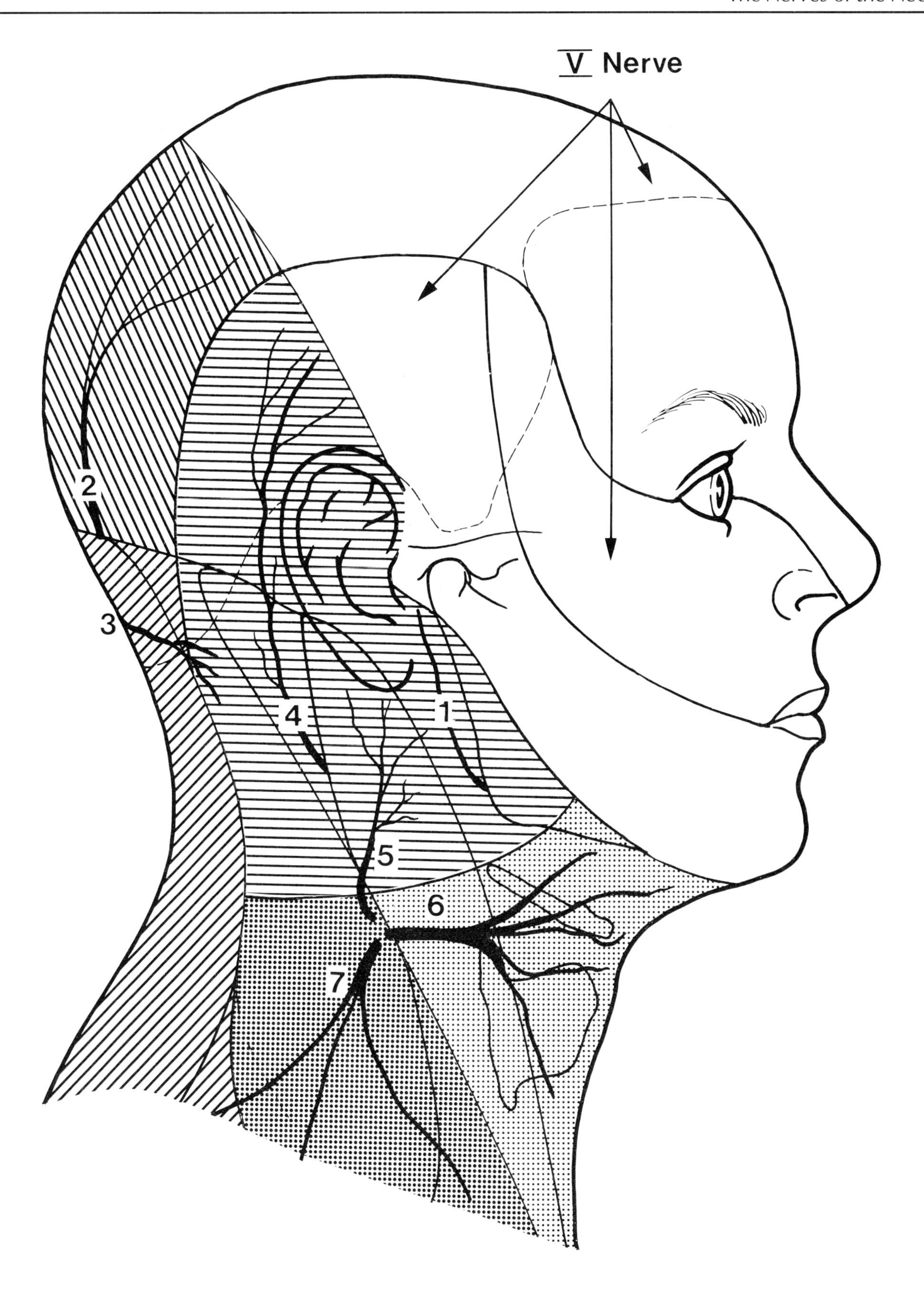
V Nerve
1
2
3
4
5
6
7

The accessory nerve [1] may be marked by a line commencing midway between the tip of the mastoid process and the neck of the mandible which runs backwards and downwards crossing the posterior border of the sternocleidomastoid muscle at the junction of its upper and middle third. Thereafter, the line crosses the posterior triangle of the neck and meets the anterior border of the trapezius at the junction of the middle and lower thirds.

In its course through the posterior triangle of the neck, the nerve is surrounded by deep cervical lymph nodes, and this should be borne in mind when cervical lymph nodes are being biopsied or removed.

The brachial plexus lies in three regions: the base of the posterior triangle of the neck, behind the clavicle and in the upper part of the axilla.

The origin of the plexus is at the lateral border of the scalenus anterior which reasonably corresponds with the posterior border of the sternocleidomastoid. Thus, the surface marking for the origin of the plexus is a line drawn down the lower third of the posterior border of the sternocleidomastoid.

The upper or lateral margin may be represented by a line from the junction of the lower and middle thirds of the sternocleidomastoid to the medial side of the tip of the coracoid process, crossing the clavicle just lateral to its midpoint.

The lower or medial margin is represented by a line from the junction of the posterior border of the sternocleidomastoid with the clavicle to a point one finger's breadth below and medial to the tip of the coracoid process.

The roots of the brachial plexus lie between the scalenus anterior and scalenus medius muscles.

The trunks lie in the posterior triangle of the neck where the upper trunk can be easily palpated in the thin subject with his arm hanging by his side. Here the upper trunk may be subject to damage during forceps delivery, from a karate chop, or from violent falls onto the shoulder. The upper trunk is easily confused with the inferior belly of the omohyoid muscle on palpation, the muscle lying below the upper trunk.

The divisions occur behind the clavicle just proximal to the lateral border of the first rib.

The cords are formed behind the clavicle at the outer border of the first rib where they enter the axilla above the first part of the axillary artery. The cords then embrace the second part of the artery in the relationship implied by their names, i.e. medial cord, medial to the artery, posterior cord, posterior to the artery and the lateral cord, lateral to the artery.

The branches arise around the third part of the artery.

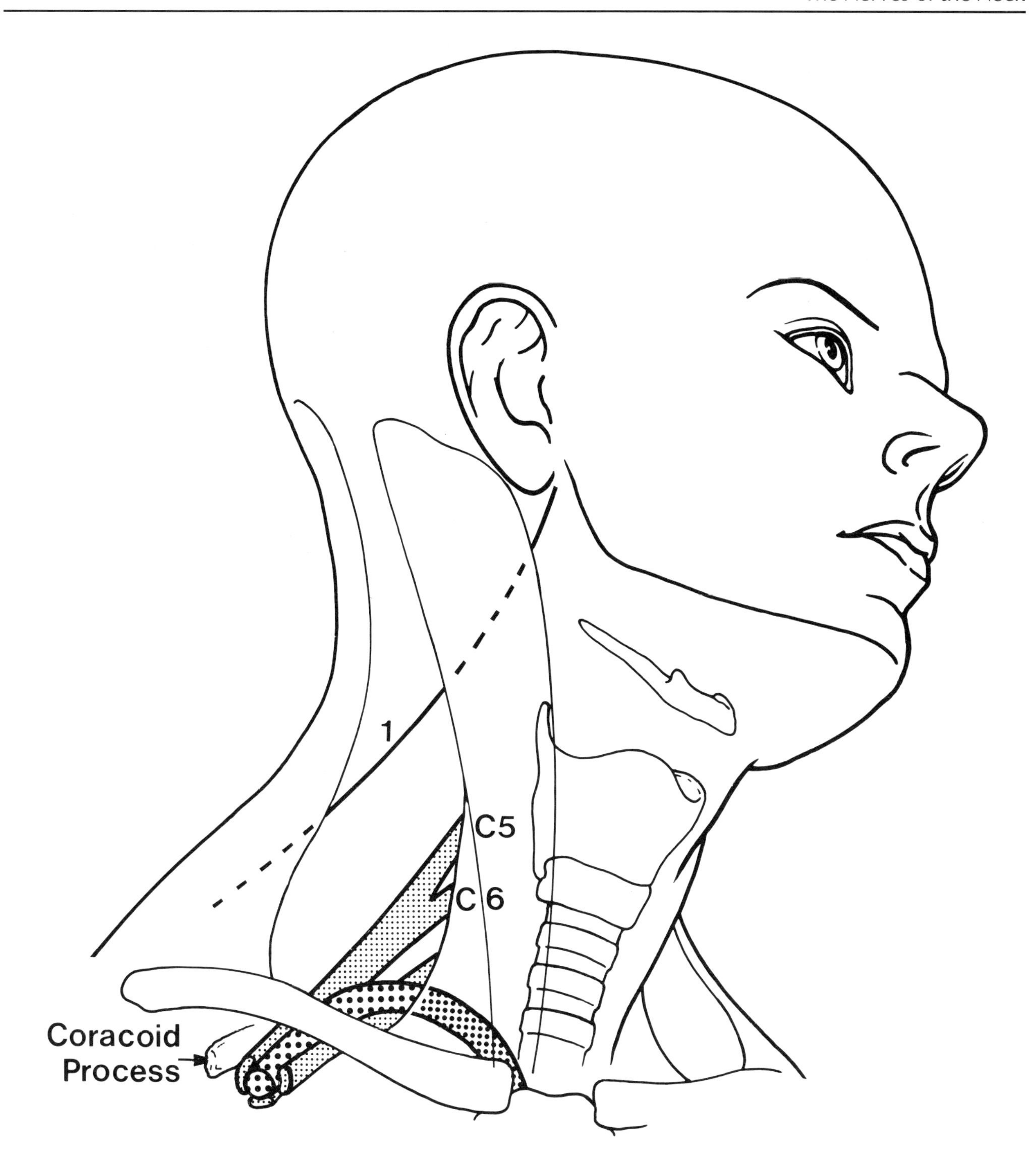
1
C5
C 6
Coracoid
Process

THE EXTERNAL EAR

The helix [**1**]
The auricular tubercle (of Darwin) [**2**]
The antihelix [**3**]
Triangular fossa [**4**]—this lies between the two crura of the antihelix
Scaphoid fossa [**5**]
Concha [**6**]
Cymba conchae [**7**]—this is the part of the concha above the crus of the helix
Tragus [**8**]
Antitragus [**9**]
Intertragic incisure [**10**]
Lobule [**11**]

On the cranial surface of the pinna there are elevations corresponding to the depressions on the lateral surface. They are named in the genitive after the lateral depressions, e.g. eminentia conchae, eminentia fossae triangularis.

The skin of the auricle is thin and covered with fine hair and sebaceous glands, the latter are more numerous in the concha and scaphoid fossa. In elderly males, the hair around the tragus and antitragus may become thicker.

Cutaneous Nerve Supply

The auricle is supplied as indicated in the diagram. The cranial surface is supplied mainly by the great auricular nerve, except for the upper part which is supplied by the lesser occipital nerve.

The concha receives its supply from the auricular branch of the vagus and the facial nerve. These also supply the eminentia conchae.

The meatus is supplied by two nerves. The auriculotemporal branch of the mandibular nerve supplies the antero-superior half of the wall and the auricular branch of the vagus supplies the postero-inferior half.

Clinical Relevance

Facial nerve. In geniculate herpes (shingles of the facial nerve), small vesicles appear over the areas of sensory supply of the facial nerve. These areas are the concha and the anterior two-thirds of the tongue which are supplied by the pretrematic branch of the facial nerve—the chorda tympani.

The vagus. The auricular branch of the vagus is probably a pretrematic branch and certain vagal reflexes may be elicited when the meatus is stimulated or irritated. If the ears are syringed with cold water, rather than warm water, young children may vomit and the elderly may be precipitated into heart failure by vagal inhibition. This latter reflex may be employed to advantage in the treatment of spontaneous supraventricular tachycardias. Sneezing and coughing are lesser manifestations of irritant reflexes.

The auriculotemporal nerve. A branch of the mandibular division of the trigeminal nerve which supplies the tongue and teeth. Thus, pain from a carious tooth or from carcinoma of the tongue may be referred to the ear resulting in a false localising sign.

Surgical Anatomy

The cymba conchae overlies the suprameatal triangle which may be palpated through it (*See* Bony Landmarks of the Head—lateral aspect, p. 184). Deep to this lies the mastoid antrum.

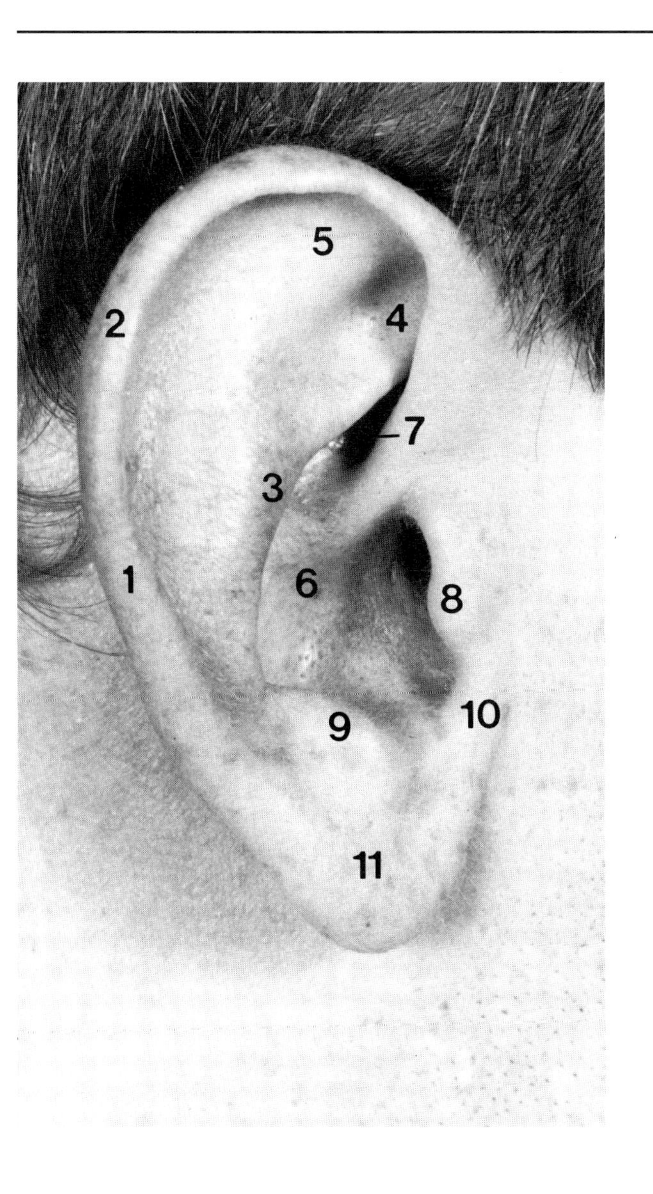
1
2
3
4
5
6
7
8
9
10
11

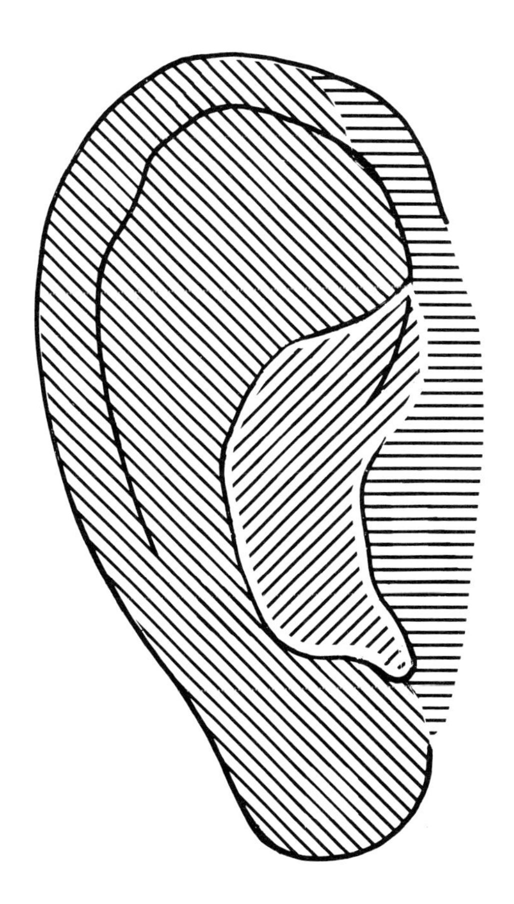
Great Auricular Nerve
Vagus & Facial Nerves
Auriculotemporal Nerve

THE EXTERNAL ACOUSTIC MEATUS

The external acoustic meatus and the tympanic membrane are examined with the aid of an electric auroscope. For best results, it is wise to ensure that the batteries are not half flat, that the beam of light is directed out of the end of the speculum, and that the speculum employed is of the largest size that can be inserted without causing pain.

The external acoustic meatus is a sinuous tube about 3 cm long. Its outer one-third is cartilaginous and the inner two-thirds lie in the petrous temporal bone. To facilitate auroscopic examination, the outer part of the tube may be straightened in the adult by pulling the auricle upwards, backwards and a little laterally. In the infant the auricle should be pulled backwards and *downwards*.

The speculum is inserted as far as the osseo-cartilaginous junction where there is a constriction. The speculum should not be introduced beyond this constriction for fear of damaging the tympanic membrane, particularly in children in whom the meatus is deceptively short.

The constriction at the osseo-cartilaginous junction should be borne in mind when contemplating the removal of foreign objects from the ear. In addition, it should be remembered that there is a depression in the floor of the meatus immediately antero-lateral to the tympanic membrane where foreign objects may become lodged.

Once inserted, the angle of the speculum in the meatus should be varied as much as possible in order to view all the tympanic membrane.

THE TYMPANIC MEMBRANE

The ear drum appears as a pearly-grey glistening membrane which is placed obliquely such that the antero-inferior wall of the meatus is longer than the postero-superior wall. The degree of obliquity varies from individual to individual, and during the growth of a single individual; at birth the membrane is far more oblique lying almost horizontally in the plane of the base of the skull.

The membrane is not flat but is concave, being pulled taught by the handle of the malleus. The resultant large tense part of the membrane is sometimes referred to as the pars tensa.

The overall appearance is as shown in the diagram.

Handle of malleus [**1**]—a yellow-red streak
Short lateral process of malleus [**2**]—a rounded white prominence
Anterior malleolar fold [**3**]
Posterior malleolar fold [**4**]
Pars flaccida [**5**]—between the malleolar folds
Long process of incus [**6**]
Cone of reflected light [**7**]

Surgical Anatomy

The upper half of the tympanic membrane is more vascular than the lower half. In addition, the ossicles and chorda tympani nerve are concealed behind the upper part of the membrane, therefore incisions through the membrane should be in the postero-inferior quadrant.

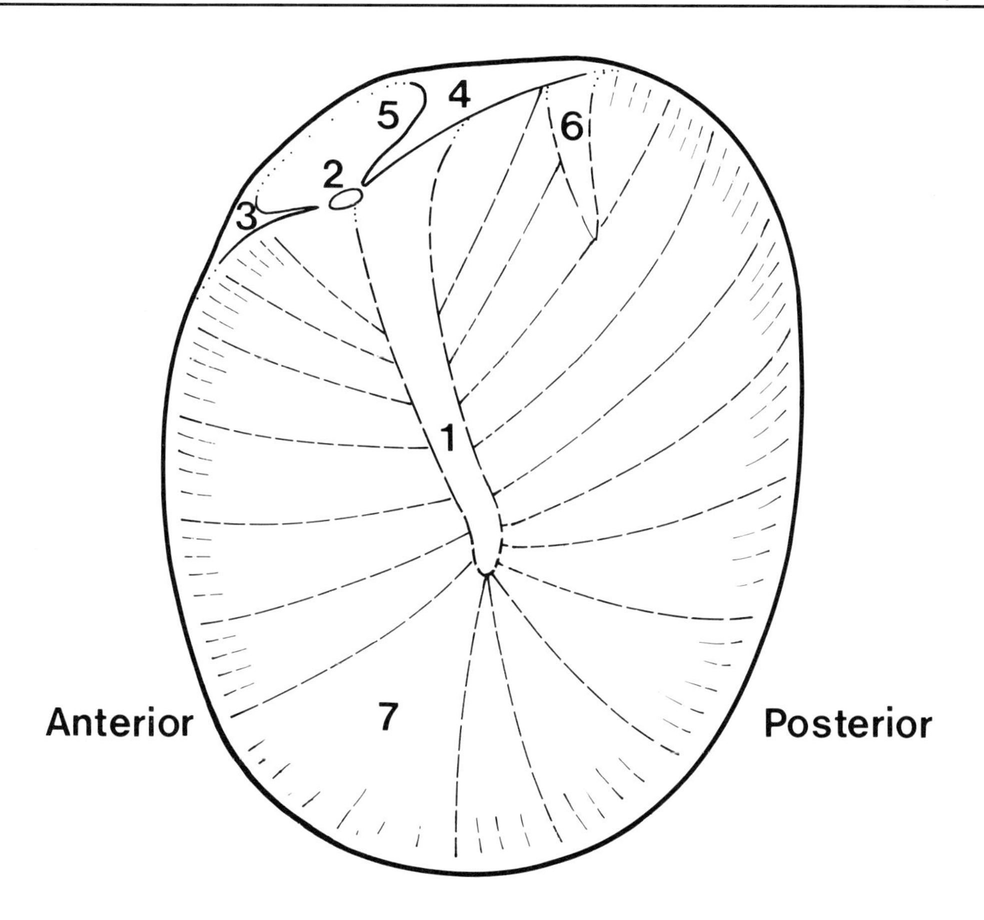
5
4
2
6
3
1
Anterior
7
Posterior

THE EYE AND LACRIMAL APPARATUS

Bony Landmarks

Supraorbital margin [**1**]
Supraorbital notch and nerve [**2**]
Zygomaticofrontal suture [**3**]
Infraorbital margin [**4**]
Frontomaxillary suture [**5**]

Soft Tissues

The palpebral fissure is the interval between the upper and lower lids. When the eye is closed it lies just below the level of the pupil.

The canthus is the angle where the upper and lower lids meet. There is, therefore, a medial canthus and a lateral canthus.

The upper eyelid is much bigger than the lower and its deep surface is best seen by turning the lid inside out. This may be achieved by everting the lid over a rod or pencil held horizontally over the lid, above the level of the eyeball, and exerting upward and outward traction on the eyelashes. In this case the superior fornix of the conjunctiva is still not seen even when the eye is rotated downwards.

The palpebral ligaments are fibrous bands stretching from the junction of the lids on either side to the corresponding margin of the orbit. In thin people, they may raise up a fold of skin especially on the medial side. They are known as the medial [**6**] and lateral [**7**] palpebral ligaments, respectively.

The conjunctiva covers the eyeball and lines the deep surface of the eyelids to form the conjunctival sac which is sealed when the eyelids are shut. The blood vessels of the ocular conjunctiva are so small that it appears almost clear; but the palpebral conjunctiva is more vascular and appears red or pink. The palpebral conjunctiva of the lower lid is readily visible and provides a useful guide to the level of haemoglobin in the blood—in the anaemic patient it appears pale.

The tarsal glands are visible through the conjunctiva when the upper lid is everted. They appear as parallel yellow streaks at right angles to the palpebral fissure.

The tarsal plate is the plate which makes the upper lid rigid and becomes obvious when the upper lid is everted.

The eyelashes are absent at the medial margins of the palpebral fissure.

The plica semilunaris [**8**] is the small fold of conjunctiva immediately lateral to the lacrimal caruncle. It represents the vestigial nictitating membrane.

The lacrimal caruncle [**9**] is the pink fleshy body in the medial canthus.

The lacrimal papilla [**10**] is the small elevation seen at the medial end of each lid where the eyelashes stop. It is more easily seen on the lower lid.

The punctum lacrimale is the small opening on the summit of the lacrimal papilla. It represents the opening of the lacrimal duct.

The lacrimal duct [**11**] conveys lacrimal fluid (tears) to the lacrimal sac. Each is 10 mm long.

The lacrimal sac [**12**] lies behind the medial palpebral ligament and projects above and below it. The infratrochlear nerve passes above the sac.

The nasolacrimal duct [**13**] passes downwards and backwards from the sac in the lateral wall of the nose for about 18 mm before opening into the inferior meatus of the nose.

The lacrimal gland [**14**] is lodged in a fossa behind the supero-lateral part of the orbital margin under cover of the zygomatic process of the frontal bone. It is divided into a large orbital portion and a smaller palpebral portion by the lateral part of the aponeurosis of the levator palpebrae superioris. The gland drains by multiple ducts into the superior temporal fornix.

Supratrochlear nerve [**15**]

Trochlear and superior oblique muscle [**16**]

Infratrochlear nerve [**17**]

Infraorbital nerve [**18**]

Zygomaticofacial nerve [**19**]

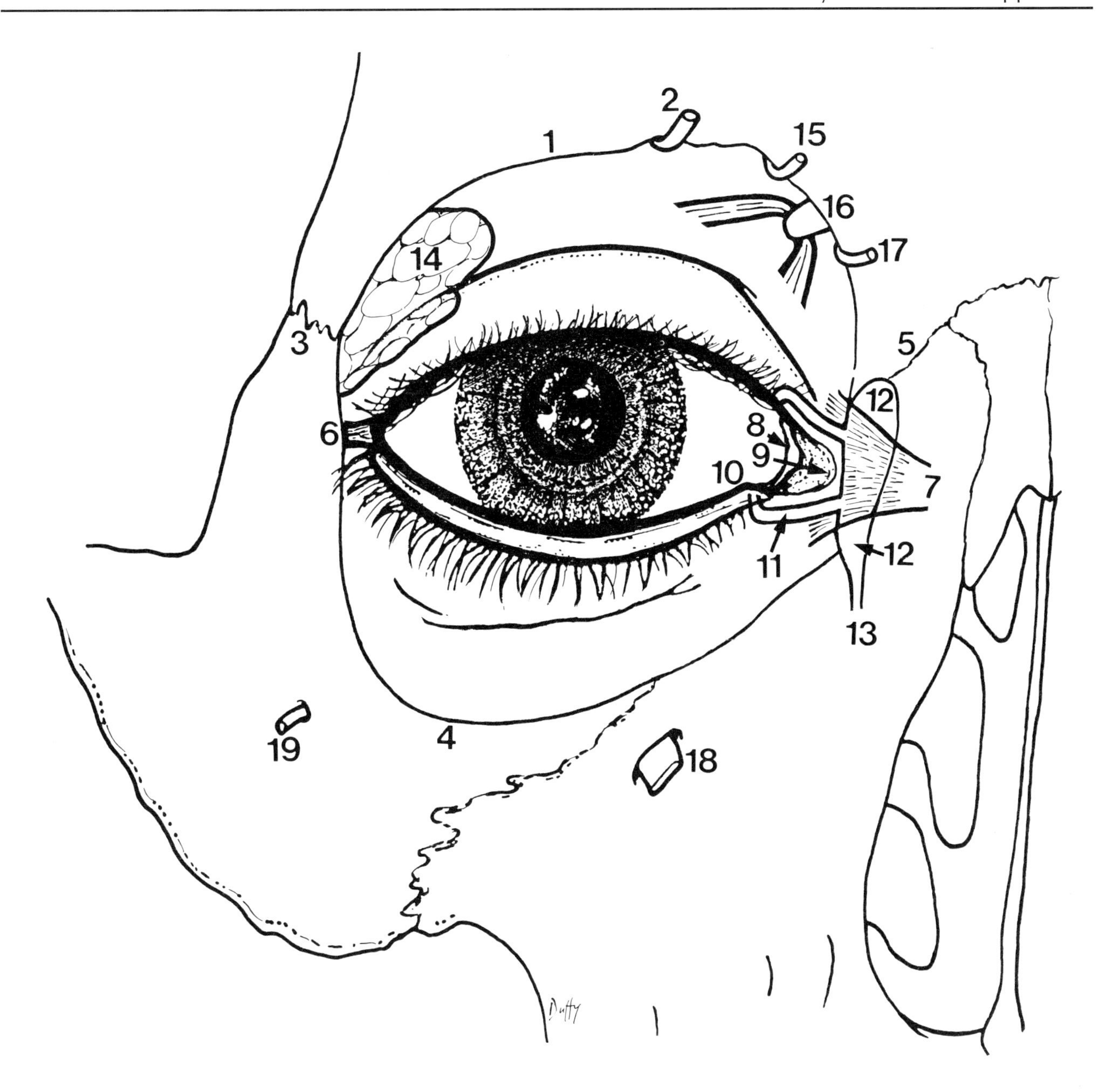
1
2
3
4
5
6
7
8
9
10
11
12
12
13
14
15
16
17
18
19
Duffy

The Eyeball

Sclera [**1**]
Iris [**2**]
Pupils [**3**]. These tend to be larger in dark eyes and smaller in light-coloured eyes and the elderly. They may occasionally be of slightly different size in the healthy individual, but this should always be assumed to be abnormal until proven otherwise.
Cornea [**4**]. In the elderly, there may be a whitish ring around the periphery known as the *arcus senilis*, in younger people it may be associated with hypercholesterolaemia.
Lens [**5**].
Optic nerve [**6**]. The site of emergence of the optic nerve through the choroid is the optic disc seen on ophthalmoscopy.
Fovea [**7**]

Eye Movements

Eye movements are controlled by the third, fourth and sixth cranial nerves which supply the extra-ocular muscles. The nuclei of these nerves are connected together, and to the vestibular nuclei and cerebellum, by the medial longitudinal bundle of the brain stem which provides a pathway for coordinating the eye movements. Any defect of either the extra-ocular muscles or the nerves that supply them, will upset the delicate balance and may result in a squint and, in many cases, double vision.

Although no individual muscle acts in isolation, each muscle does have a prime action and these actions may easily be tested by asking the patient to look in six different directions which correspond approximately to the prime actions of each of the six extra-ocular muscles.

Nerve supply
Fourth nerve—superior oblique
Sixth nerve—lateral rectus
Third nerve—all other extra-ocular muscles, including levator palpebrae superioris and parasympathetic fibres to sphincter pupillae.

Pupillary Reflexes

There are two pupillary reflexes:
(i) the light reflex
(ii) the accommodation reflex

The light reflex. When a light is shone in one eye both pupils constrict. The constriction of the pupil of the stimulated eye is known as the *direct reaction to light*, whereas the constriction of the contra-lateral pupil is known as the *consensual reaction to light*.

The accommodation reflex. This reflex consists of ocular convergence, pupillary constriction and thickening of the lens. The subject is asked to focus on a distant object and then on an object about 30 cm in front of his eyes. His eyes converge on the closer object, his lens thickens to adjust the focus and the pupil constricts to sharpen the image on the retina. The observer should be looking for the pupillary constriction.

The pathways for the light and accommodation reflexes are different and in certain disease states they may become dissociated. Since they both form part of the standard neurological examination, it is imperative that their respective pathways are understood, although such a discussion is outside the scope of this book.

Hippus. If a light is shone continuously in the eye of a subject the pupil is sometimes seen to constrict rhythmically. It may be prominent in some neurological conditions.

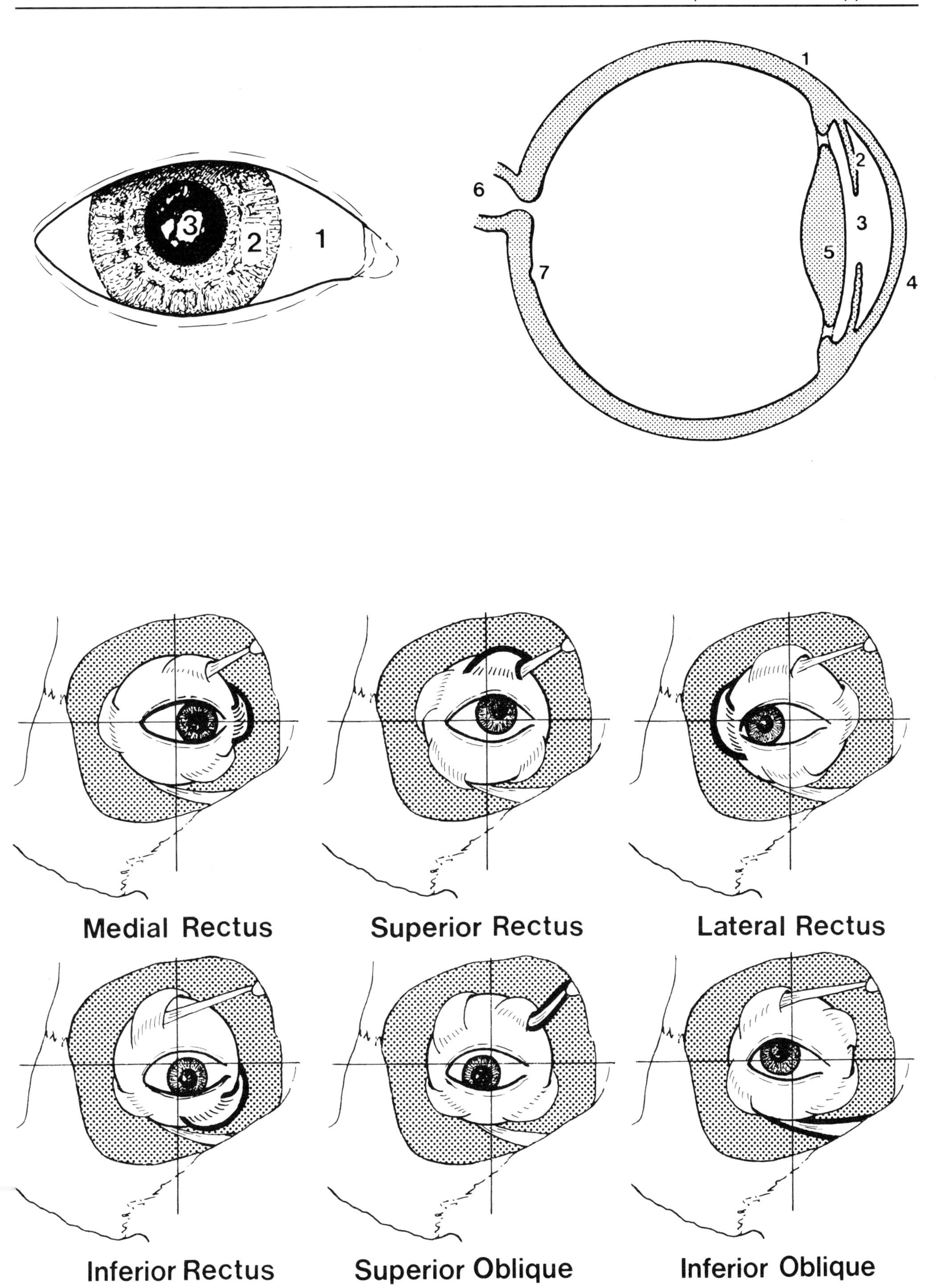
3
2
1
1
2
3
4
5
6
7
Medial Rectus
Superior Rectus
Lateral Rectus
Inferior Rectus
Superior Oblique
Inferior Oblique

OPHTHALMOSCOPY AND THE FUNDUS OF THE EYE

Ophthalmoscopy

Ophthalmoscopy is the visual examination of the fundus of the eye. It provides us with an opportunity to examine, non-invasively, certain otherwise inaccessible elements of the cardiovascular and central nervous systems.

In everyday practice the medical practitioner employs a direct ophthalmoscope to visualise the fundus. This instrument is structured such that light from a light source [**a**] is reflected by an angled mirror [**m**] into the patient's eye through the pupil [**p**]. The observer [**o**] views the fundus through a hole in the centre of the mirror.

Reflected light from the fundus forms an image which may be clearly focussed by one of a series of lenses mounted in a moveable circular disc in the head of the ophthalmoscope. These lenses compensate for any refractive errors in the eye of the subject or observer. They also assist in providing a magnified view of the fundus—*circa* ×15.

Procedure. The use of the ophthalmoscope is an acquired but essential skill which requires much practice.

The ophthalmoscope is held as close as possible to the observer's eye and within 5 cm of the patient's eye. The eye is best approached from the temporal side to obtain a good view of the fundus before the pupil constricts when the light is shone on the macula. Ideally, both the nasal and temporal vessels should be examined before the macula because the macula is so sensitive to light that the pupil constricts almost immediately, making a good examination sometimes only possible with the aid of mydriatic eyedrops to dilate the pupil.

To reduce pupillary constriction due to the light reflex, the examination is best performed in subdued lighting. In addition, pupillary constriction due to accommodation may be reduced by asking the patient to focus on a fixed point in the distance. This also fixes the focal point of the lens on the retina in the normal eye.

The left eye should be examined by the observer's left eye and holding the instrument in the left hand. This is difficult (except for the left-handed) and is perfected only with a great deal of practice.

The Fundus

The background is red because of the choroidal vessels and the pigment layer. In darker races the fundus may appear almost yellowy-grey.

Upper temporal quadrant [**1**]
Upper nasal quadrant [**2**]
Lower temporal quadrant [**3**]
Lower nasal quadrant [**4**]
Artery [**5**]—narrower than veins. Each quadrant has its own vessels, e.g. superior temporal vessels, inferior nasal vessels etc.
Vein [**6**]
Arteriovenous crossing [**7**]—arteries cross anterior to veins
Macular area [**8**]—normally darker than rest of fundus
Fovea centralis [**9**]
Optic disc [**10**]—Pink, temporal side paler. The margin should be sharp and flat. There is a cup in the centre from which the retinal vessels emerge. It varies in size and depth and usually slopes to the temporal side.

The red reflex. If the ophthalmoscope is turned to zero dioptres and the patient's eyes are observed from a distance of about 1 m, the pupils appear red. Alternatively, the ophthalmoscope may be turned to +5 dioptres and the eyes examined from 10 cm away. The red colour is due to reflection of the light from the choroidal vessels. Any opacity in the way of the light path will be seen as a dark area.

The corneal light reflex. If a light is shone at the eyes, the cornea gives a bright shiny reflection. This is called the *corneal light reflex*, and it should be seen at the centre of the pupil. If one of the eyes is misaligned, the reflection (reflex) will not be at the centre of the pupil.

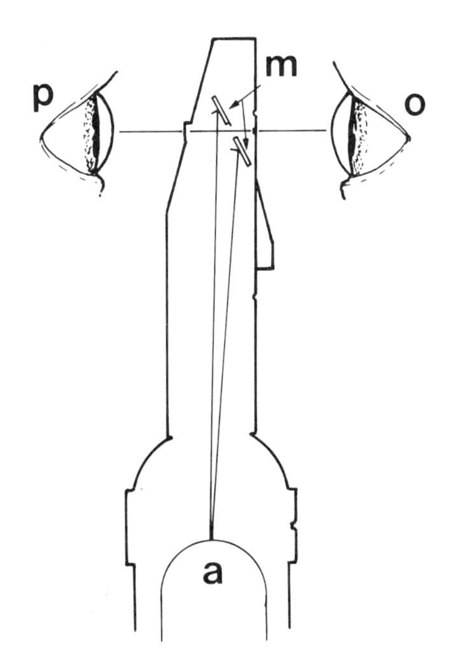

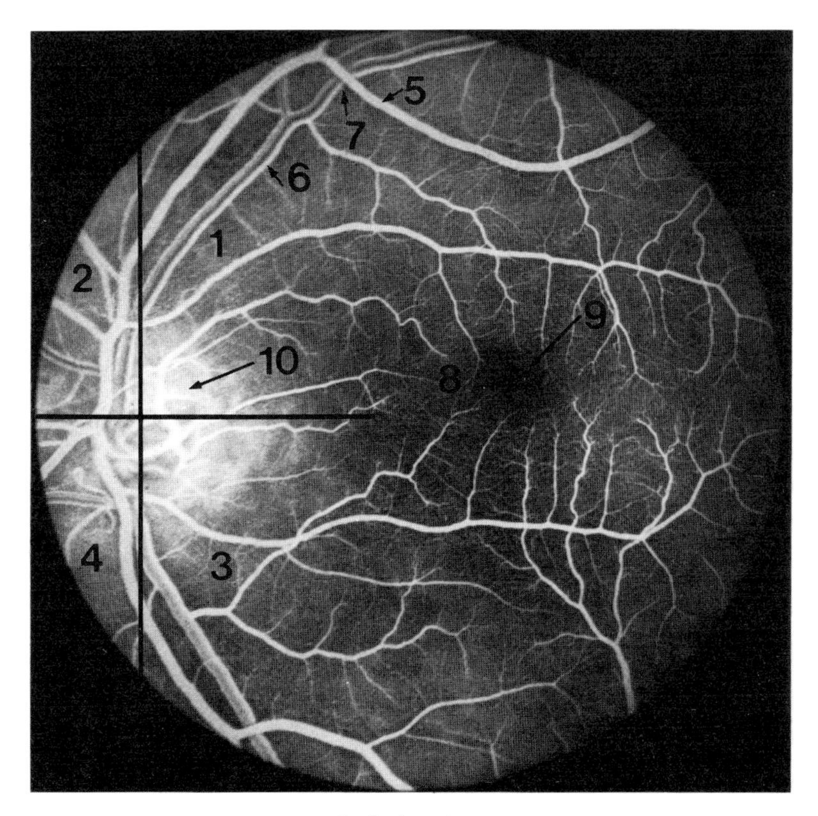

Left fundus

THE NOSE AND NASAL CAVITIES

The nose consists of two nasal cavaties separated by a midline septum. Anteriorly, the cavities open to the exterior via the nostrils, and posteriorly they are continuous with the upper part of the pharynx—the nasopharynx.

Anterior Rhinoscopy (A)

Rhinoscopy means visualisation of the nose, and anterior rhinoscopy means inspection of the nasal cavities through the nostrils using suitable illumination. This requires that the nostrils be dilated with an appropriate speculum (Thudicum's speculum), while the tip of the nose is tilted upwards. Care should be taken not to touch the sensitive nasal septum with the speculum.

Immediately within each nostril is a chamber known as the *vestibule.* This is separated from the nasal cavity by a lateral constriction or ridge called the *limen nasi.* The skin of the vestibule is lined with hairy skin containing many sebaceous glands. However, beyond the limen nasi the nasal cavities are lined with a vascular mucoperiosteum.

On anterior rhinoscopy, a pink fleshy structure is seen projecting into the nasal cavity from the lateral side just below the level of the infraorbital margin, this is the *inferior concha.*

The light coloured *middle nasal concha* is also seen opposite the medial angle of the eye; sometimes the cleft-like *hiatus semilunaris* is visible below it. The superior concha lies about 0.5 cm above the middle concha and is not visible.

POSTERIOR RHINOSCOPY (B)

This procedure is carried out with the aid of a long post-nasal mirror, about 12 mm in diameter. Illumination is provided by a mirror or lamp on the observer's forehead.

The subject is asked to open his mouth wide, but to breathe through the nose; this brings the soft palate forwards and away from the posterior pharyngeal wall. The mirror is introduced to one side of the uvula with great care being taken to avoid contact with the sensitive mucosa of the uvula, tongue, fauces or pharyngeal wall.

The structures seen are indicated in the diagram.

Nasal septum [**1**]
Inferior concha [**2**]
Middle concha [**3**]
Superior concha [**4**]
Nasopharyngeal tonsil [**5**]
Pharyngeal recess [**6**]
Opening into sphenoidal air sinus [**7**]
Hiatus semilunaris [**8**] which receives the opening of the frontal, anterior ethmoidal and maxillary air sinuses
Auditory tube orifice [**9**] (Eustachian orifice)
Tubal elevation [**10**]

The space under the overhang of each concha is known as a meatus. Thus, there are three meati in each nasal cavity. Several structures open through the lateral wall of the nose into the meati. They are listed below:

The superior meatus
The posterior ethmoidal air cells

The middle meatus
The hiatus semilunaris (this includes the frontal, anterior ethmoidal and the maxillary air sinuses), and the middle ethmoidal air cells.

The inferior meatus
The nasolacrimal duct.

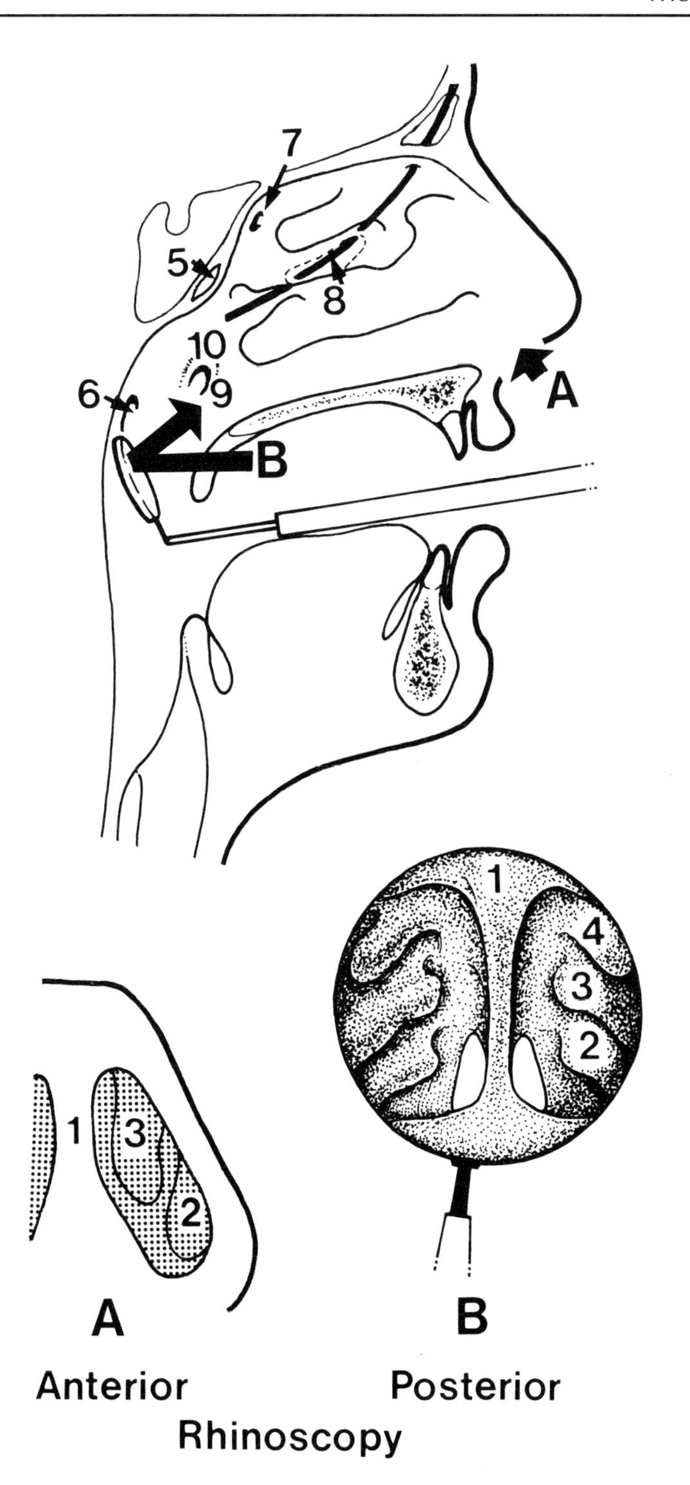
7
5
8
10
9
6
A
B
1
4
3
2
1
3
2
A
B
Anterior
Posterior
Rhinoscopy

The Paranasal Sinuses

Paired paranasal sinuses develop as diverticula from each nasal cavity into the frontal, maxillary, ethmoid and sphenoid bones. They are lined with mucoperiosteum and are small at birth. They enlarge mainly during the eruption of the permanent dentition and reach adult size shortly after puberty. Their drainage is poor and they are, therefore, a common site of infection.

The maxillary sinus [**M**] is a pyramidal-shaped sinus enclosed in the maxilla. It is the largest of the paranasal sinuses with an adult volume of some 15 ml, although the size is very variable and may even vary from side to side in the same individual. Its apex projects laterally into the zygomatic process of the maxilla, while its base forms part of the lateral wall of the nose. Its ostium opens from the postero-superior part of its base into the hiatus semilunaris in the middle meatus. The roof of the sinus forms the floor of the orbit and contains the infraorbital vessels and nerve. Inferiorly, it is related to the lateral part of the hard palate and the upper premolar and first two molar teeth. Posteriorly, it is related to the sphenopalatine fossa and the posterior superior alveolar nerves.

Surgically, the sinus may be approached through the nose via the inferior meatus or via the mouth, making an incision in the canine fossa above the premolar teeth.

The frontal sinus [**F**] lies posterior to the superciliary arch, between the two tables of the frontal bone above the orbit and in front of the anterior cranial fossa. It may be marked on the surface by a triangle between the following three points:
The nasion [**1**]
A point 3 cm above the nasion [**2**]
The junction of the medial and middle thirds of the supraorbital margin [**3**].

Rarely, one or both sinuses may be absent but this does not affect the prominence of the superciliary ridge. The frontal sinuses are usually more developed in males giving the profile of the forehead an obliquity which contrasts with the vertical or convex profile of the woman or child.

Each sinus opens into the middle meatus beneath the middle concha via the long frontonasal duct which transgresses the anterior ethmoidal air cells.

The ethmoidal sinuses are also bilateral, but consist of a honeycomb of air cells lying between the medial wall of the orbit and the upper lateral wall of the nose. The air cells are divided into anterior, middle and posterior groups which drain separately—the anterior into the anterior part of the hiatus semilunaris, often via the frontonasal duct; the middle into the middle meatus via the bulla ethmoidalis; and the posterior into the superior meatus. The posterior group is closely related to the optic canal and nerve.

The sphenoidal sinus lies postero-superior to the nasopharynx enclosed in the body of the sphenoid bone, opposite a level just above the zygomatic arch. It is related laterally to the cavernous sinus, the internal carotid artery and the nerves lying in the cavernous sinus. Superiorly, the sinus is related to the sella turcica and the pituitary and hence it is possible to operate on the pituitary through the nose. Medially, the sinus is separated from the sinus of the opposite side by a thin bony septum. Each sinus opens into the sphenoethmoidal recess of the nasopharynx about 6 cm behind the upper margin of the apertura piriformis.

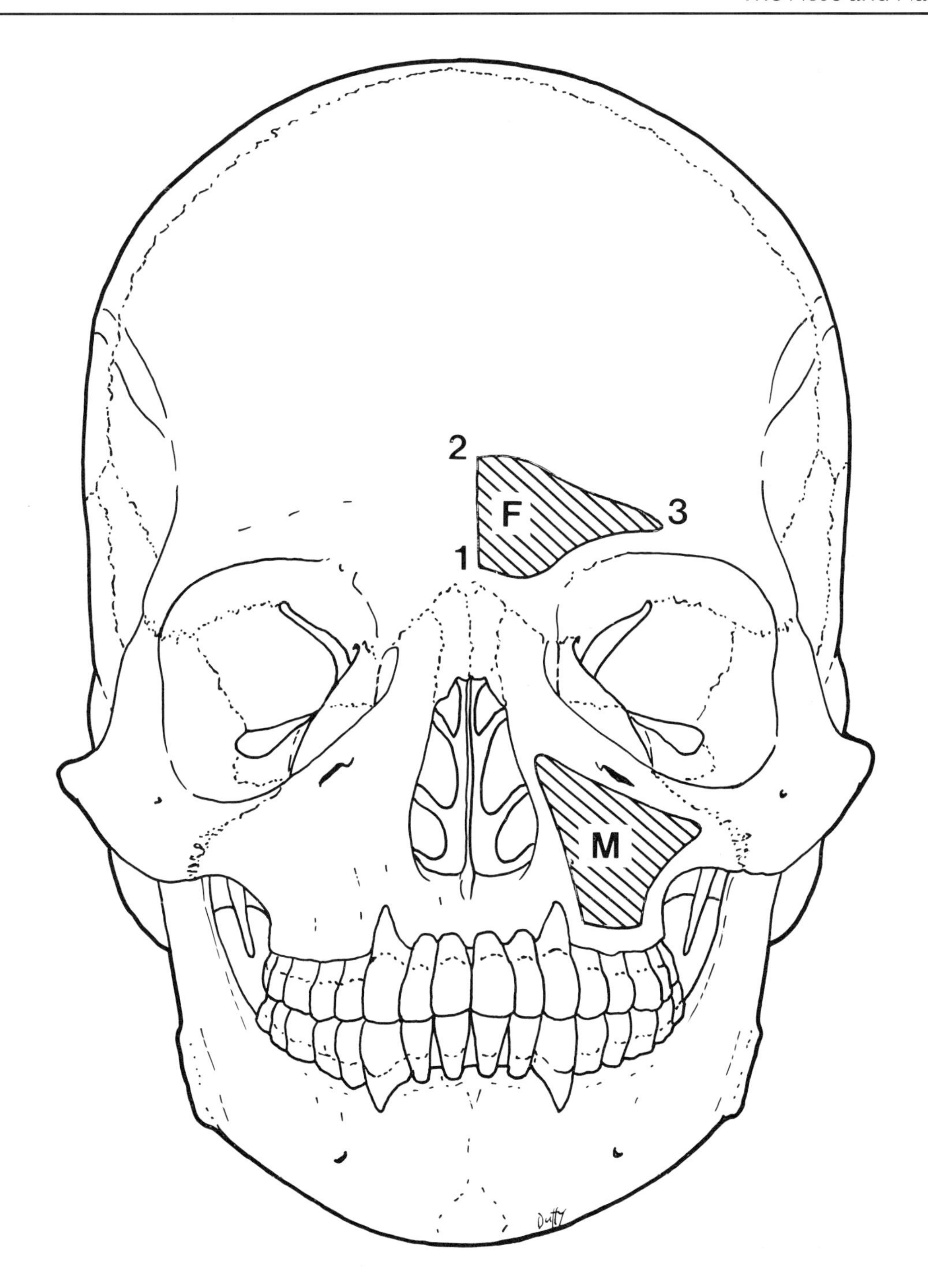
2
F
3
1
M

THE SALIVARY GLANDS

The Parotid Gland

The parotid gland lies in the hollow between, and overlaps, the mandible and masseter in front and the mastoid process and sterno-cleidomastoid behind. It may be marked on the surface by a series of lines joining the following points:
A point immediately behind the head of the mandible [**1**]
The centre of the masseter [**2**] (*See also* The Parotid Duct)
A point one finger's breadth below and behind the mandible [**3**]
The tip of the mastoid process [**4**], and back to [**1**]

The gland is enclosed in the deep cervical fascia and it is the stretching of this fascia by the swollen gland that causes the discomfort and pain in mumps.

The Parotid Duct [5]

This may be represented by the middle third of a line drawn from the lower margin of the tragus to a point midway between the ala of the nose and the red margin of the upper lip. It arises from the anterior border of the gland and runs forwards on the masseter, diving round its anterior border to pierce the buccinator. Then, it runs forwards for 1 cm between the buccinator and the mucus membrane of the cheek to enter the oral cavity opposite the second upper molar tooth [**6**]. It may be rolled between the finger and the tensed masseter, but despite its size, its lumen is usually very fine. It may be extremely tortuous.

The patency of the parotid duct may be tested radiologically by introducing dye into the duct by means of a fine cannula. This procedure is known as *sialography* and is used mainly when a calculus is suspected in the parotid duct.

The Submandibular Gland [7]

This gland may be represented by an oval extending from the angle of the mandible post-eriorly, to a point on the lower border of the mandible two finger breadths from the angle. Superiorly, it reaches a level one finger's breadth above the lower border of the mandible and inferiorly it extends to the hyoid bone. It is very variable in shape and size.

The facial artery runs in a groove on its posterior surface before looping round the lower border of the mandible at the anterior border of the masseter where its pulse can be felt.

The Submandibular Duct

This runs deeply on the oral side of the mylohyoid muscle to enter the mouth on the summit of the sublingual papilla just to the side of the frenulum of the tongue.

Its course may be marked by a line, about 5 cm long, beginning at the angle of the mandible, and running to a point two finger breadths from the symphysis menti and one finger's breadth above the lower border of the mandible.

The Sublingual Gland [8]

This is the smallest of the three glands and it lies beneath the mucosa of the floor of the mouth in contact with the anterior part of the mandible. Its upper border may be felt by the tongue because, with the tongue bent downwards, the gland bulges upwards to produce the distinctive *plica sublingualis* on the floor of the mouth. The multiple ducts of this gland open onto the crest of this fold.

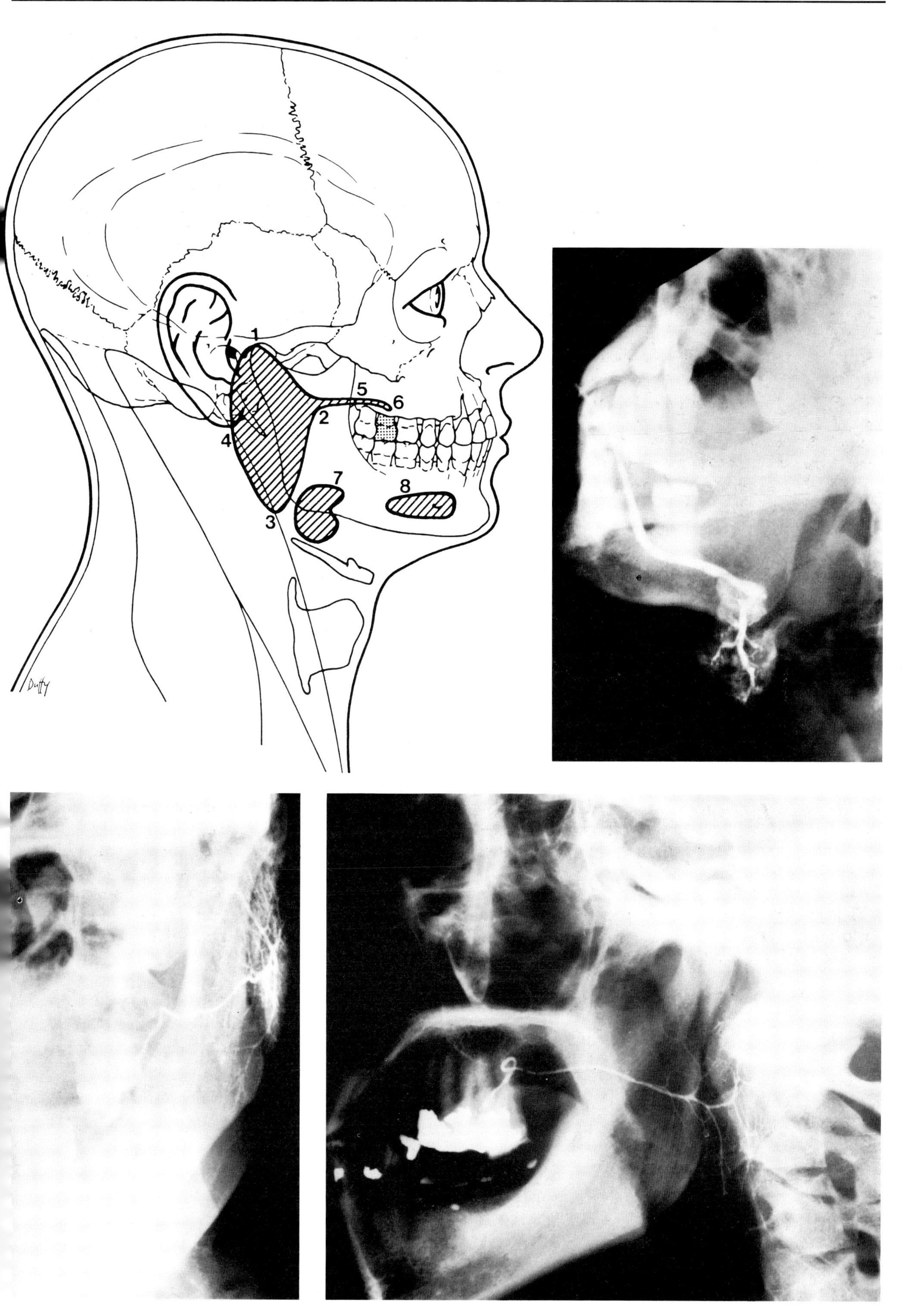
1
2
3
4
5
6
7
8
Duffy

THE MOUTH AND THE TONGUE

The lips are covered externally with skin and superficial fascia bound down tightly to the underlying orbicularis oris muscle and the muscles which blend with it. Internally, they are lined with mucus membrane. Each lip is attached to the front of the corresponding gum by the *frenulum of the lip*.

The lateral angle of the mouth is usually opposite the interval between the premolar teeth. The opening between the lips is sometimes referred to as the *oral fissure*.

The oral cavity (or mouth) may be considered in two parts:

i *The vestibule*, which is the slit-like space between the teeth and cheeks.

ii *The oral cavity proper* which is bounded anteriorly and laterally by the alveolar arches and teeth, and which opens posteriorly into the oropharynx via the pharyngeal isthmus between the two palatoglossal folds.

On the inner surface of the cheek, opposite the crown of the second upper molar tooth, a small papilla marks the opening of the parotid duct.

Hard palate showing palatine folds **[1]**
Soft palate **[2]**
Uvula **[3]**
Posterior wall of pharynx **[4]**
Palatopharyngeal arch formed by the muscle of the same name **[5]**
Palatoglossal arch **[6]** formed by the muscle of the same name; isthmus into oropharynx
Palatine tonsil **[7]**

The Tongue

The tongue is divided into oral and pharyngeal parts by the V-shaped *sulcus terminalis* **[ST]** which extends on the dorsal surface of the tongue from the base of the palatoglossal fold on either side to the *foramen caecum* **[F]**.

The oral part of the tongue fills in the floor of the mouth with its apex resting on the incisor teeth and its free lateral border resting on the cheek, teeth and gums. Its superior or dorsal surface lies in contact with the hard and soft palates and is marked by a median furrow extending to the foramen caecum. It represents the embryological anterior two-thirds of the tongue and has a rough, shaggy appearance produced mainly by the filiform papillae which are distributed widely over its superior surface. There are several sorts of papillae distributed as shown.

Filiform papillae **[8]** are conical papillae which cover the whole of the oral surface of the tongue. They lie in transverse rows anteriorly, but come to lie parallel with the sulcus terminalis posteriorly.

Fungiform papillae **[9]** are scattered irregularly over the whole of the oral surface but are concentrated mainly at the sides and apex of the tongue.

Vallate papillae **[10]** are the most conspicuous papillae. They are 8–12 in number and lie just in front of the sulcus terminalis.

Foliate papillae **[11]** are 4–5 vertical folds lying on the lateral border of the tongue just in front of the palatoglossal arch.

The pharyngeal part of the tongue lies behind the palatoglossal arches and forms the anterior wall of the oropharynx. It has *no* papillae, but it is studded with multiple low elevations (superficially resembling vallate papillae) which are formed by underlying nodules of lymphoid tissue which collectively form the *lingual tonsil*.

The inferior surface
Frenulum **[12]**—a median fold of mucus membrane
Lingual artery **[13]**
Lingual nerve **[14]**
Profunda linguae vein (lingual vein) **[15]**
The anterior lingual gland **[16]**
Plica fimbriata (fimbriated fold) **[17]**
Plica sublingualis (sublingual fold) **[18]** covers the sublingual gland. The ducts of the sublingual gland open onto the fold
The sublingual papilla **[19]** marks the opening of the submandibular duct

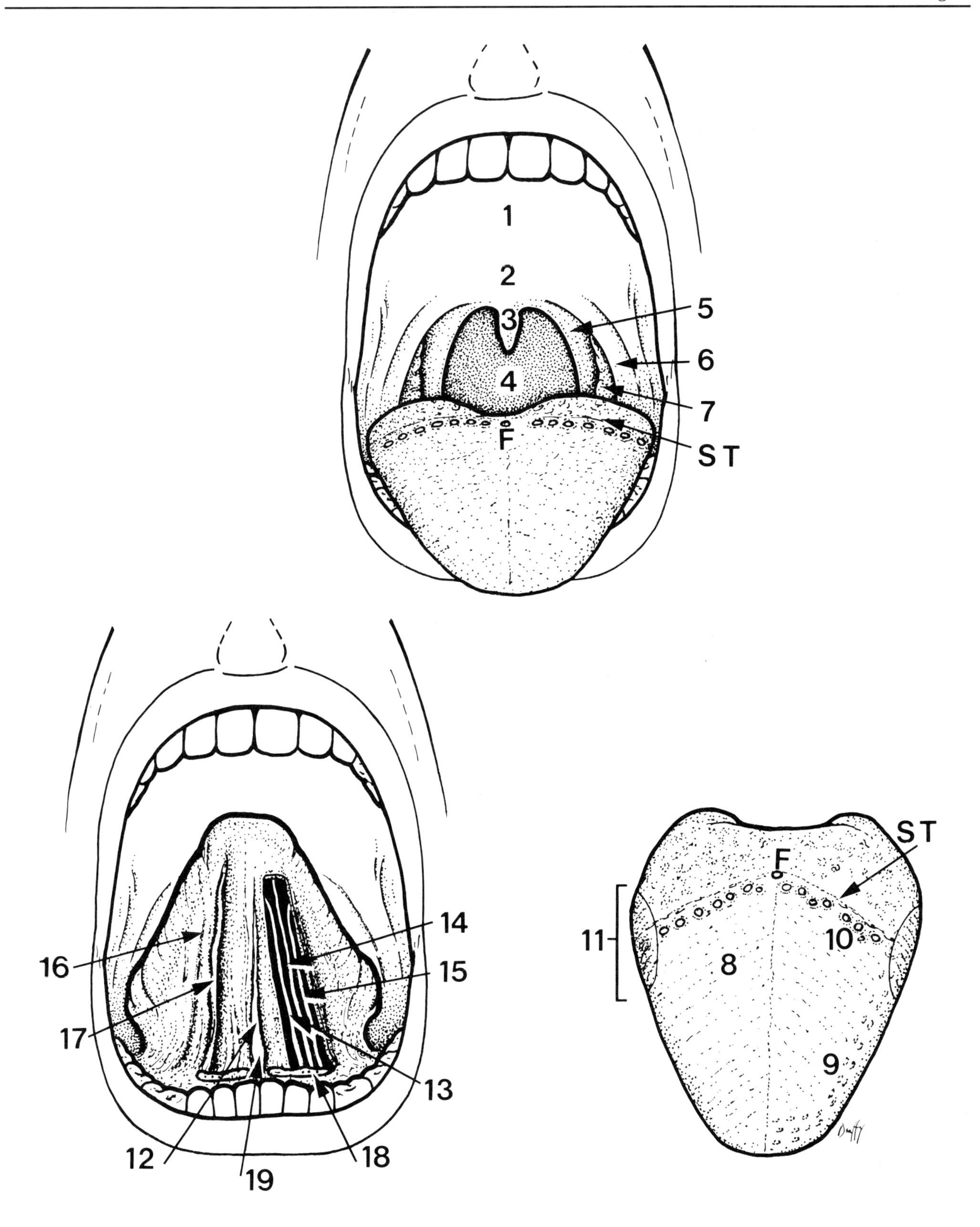
1
2
3
4
5
6
7
F
S T
16
17
14
15
13
12
19
18
S T
F
11
10
8
9

THE LARYNX

The larynx lies behind the thyroid cartilage (the Adam's apple). The vocal cords lie in an antero-posterior direction opposite the junction of the upper and middle thirds of the anterior border.

The larynx is continuous with the trachea at the upper border of the cricoid cartilage which lies at the level of C6.

The larynx may be examined in the conscious patient by the use of a mirror in the mouth. This procedure is known as *indirect laryngoscopy*. In the unconscious patient, the larynx may be visualised directly—*direct laryngoscopy*.

Indirect Laryngoscopy

The patient is asked to sit comfortably and to open his mouth as wide as possible. He is then invited to protrude his tongue while breathing with short panting breaths through his mouth. The tongue is now gently pulled forward by the examiner. This is facilitated by wrapping the anterior part of the tongue in dry gauze to aid gripping, but the gauze also serves to protect the frenulum of the protruded tongue from the lower incisor teeth. A light is now shone onto the uvula and a warmed laryngeal mirror is passed back into the mouth and pressed firmly upwards and backwards against the base of the uvula. The mirror should be kept in the midline because any contact with the fauces will induce gagging.

The first structures to be seen are the pharyngeal surface of the tongue and the lingual surface of the epiglottis. Below the epiglottis are the two pearly white vocal cords.

To inspect the vocal cords, the subject is asked to produce a high pitched 'ee-ee'. This causes the larynx to rise closer to the mirror and the cords adduct. Normally, each cord is slightly convex laterally, but a paralysed cord has a taut, straight edge because of its intrinsic elasticity. When phonation ceases the cords abduct to reveal the upper rings of the trachea through the rima glottidis. The cords are also seen to move rhythmically with respiration—abduction becoming more prominent during inspiration.

Several potential pitfalls exist in this examination which should always be borne in mind by the examiner. The laryngeal sinuses which open between the vestibular folds (false cords) and the true vocal cords cannot be inspected. In addition, the overhang of the true cords may easily conceal a small lesion in the subglottic region.

The valleculae and piriform fossae, which are well recognised traps for foreign bodies, may be inspected easily.

Pharyngeal surface of tongue in valleculae [**1**]
Epiglottis [**2**]
Vestibular folds [**3**]—pink and fleshy
True vocal cords [**4**]—pearly white, slightly convex laterally
Corniculate cartilage [**5**]
Cuneiform cartilage [**6**]
Interarytenoid region [**7**]
Piriform fossa [**8**]
Laryngeal sinus [**9**]

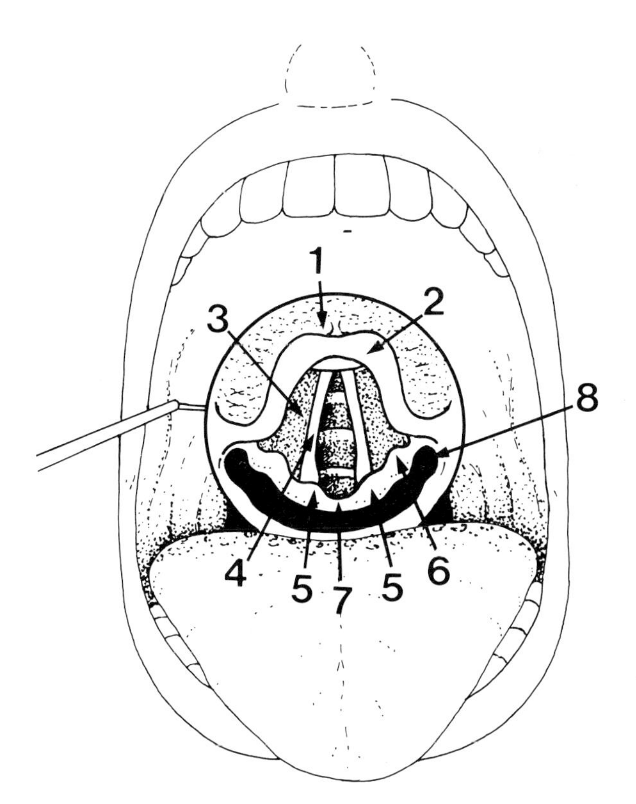

Indirect laryngoscopy

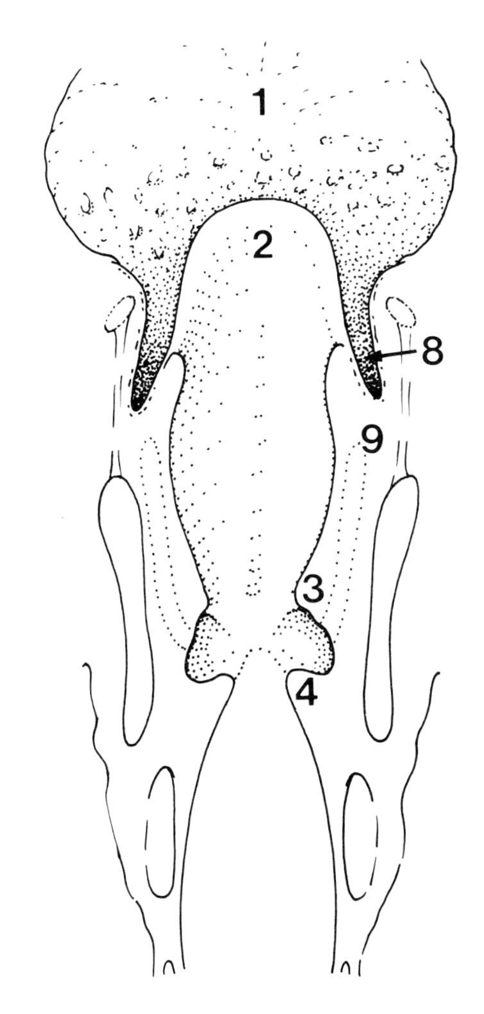

Coronal section of the larynx viewed from behind

LYMPH NODES OF THE HEAD AND NECK

All lymph from the head and neck ultimately drains into the deep cervical lymph nodes which lie as a chain embedded in the carotid sheath adjacent to the internal jugular vein. Most of these nodes lie deep to the sternocleidomastoid muscle but, superiorly, some nodes extend above it and lie in the triangle between the posterior belly of the digastric muscle, the facial vein and the internal jugular vein. This group is known as the *jugulodigastric group* and is concerned primarily with the lymph drainage of the tonsils and tongue. Sometimes, a single large node in this region is referred to as the jugulodigastric node. Inferiorly, the chain spreads into the subclavian triangle from under the posterior border of the sternocleidomastoid muscle. Often one of the nodes of this group lies in the region of the omohyoid tendon and is known as the *jugulo-omohyoid node*. This node is also concerned with the lymph drainage of the tongue. Thus, enlarged nodes in the neck, especially the jugulo-omohyoid and the jugulodigastric nodes, should alert the clinician to the possibility of carcinoma of the tongue. Enlarged nodes in the neck may be the *only* early presenting feature of this disease. The efferent lymph vessels from the deep cervical chain unite to form the jugular lymph trunk which usually drains into the right lymphatic duct or the thoracic duct and, occasionally, into the subclavian lymph trunk or directly into the brachiocephalic vein.

Regional Nodes

There are two main series of regional nodes both of which drain to the deep cervical nodes.

The outer collar of lymph nodes consists of several distinct groups each draining a particular area of the scalp or face. When enlarged, the nodes of each of these groups, with the exception of the parotid nodes, are easily palpable.

The occipital group **[1]** lies over the occipital bone at the apex of the posterior triangle of the neck. They drain the back of the scalp.

The retro-auricular (mastoid) group **[2]** lies over the lateral surface of the mastoid process. They drain the scalp above the auricle and the posterior wall of the external auditory meatus.

The superficial cervical lymph nodes **[3]** extend along the external jugular vein and receive lymph from the angle of the jaw, the skin over the apex of the parotid and the lobe of the ear.

The parotid nodes **[4]** may be divided into the superficial (pre-auricular) nodes and the deep parotid nodes. The superficial nodes receive lymph from the auricle, the anterior part of the temple and scalp, and the upper part of the face.

The buccal nodes **[5]** are few or single in number and lie over the buccinator muscle close to the facial vein. They lie along the course of the vessels taking lymph to the submandibular glands.

The submandibular nodes **[6]** lie on the surface of the submandibular salivary gland deep to the deep cervical fascia and along the lower border of the mandible, against which they can be palpated as far as the anterior border of the masseter. They are therefore in close relation to the mandibular branch of the facial nerve—a point to be borne in mind when attempting a lymph node biopsy. They receive lymph from the front of the scalp, the nose, the cheek, the frontal, ethmoid and the maxillary air sinuses, the floor of the mouth and the vestibule, the anterior two-thirds of the tongue (except the tip), the gums, the teeth and lips, except the lower two incisors and the lip opposite them.

The submental nodes **[7]** lie in the submental triangle between the anterior bellies of the digastric muscles. They drain a wedge of tissue opposite the premaxilla. They receive lymph from the tip of the tongue, the floor of the mouth beneath the tip of the tongue, the lip and the lingual and labial gums of the two lower incisors and the incisors themselves. They drain *bilaterally* to the jugulo-omohyoid nodes.

The Inner Circle of Lymph Nodes

The viscerotubular structures of the neck are each surrounded by an extensive lymphatic plexus which drains via regional nodes into the deep cervical nodes. The regional nodes lie in the positions indicated by their names:
Retropharyngeal
Infrahyoid
Prelaryngeal
Pre- and paratracheal

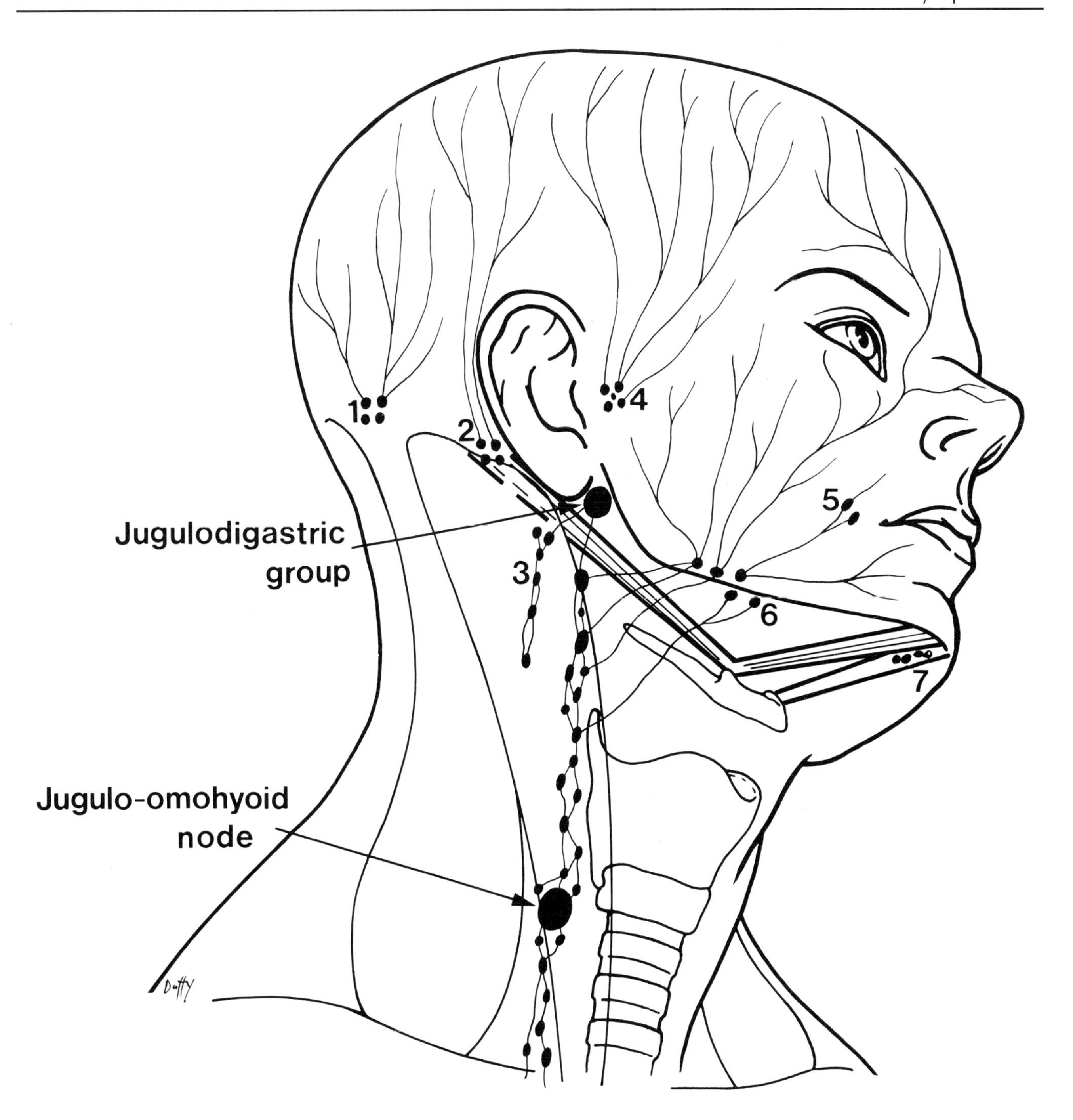
1
2
3
4
5
6
7
Jugulodigastric group
Jugulo-omohyoid node
Duffy

Waldeyer's Ring

This is a circular band of lymphoid tissue surrounding the common entrance to the respiratory and digestive tracts. The ring is formed inferiorly by the lingual tonsil, laterally by palatine tonsils, and superiorly by the tubal and pharyngeal tonsils.

The lingual tonsil [**1**]. *See* 'The Mouth and Tongue' (p. 252).

The palatine tonsils [**2**]. Each lies in the triangular tonsillar recess between the diverging palatoglossal and palatopharyngeal arches. In early childhood they form conspicuous projections which recede by puberty; 12–15 orifices may be seen on the medial surface of the gland. These are the openings of the tonsillar crypts.

The pharyngeal tonsil [**3**]. This forms a pyramidal prominence on the roof of the nasopharynx with its apex pointing towards the basal septum. It is particularly prominent in children under the age of seven years, after which it begins to atrophy, often disappearaing completely by adulthood. The lateral prolongation of the pharyngeal tonsil behind the Eustachian tube is known as the tubal tonsil [**4**].

THE THYROID

The thyroid gland consists of two piriform (pear-shaped) lobes on either side of the neck connected by an isthmus in front of the trachea.

The upper poles of each lobe reach the level of the middle of the thyroid cartilage, whereas the lower poles may extend behind the sternal end of the clavicle.

The isthmus of the gland is a 2 cm broad band of glandular tissue connecting the two lobes across the midline over the trachea. Its upper border lies about 1 cm below the cricoid cartilage.

Ectopic thyroid may occur anywhere along the course of the thyroglossal duct, in which case there may be no functioning thyroid at the normal site. The site of the abnormal thyroid and the absence of functioning thyroid at the normal site may be demonstrated by radioactive scanning techniques.

Radiology of the Thyroid

Conventional x-rays in the study of the thyroid have largely been superseded by newer methods of nuclear scanning.

The thyroid has a strong affinity for iodine which it uses for the manufacture of thyroid hormones. This affinity is exploited in nuclear scanning. The subject is given an oral dose of the radioactive isotope ^{131}I. Twenty-four hours later the subject's neck is scanned with a gamma camera connected to a scintillation counter. A plot is made of the radioactivity over the neck to produce a map of the ^{131}I uptake—the resultant plot is called a 'scintigram'. Most of the uptake is in the lateral lobes, the right often taking up more than the left. There is much less uptake by the upper poles and isthmus.

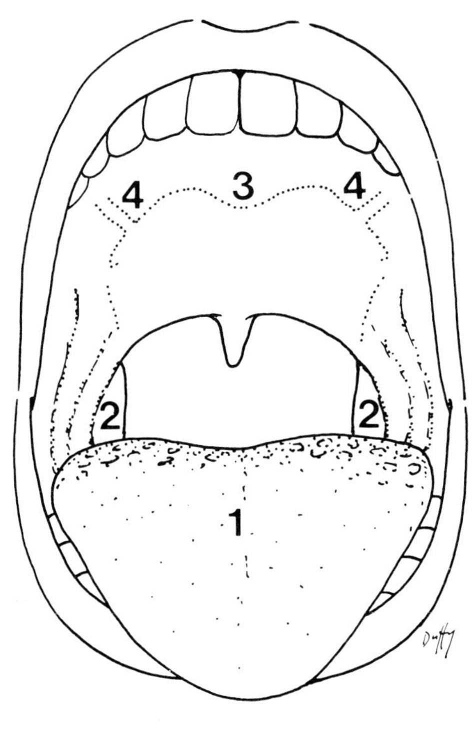

Waldeyer's ring

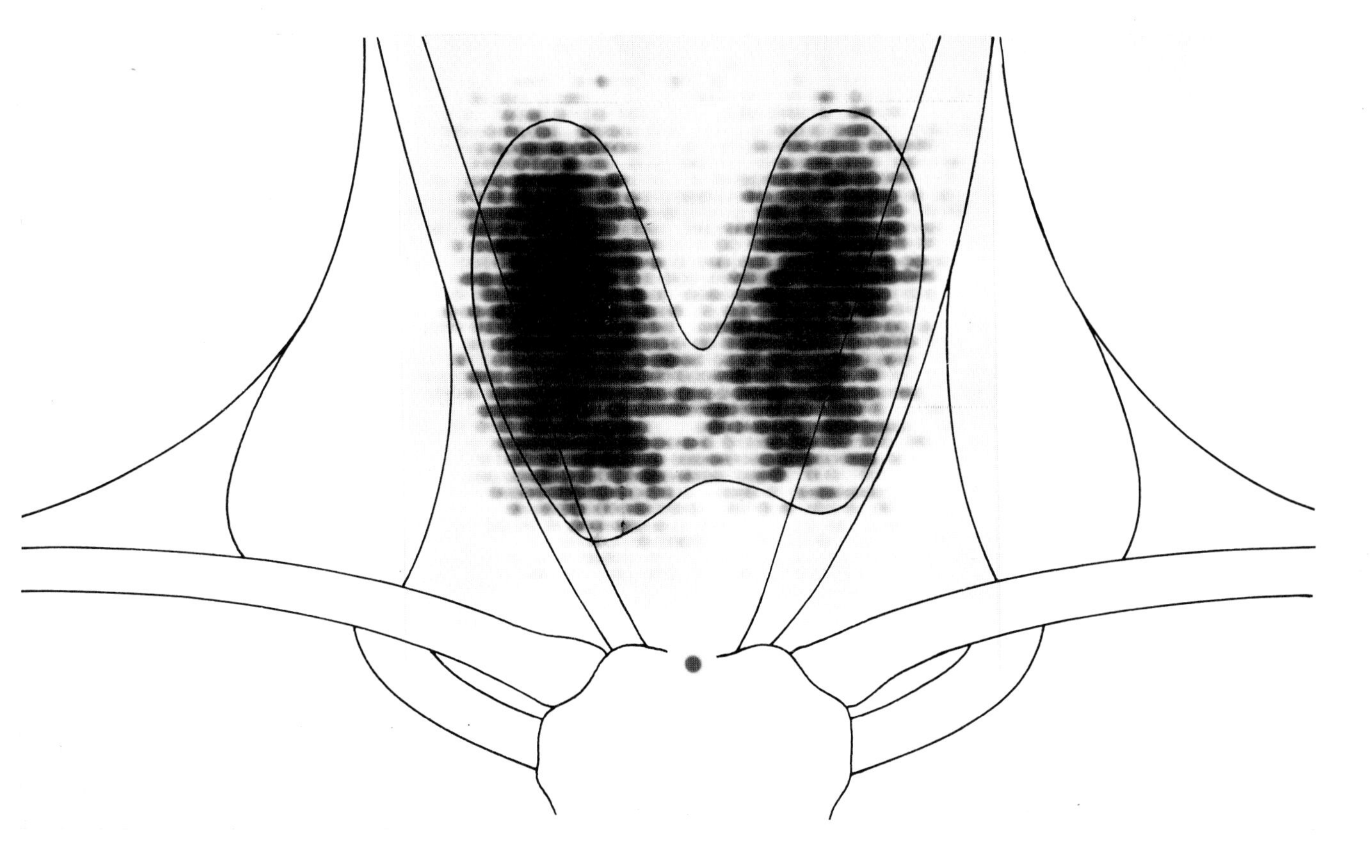

Thyroid scintigram

11. Motor Points

A motor point is a point on the body surface where an appropriate electrical stimulus will stimulate a single muscle or group of muscles to contract. It is therefore useful not only for demonstrating the action of muscles but also for stimulating particular muscles to prevent atrophy during a period of motor nerve regeneration. In order to stimulate a *single* muscle, that muscle must be relatively superficial and the nerves to adjacent muscles should not be too close to the point of stimulation. Most muscles of the arm and forearm can be stimulated satisfactorily without any appreciable contraction of adjacent muscles, but in the buttock and thigh this is not feasible, except for the rectus femoris and the tensor fasciae latae.

It is possible in some cases, by using a weak stimulus, to make a tendon stand out for demonstrative purposes without any associated joint movement. However, a stronger stimulus to the same muscle would produce movement of the joint over which the tendon passes.

THE UPPER LIMB

Anterior Aspect

Pectoralis major [1]
Deltoid [2]
Biceps [3]
Brachioradialis [4]
Flexor carpi radialis [5]
Palmaris longus [6]
Flexor digitorum superficialis [7]
Flexor pollicis longus [8]
Pronator teres [9]
Median nerve [10]
Abductor pollicis brevis [11]
Opponens pollicis [12]
Flexor pollicis brevis [13]
First lumbrical [14]
Fourth lumbrical [15]
Abductor digiti minimi [16]

Posterior Aspect

Trapezius [1]
Rhomboideus major [2]
Infraspinatus [3]
Teres major [4]
Triceps—lateral head [5]
Triceps—long head [6]
Anconeus [7]
Extensor carpi ulnaris [8]
Extensor digiti minimi [9]
Abductor pollicis longus [10]
Extensor indicis [11]
Third dorsal interosseous [12]
First dorsal interosseous [13]
Extensor pollicis brevis [14]
Extensor digitorum [15]
Extensor carpi ulnaris [16]
Extensor carpi radialis brevis [17]
Extensor carpi radialis longus [18]
Deltoid [19]
Brachialis [20]

THE LOWER LIMB

Lateral Aspect

Gluteus maximus [1]
Sciatic nerve [2]
Semitendinosus [3]
Biceps—long head [4]
Biceps—short head [5]
Gastrocnemius [6]
Common peroneal nerve [7]
Soleus [8]
Extensor digitorum brevis [9]
Extensor hallucis [10]
Peroneus brevis [11]
Extensor digitorum longus [12]
Peroneus longus [13]
Tibialis anterior [14]

Medial Aspect

Sartorius [1]
Rectus femoris [2]
Adductor longus [3]
Vastus medialis [4]
Soleus [5]
Abductor hallucis [6]
Gastrocnemius [7]
Tibial nerve [8]
Gracilis [9]
Semimembranosus [10]
Semitendinosus [11]
Adductor magnus [12]
Gluteus maximus [13]

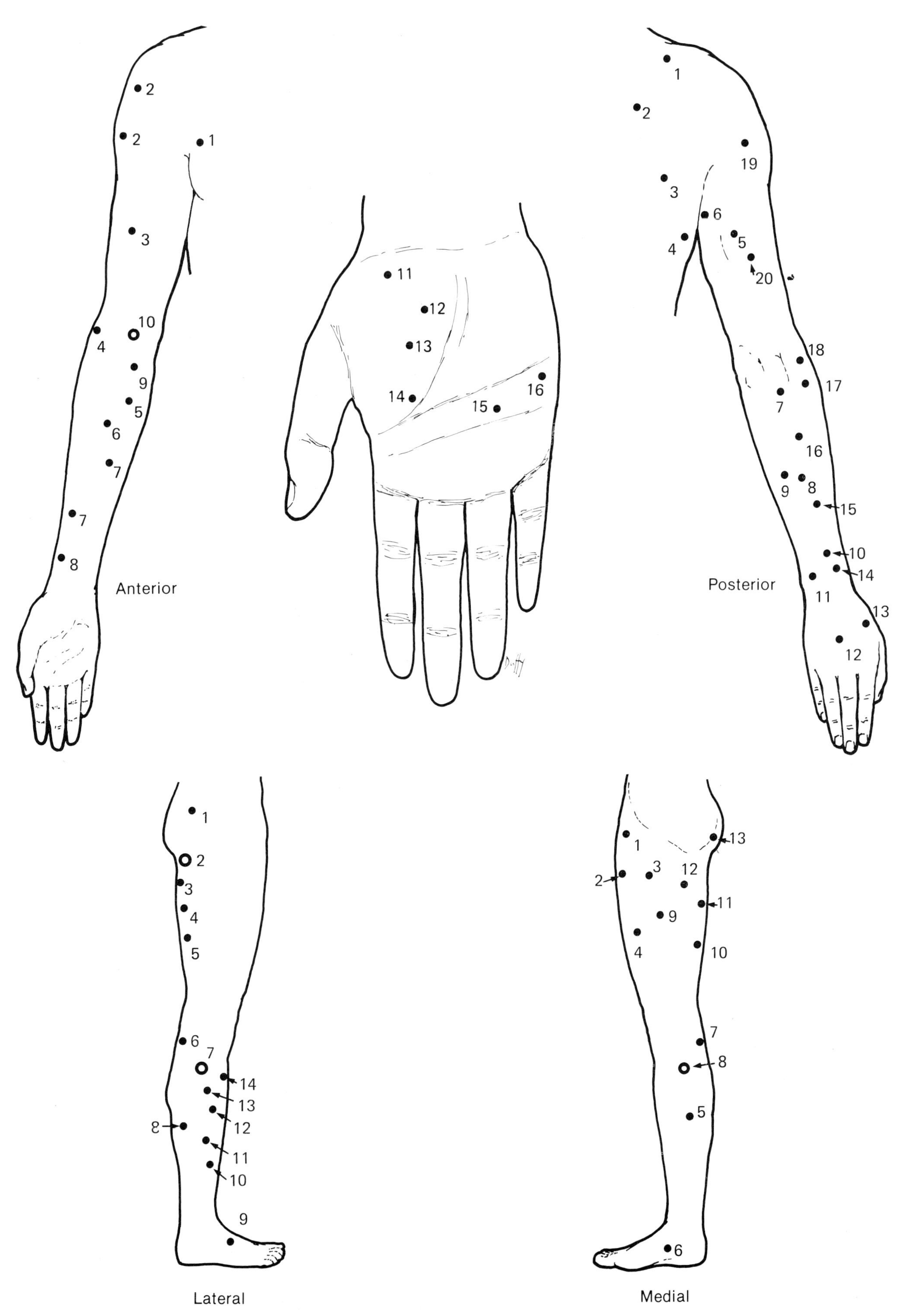
Anterior
Posterior
Lateral
Medial

12. Index

Index